Student Solutions Manual

Mathematics
A Practical Odyssey

SEVENTH EDITION

David B. Johnson
Diablo Valley College

Thomas A. Mowry
Diablo Valley College

Prepared by

Ann Ostberg
Grace University

BROOKS/COLE
CENGAGE Learning

Australia • Brazil • Japan • Korea • Mexico • Singapore • Spain • United Kingdom • United States

ISBN-13: 978-0-8400-5387-9
ISBN-10: 0-8400-5387-8

Brooks/Cole
20 Davis Drive
Belmont, CA 94002-3098
USA

Cengage Learning is a leading provider of customized learning solutions with office locations around the globe, including Singapore, the United Kingdom, Australia, Mexico, Brazil, and Japan. Locate your local office at: **www.cengage.com/global**

Cengage Learning products are represented in Canada by Nelson Education, Ltd.

To learn more about Brooks/Cole, visit **www.cengage.com/brookscole**

Purchase any of our products at your local college store or at our preferred online store **www.cengagebrain.com**

For product information and technology assistance, contact us at **Cengage Learning Customer & Sales Support, 1-800-354-9706**

For permission to use material from this text or product, submit all requests online at **www.cengage.com/permissions** Further permissions questions can be emailed to **permissionrequest@cengage.com**

Printed in the United States of America
1 2 3 4 5 6 7 14 13 12 11 10

Table of Contents

1.1 Deductive vs. Inductive Reasoning

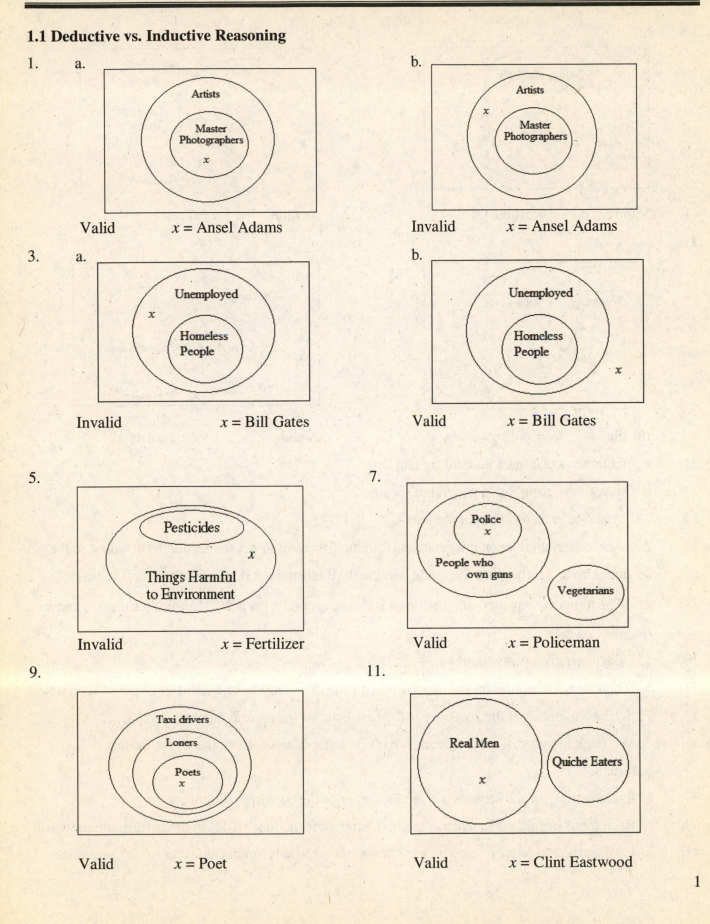

1.
   a.
   Valid x = Ansel Adams

   b.
   Invalid x = Ansel Adams

3.
   a.
   Invalid x = Bill Gates

   b.
   Valid x = Bill Gates

5.
   Invalid x = Fertilizer

7.
   Valid x = Policeman

9.
   Valid x = Poet

11.
   Valid x = Clint Eastwood

13.   15.

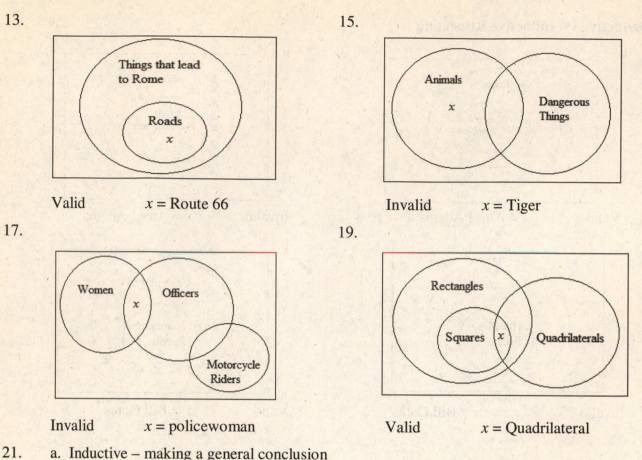

Valid x = Route 66   Invalid x = Tiger

17.   19.

Invalid x = policewoman   Valid x = Quadrilateral

21.   a. Inductive – making a general conclusion

b. Deductive – going from general to specific

23.   23. Add 5 to each term to get the next.

25.   20. Add consecutive even integers: add 2 to the first term to get the second term, add 4 to the

second term to get the third term, add 6 to the third term to get the fourth term, and so on.

27.   25. The terms are squares: the first term is 1^2, the second term is 2^2, the third term is 3^2, and so

on.

29.   13. Each term is a prime number.

31.   5. Think of each number in the sequence as a time (5:00, 8:00, etc.). Add three hours each time.

33.   F, S. Each member of the sequence is the first letter of the natural numbers 1, 2, 3, …

35.   T, W. Each member of the sequence is the first letter of the days of the week: Friday,

Saturday, …

37.   r. Each member of the sequence is the first letter of the corresponding word.

39.   f. Each member of the sequence is the first letter after the first vowel in the corresponding word.

41.   f. Look at the last letter in each word. Choose the next letter in the alphabet for the sequence.

2

43. 13. Add 3 each time.

1. Think of the numbers as times and add 3 hours each time.

45. $x^2 = \dfrac{\left(-b+\sqrt{b^2-4ac}\right)^2}{(2a)^2} = \dfrac{b^2-2b\sqrt{b^2-4ac}+b^2-4ac}{4a^2}$ So,

$$
\begin{aligned}
ax^2+bx+c &= a\cdot\dfrac{b^2-2b\sqrt{b^2-4ac}+b^2-4ac}{4a^2}+\dfrac{b\left(-b+\sqrt{b^2-4ac}\right)}{2a}+c \\[2mm]
&= \dfrac{2b^2-2b\sqrt{b^2-4ac}-4ac}{4a}+\dfrac{-b^2+b\sqrt{b^2-4ac}}{2a}+c \\[2mm]
&= \dfrac{2\left(b^2-b\sqrt{b^2-4ac}-2ac\right)}{4a}+\dfrac{-b^2+b\sqrt{b^2-4ac}}{2a}+c \\[2mm]
&= \dfrac{b^2-b\sqrt{b^2-4ac}-2ac}{2a}+\dfrac{-b^2+b\sqrt{b^2-4ac}}{2a}+c \\[2mm]
&= \dfrac{-2ac}{2a}+c = -c+c = 0
\end{aligned}
$$

47. $a=1,\ b=-6,\ c=7$

$$
x = \dfrac{6\pm\sqrt{(-6)^2-4\cdot 1\cdot 7}}{2\cdot 1} = \dfrac{6\pm\sqrt{8}}{2} = \dfrac{6\pm 2\sqrt{2}}{2} = 3\pm\sqrt{2}
$$

49. Each row, each column, and each box must contain the digits 1 through 9. For reference, number the boxes 1 through 9, and label the 9 rows and 9 columns as in Example 8. Consider starting with the digit 1. Concentrate, for example, on boxes 3, 6, and 9. We see that boxes 3 and 9 each contain the digit 1, whereas box 6 does not. Consequently, the 1 in box 6 must be placed in column 9 because columns 7 and 8 already have a 1. However, row 4 already has a 1, so we can deduce that 1 must be placed in row 5, column 9, that is, in square $(5, 9)$. Concentrating on boxes 2, 5, and 8, the method can be used to place a 1 in box 5, namely in square $(6, 6)$. Then concentrating on boxes 1, 4, and 7, the method can be used to place a 1 in box $(8, 1)$. Use logic and the three consecutive box strategy to complete the grid.

3

3	5	6	4	1	8	2	7	9
2	1	7	6	5	9	3	4	8
9	8	4	3	7	2	1	5	6
6	2	1	8	9	5	4	3	7
8	3	5	7	4	6	9	2	1
7	4	9	2	3	1	6	8	5
5	9	2	1	8	4	7	6	3
1	6	3	5	2	7	8	9	4
4	7	8	9	6	3	5	1	2

51. Each row, each column, and each box must contain the digits 1 through 9. For reference, number the boxes 1 through 9, and label the 9 rows and 9 columns as in Example 8. Consider starting with the digit 9. Concentrate, for example, on boxes 3, 6, and 9. There is a 9 in box 3, column 9. Boxes 6 and 9 need a 9 and it must be placed in columns 7 and 8. Notice that rows 7 and 9 already contain a 9, so the only square left for a 9 in box 9 is square (8, 8). Now we see the 9 in box 6 must be placed in column 7. Since there is already a 9 in row 5, the 9 must be placed in square (4, 7). This same method can be used concentrating on boxes 2, 5, and 8. A 9 must be placed in square (6, 5). Then again the method can be used concentrating on boxes 1, 4, and 7. A 9 must be placed in square (1, 1). Use logic and the three consecutive box strategy to complete the grid.

9	5	6	3	4	1	2	8	7
8	2	3	6	5	7	1	4	9
4	7	1	9	8	2	5	3	6
6	8	7	1	3	4	9	5	2
5	9	4	2	6	8	7	1	3
3	1	2	7	9	5	4	6	8
7	6	5	8	1	9	3	2	4
2	4	8	5	7	3	6	9	1
1	3	9	4	2	6	8	7	5

53. Each row, each column, and each box must contain the digits 1 through 9. For reference, number the boxes 1 through 9, and label the 9 rows and 9 columns as in Example 8.

Consider concentrating first on boxes 1, 2, and 3. Starting with the digit 5, notice that boxes 2 and 3 each contain the digit 5, whereas box 1 does not. Consequently, the 5 in box 1 should be placed in row 2 because rows 1 and 3 already have a 5. Now notice the 5 cannot be placed in column 1 since it already has a 5. The 5 in box 1 must be placed in square (2, 2). Once again concentrating on boxes 1, 2, and 3, and using the same method, notice the digit 6 must be placed in square (2, 7). Concentrating still on boxes 1, 2, and 3, and using the same method, notice the digit 7 must be placed in square (3, 9). Use logic and the three consecutive box strategy to complete the grid.

7	6	8	5	4	2	9	1	3
1	5	2	7	3	9	6	4	8
9	3	4	6	1	8	2	5	7
8	7	1	9	6	3	5	2	4
5	2	6	1	7	4	8	3	9
4	9	3	8	2	5	7	6	1
6	8	9	4	5	1	3	7	2
3	4	7	2	9	6	1	8	5
2	1	5	3	8	7	4	9	6

1.2 Symbolic Logic

1.   a. Statement. It must be either true or false.   b. Statement. It must be either true or false.

    c. Not a statement. (question)   d. Not a statement. (opinion)

3.   a and d are negations; b and c are negations

5.   a. Her dress is red.

    b. No computer is priced under $100. (All computers are $100 or more.)

    c. Some dogs are not four-legged animals.   d. Some sleeping bags are waterproof.

7.   a. $p \wedge q$ b. $\sim p \to \sim q$ c. $\sim (p \vee q)$ d. $p \wedge \sim q$

    e. $p \to q$ f. $\sim q \to \sim p$

9.   a. $(p \vee q) \to r$ b. $p \wedge q \wedge r$ c. $r \wedge \sim (p \vee q)$ d. $q \to r$

    e. $\sim r \to \sim (p \vee q)$ f. $r \to (q \vee p)$

11. p: A shape is a square.

    q: A shape is a rectangle.

    $p \to q$

13. p: A shape is a square.

    q: A shape is a triangle.

    $p \to \sim q$

15. p: A number is a whole number.

    q: A number is even.

    r: A number is odd.

    $p \to (q \vee r)$

17. p: A number is a whole number.

    q: A number is greater than 3.

    r: A number is less than 4.

    $p \to \sim (q \wedge r)$

19. p: A person is an orthodontist.

    q: A person is a dentist.

    $p \to q$

21. p: A person knows Morse code.

    q: A person operates a telegraph.

    $q \to p$

23. p: The animal is a monkey.

    q: The animal is an ape.

    $p \to \sim q$

25. p: The animal is a monkey.

    q: The animal is an ape.

    $q \to \sim p$

27. p: I sleep soundly.

    q: I drink coffee.

    r: I eat chocolate.

    $(q \vee r) \to \sim p$

29. p: Your check is accepted.

    q: You have a driver's license.

    r: You have a credit card.

    $\sim (q \vee r) \to \sim p$

31. p: You do drink.

    q: You do drive.

    r: You are fined.

    s: You go to jail.

    $(p \wedge q) \to (r \vee s)$

33. p: You get a refund.

    q: You get a store credit.

    r: The product is defective.

    $r \to (p \vee q)$

35. a. I am an environmentalist and I recycle my aluminum cans.

    b. If I am an environmentalist, then I recycle my aluminum cans.

    c. If I do not recycle my aluminum cans, then I am not an environmentalist.

    d. I recycle my aluminum cans or I am not an environmentalist.

37. a. If I recycle my aluminum cans or newspapers, then I am an environmentalist.

    b. If I am not an environmentalist, then I do not recycle my aluminum cans or newspapers.

    c. I recycle my aluminum cans and newspapers or I am not an environmentalist.

    d. If I recycle my newspapers and do not recycle my aluminum cans, then I am not an

6

environmentalist.

39. Statement #1: Cold weather is required in order to have snow.

41. Statement #2: A month with 30 days would also indicate that the month is not February.

55. a. 57. b.

1.3 Truth Tables

1. $p \lor \sim q$

p	q	$\sim q$	$p \lor \sim q$
T	T	F	T
T	F	T	T
F	T	F	F
F	F	T	T

3. $p \lor \sim p$

p	$\sim p$	$p \lor \sim p$
T	F	T
F	T	T

5. $p \rightarrow \sim q$

p	q	$\sim q$	$p \rightarrow \sim q$
T	T	F	F
T	F	T	T
F	T	F	T
F	F	T	T

7. $\sim q \rightarrow \sim p$

p	q	$\sim q$	$\sim p$	$\sim q \rightarrow \sim p$
T	T	F	F	T
T	F	T	F	F
F	T	F	T	T
F	F	T	T	T

9. $(p \lor q) \rightarrow \sim p$

p	q	$p \lor q$	$\sim p$	$(p \lor q) \rightarrow \sim p$
T	T	T	F	F
T	F	T	F	F
F	T	T	T	T
F	F	F	T	T

11. $(p \lor q) \rightarrow (p \land q)$

p	q	$p \lor q$	$p \land q$	$(p \lor q) \rightarrow (p \land q)$
T	T	T	T	T
T	F	T	F	F
F	T	T	F	F
F	F	F	F	T

13. $p \land \sim (q \lor r)$

p	q	r	$q \lor r$	$\sim (q \lor r)$	$p \land \sim (q \lor r)$
T	T	T	T	F	F
T	T	F	T	F	F
T	F	T	T	F	F
T	F	F	F	T	T
F	T	T	T	F	F
F	T	F	T	F	F
F	F	T	T	F	F
F	F	F	F	T	F

15. $p \lor (\sim q \land r)$

p	q	r	$\sim q$	$\sim q \land r$	$p \lor (\sim q \land r)$
T	T	T	F	F	T
T	T	F	F	F	T
T	F	T	T	T	T
T	F	F	T	F	T
F	T	T	F	F	F
F	T	F	F	F	F
F	F	T	T	T	T
F	F	F	T	F	F

7

17. $(\sim r \vee p) \rightarrow (q \wedge p)$

p	q	r	$\sim r$	$\sim r \vee p$	$q \wedge p$	$(\sim r \vee p) \rightarrow (q \wedge p)$
T	T	T	F	T	T	T
T	T	F	T	T	T	T
T	F	T	F	T	F	F
T	F	F	T	T	F	F
F	T	T	F	F	F	T
F	T	F	T	T	F	F
F	F	T	F	F	F	T
F	F	F	T	T	F	F

19. $(p \vee r) \rightarrow (q \wedge \sim r)$

p	q	r	$\sim r$	$p \vee r$	$q \wedge \sim r$	$(p \vee r) \rightarrow (q \wedge \sim r)$
T	T	T	F	T	F	F
T	T	F	T	T	T	T
T	F	T	F	T	F	F
T	F	F	T	T	F	F
F	T	T	F	T	F	F
F	T	F	T	F	T	T
F	F	T	F	T	F	F
F	F	F	T	F	F	T

21. p: It is raining.

   q: The streets are wet.

   $p \rightarrow q$

p	q	$p \rightarrow q$
T	T	T
T	F	F
F	T	T
F	F	T

23. p: It rains.

   q: The water supply is rationed.

   $\sim p \rightarrow q$

p	q	$\sim p$	$\sim p \rightarrow q$
T	T	F	T
T	F	F	T
F	T	T	T
F	F	T	F

25. p: A shape is a square.

   q: A shape is a rectangle.

   $p \rightarrow q$

p	q	$p \rightarrow q$
T	T	T
T	F	F
F	T	T
F	F	T

27. p: It is a square.

   q: It is a triangle.

   $p \rightarrow \sim q$

p	q	$\sim q$	$p \rightarrow \sim q$
T	T	F	F
T	F	T	T
F	T	F	T
F	F	T	T

29. p: The animal is a monkey.

     q: The animal is an ape.

     $p \to \sim q$

p	q	$\sim q$	$p \to \sim q$
T	T	F	F
T	F	T	T
F	T	F	T
F	F	T	T

31. p: The animal is a monkey.

     q: The animal is an ape.

     $q \to \sim p$

p	q	$\sim p$	$q \to \sim p$
T	T	F	F
T	F	F	T
F	T	T	T
F	F	T	T

33. p: You have a driver's license.

     q: You have a credit card.

     r: Your check is approved.

     $(p \vee q) \to r$

p	q	r	$p \vee q$	$(p \vee q) \to r$
T	T	T	T	T
T	T	F	T	F
T	F	T	T	T
T	F	F	T	F
F	T	T	T	T
F	T	F	T	F
F	F	T	F	T
F	F	F	F	T

35. p: Leaded gas is used.

     q: The catalytic converter is damaged.

     r: The air is polluted.

     $p \to (q \wedge r)$

p	q	r	$q \wedge r$	$p \to (q \wedge r)$
T	T	T	T	T
T	T	F	F	F
T	F	T	F	F
T	F	F	F	F
F	T	T	T	T
F	T	F	F	T
F	F	T	F	T
F	F	F	F	T

37. p: I have a college degree.

     q: I have a job.

     r: I own a house.

     $p \wedge \sim (q \vee r)$

p	q	r	$q \vee r$	$\sim (q \vee r)$	$p \wedge \sim (q \vee r)$
T	T	T	T	F	F
T	T	F	T	F	F
T	F	T	T	F	F
T	F	F	F	T	T
F	T	T	T	F	F
F	T	F	T	F	F
F	F	T	T	F	F
F	F	F	F	T	F

39.   *p*: Proposition A passes.

        q: Proposition B passes.

        r: Jobs are lost.

        s: New taxes are imposed.

$(p \wedge \sim q) \rightarrow (r \vee s)$

p	*q*	*r*	*s*	$\sim q$	$p \wedge \sim q$	$r \vee s$	$(p \wedge \sim q) \rightarrow (r \vee s)$
T	T	T	T	F	F	T	T
T	T	T	F	F	F	T	T
T	T	F	T	F	F	T	T
T	T	F	F	F	F	F	T
T	F	T	T	T	T	T	T
T	F	T	F	T	T	T	T
T	F	F	T	T	T	T	T
T	F	F	F	T	T	F	F
F	T	T	T	F	F	T	T
F	T	T	F	F	F	T	T
F	T	F	T	F	F	T	T
F	T	F	F	F	F	F	T
F	F	T	T	T	F	T	T
F	F	T	F	T	F	T	T
F	F	F	T	T	F	T	T
F	F	F	F	T	F	F	T

41.   *p*: The streets are wet.

        q: It is raining.

p	*q*	$\sim q$	$p \vee \sim q$	$q \rightarrow p$
T	T	F	T	T
T	F	T	T	T
F	T	F	F	F
F	F	T	T	T

The statements are equivalent.

45.   *p*: Handguns are outlawed.

        q: Outlaws have handguns.

p	*q*	$p \rightarrow q$	$q \rightarrow p$
T	T	T	T
T	F	F	T
F	T	T	F
F	F	T	T

The statements are not equivalent.

43.   *p*: He has a high school diploma.

        q: He is unemployed.

p	*q*	$\sim p$	$p \vee q$	$\sim p \rightarrow q$
T	T	F	T	T
T	F	F	T	T
F	T	T	T	T
F	F	T	F	F

The statements are equivalent.

47.   *p*: The spotted owl is on the endangered species list.

   q: Lumber jobs are lost.

p	*q*	~*p*	~*q*	*p* → *q*	~*q* → ~*p*
T	T	F	F	T	T
T	F	F	T	F	F
F	T	T	F	T	T
F	F	T	T	T	T

The statements are equivalent.

49.   *p*: The plaintiff is innocent.

   q: The insurance company settles out of court.

p	*q*	~*p*	~*q*	*p* ∨ ~*q*	*q* ∧ ~*p*
T	T	F	F	T	F
T	F	F	T	T	F
F	T	T	F	F	T
F	F	T	T	T	F

The statements are not equivalent.

51.   *p*: The person knows Morse code.   *q*: A person can operate a telegraph.

   i.   ii.   iii.

p	*q*	*p* → *q*
T	T	T
T	F	F
F	T	T
F	F	T

p	*q*	*q* → *p*
T	T	T
T	F	T
F	T	F
F	F	T

p	*q*	~*p*	~*q*	~*p* → ~*q*
T	T	F	F	T
T	F	F	T	T
F	T	T	F	F
F	F	T	T	T

   iv.

p	*q*	~*p*	~*q*	~*q* → ~*p*
T	T	F	F	T
T	F	F	T	F
F	T	T	F	T
F	F	T	T	T

i. and iv. are equivalent.

ii. and iii. are equivalent.

53.   *p*: The water is cold.   *q*: A person going swimming.

   i.   ii.

p	*q*	~*q*	~*q* → *p*
T	T	F	T
T	F	T	T
F	T	F	T
F	F	T	F

p	*q*	~*p*	*q* → ~*p*
T	T	F	F
T	F	F	T
F	T	T	T
F	F	T	T

iii.

p	q	$\sim q$	$p \rightarrow \sim q$
T	T	F	F
T	F	T	T
F	T	F	T
F	F	T	T

iv.

p	q	$\sim p$	$\sim p \rightarrow q$
T	T	F	T
T	F	F	T
F	T	T	T
F	F	T	F

i. and iv. are equivalent.   ii. and iii. are equivalent.

55. $\sim(p \wedge q)$ $\sim p \vee \sim q$

p	q	$p \wedge q$	$\sim(p \wedge q)$
T	T	T	F
T	F	F	T
F	T	F	T
F	F	F	T

p	q	$\sim p$	$\sim q$	$\sim p \vee \sim q$
T	T	F	F	F
T	F	F	T	T
F	T	T	F	T
F	F	T	T	T

They are equivalent.

57. p: I have a college degree. q: I am employed.

$p \wedge \sim q$ Negation: $\sim(p \wedge \sim q) \equiv \sim p \vee \sim(\sim q) \equiv \sim p \vee q$

I do not have a college degree or I am employed.

59. p: The television set is broken. q: There is a power outage.

$p \vee q$ Negation: $\sim(p \vee q) \equiv \sim p \wedge \sim q$

The television set is not broken and there is not a power outage.

61. p: The building contains asbestos. q: The original contractor is responsible.

$p \rightarrow q$ Negation: $\sim(p \rightarrow q) \equiv p \wedge \sim q$

The building contains asbestos and the original contractor is not responsible.

63. p: The lyrics are censored. q: The First Amendment has been violated.

$p \rightarrow q$ Negation: $\sim(p \rightarrow q) \equiv p \wedge \sim q$

The lyrics are censored and the First Amendment has not been violated.

65. p: It is rainy weather. q: I am washing my car.

$p \rightarrow \sim q$ Negation: $\sim(p \rightarrow \sim q) \equiv p \wedge \sim(\sim q) \equiv p \wedge q$

It is rainy weather and I am washing my car.

67. p: The person is talking. q: The person is listening.

$q \rightarrow \sim p$ Negation: $\sim (q \rightarrow \sim p) \equiv q \wedge \sim (\sim p) \equiv q \wedge p$

The person is listening and talking.

1.4 More on Conditionals

1. a. If she is a police officer, then she carries a gun.

b. If she carries a gun, then she is a police officer.

c. If she is not a police officer, then she does not carry a gun.

d. If she does not carry a gun, then she is not a police officer.

e. Parts (a) and (d) are equivalent; parts (b) and (c) are equivalent. The contrapositive statement is always equivalent to the original.

3. a. If I watch television, then I do not do my homework.

b. If I do not do my homework, then I watch television.

c. If I do not watch television, then I do my homework.

d. If I do my homework, then I do not watch television.

e. Parts (a) and (d) are equivalent; parts (b) and (c) are equivalent. The contrapositive statement is always equivalent to the original.

5. a. If you do not pass this mathematics course, then you do not fulfill a graduation requirement.

b. If you fulfill a graduation requirement, then you pass this mathematics course.

c. If you do not fulfill a graduation requirement, then you do not pass this mathematics course.

7. a. If the electricity is turned on, then the television set does work.

b. If the television set does not work, then the electricity is turned off.

c. If the television set does work, then the electricity is turned on.

9. a. If you eat meat, then you are not a vegetarian.

b. If you are a vegetarian, then you do not eat meat.

c. If you are not a vegetarian, then you do eat meat.

11. a. The person not being a dentist is sufficient to not being an orthodontist.

b. Not being an orthodontist is necessary for not being a dentist.

13. a. Not knowing Morse code is sufficient to not operating a telegraph.

b. Not being able to operate a telegraph is necessary to not knowing Morse code.

15.   a. *Premise*: I take public transportation. *Conclusion*: Public transportation is convenient.

    b. If I take public transportation, then it is convenient.

    c. False statement when I take public transportation and it is not convenient.

17.   a. *Premise*: I buy foreign products. *Conclusion*: Domestic products are not available.

    b. If I buy foreign products, then domestic products are not available.

    c. False statement when I buy foreign products and domestic products are available.

19.   a. *Premise*: You may become a U. S. senator. *Conclusion*: You are at least 30 years old and have been a citizen for nine years.

    b. If you become a U. S. senator, then you are at least 30 years old and have been a citizen for nine years.

    c. False statement when you become a U. S. senator and you are not at least 30 years old or have not been a citizen for nine years, or both.

21.   If you obtain a refund, then you have a receipt and if you have a receipt, then you obtain a refund.

23.   If the quadratic equation $ax^2 + bx + c = 0$ has two distinct real solutions, then $b^2 - 4ac > 0$ and if $b^2 - 4ac > 0$, then the quadratic equation $ax^2 + bx + c = 0$ has two distinct real solutions.

25.   If a polygon is a triangle, then the polygon has three sides and if the polygon has three sides, then the polygon is a triangle.

27.   If $a^2 + b^2 = c^2$, then the triangle has a $90°$ angle, and if the triangle has a $90°$ angle, then $a^2 + b^2 = c^2$.

29. p: I can have surgery.

    q: I have health insurance.

    $\sim q \rightarrow \sim p$ and $p \rightarrow q$

p	q	$\sim p$	$\sim q$	$\sim q \rightarrow \sim p$	$p \rightarrow q$
T	T	F	F	T	T
T	F	F	T	F	F
F	T	T	F	T	T
F	F	T	T	T	T

The statements are equivalent.

31.	*p*: You earn less than $12,000 per year.

q: You are eligible for assistance.

$p \rightarrow q$ and $\sim q \rightarrow \sim p$

p	*q*	$\sim p$	$\sim q$	$p \rightarrow q$	$\sim q \rightarrow \sim p$
T	T	F	F	T	T
T	F	F	T	F	F
F	T	T	F	T	T
F	F	T	T	T	T

The statements are equivalent.

33.	*p*: I watch television.

q: The program is educational.

$p \rightarrow q$ and $\sim q \rightarrow \sim p$

p	*q*	$\sim p$	$\sim q$	$p \rightarrow q$	$\sim q \rightarrow \sim p$
T	T	F	F	T	T
T	F	F	T	F	F
F	T	T	F	T	T
F	F	T	T	T	T

The statements are equivalent.

35.	*p*: The automobile is American-made.

q: The automobile hardware is metric.

$p \rightarrow \sim q$ and $q \rightarrow \sim p$

p	*q*	$\sim p$	$\sim q$	$p \rightarrow \sim q$	$q \rightarrow \sim p$
T	T	F	F	F	F
T	F	F	T	T	T
F	T	T	F	T	T
F	F	T	T	T	T

The statements are equivalent.

37.	If I do not walk to work, then it is raining.

39.	If it is not cold, then it is not snowing.

41.	If you are a vegetarian, then you do not eat meat.

43.	If the person does not own guns, then the person is not a policeman.

45.   If the person is eligible to vote, then the person is not a convicted felon.

47. p: Proposition 111 passes.

       q: Freeways are improved.

       i. $p \rightarrow q$ ii. $\sim p \rightarrow \sim q$ iii. $q \rightarrow p$ iv. $\sim q \rightarrow \sim p$

       ii and iii are equivalent and i and iv are equivalent.

49. p: It is Sunday.

       q: I go to church.

       i. $p \rightarrow q$ ii. $q \rightarrow p$ iii. $\sim q \rightarrow \sim p$ iv. $\sim p \rightarrow \sim q$

       i and iii are equivalent and ii and iv are equivalent.

51. p: Line 34 is greater than Line 29.

       q: I use schedule X.

       i. $p \rightarrow q$ ii. $q \rightarrow p$ iii. $\sim q \rightarrow \sim p$ iv. $\sim p \rightarrow \sim q$

       i and iii are equivalent and ii and iv are equivalent.

59.   b.   61.   c.

1.5 Analyzing Arguments

1. $p \rightarrow q$ 3. $p \rightarrow q$ 5. $p \rightarrow q$

      $\dfrac{p}{q}$ $\dfrac{\sim q}{\sim p}$ $\dfrac{\sim p}{\sim q}$

7. $q \rightarrow p$ 9. $p \rightarrow \sim q$

      $\dfrac{r \wedge p}{r \wedge q}$ $\dfrac{r \wedge q}{r \wedge \sim p}$

11. $\left[(p \rightarrow q) \wedge p \right] \rightarrow q$ The argument is valid.

p	q	$p \rightarrow q$	$(p \rightarrow q) \wedge p$	$\left[(p \rightarrow q) \wedge p \right] \rightarrow q$
T	T	T	T	T
T	F	F	F	T
F	T	T	F	T
F	F	T	F	T

13. $[(p \to q) \wedge \sim q] \to \sim p$ The argument is valid.

p	q	$p \to q$	$\sim q$	$(p \to q) \wedge \sim q$	$\sim p$	$[(p \to q) \wedge \sim q] \to \sim p$
T	T	T	F	F	F	T
T	F	F	T	F	F	T
F	T	T	F	F	T	T
F	F	T	T	T	T	T

15. $[(p \to q) \wedge \sim p] \to \sim q$ The argument is invalid if you don't exercise regularly and you are healthy.

p	q	$\sim q$	$p \to q$	$\sim p$	$(p \to q) \wedge \sim p$	$[(p \to q) \wedge \sim p] \to \sim q$
T	T	F	T	F	F	T
T	F	T	F	F	F	T
F	T	F	T	T	T	F
F	F	T	T	T	T	T

17. $[(q \to p) \wedge (r \wedge p)] \to (r \wedge q)$ The argument is invalid if the person knows Morse code, doesn't operate a telegraph, and is Nikola Tesla.

p	q	r	$q \to p$	$r \wedge p$	$(q \to p) \wedge (r \wedge p)$	$r \wedge q$	$[(q \to p) \wedge (r \wedge p)] \to (r \wedge q)$
T	T	T	T	T	T	T	T
T	T	F	T	F	F	F	T
T	F	T	T	T	T	F	F
T	F	F	T	F	F	F	T
F	T	T	F	F	F	T	T
F	T	F	F	F	F	F	T
F	F	T	T	F	F	F	T
F	F	F	T	F	F	F	T

19. $[(p \to \sim q) \wedge (r \wedge q)] \to (r \wedge \sim p)$ The argument is valid.

p	q	r	$\sim p$	$\sim q$	$p \to \sim q$	$r \wedge q$	$(p \to \sim q) \wedge (r \wedge q)$	$r \wedge \sim p$	$[(p \to \sim q) \wedge (r \wedge q)] \to (r \wedge \sim p)$
T	T	T	F	F	F	T	F	F	T
T	T	F	F	F	F	F	F	F	T
T	F	T	F	T	T	F	F	F	T
T	F	F	F	T	T	F	F	F	T
F	T	T	T	F	T	T	T	T	T
F	T	F	T	F	T	F	F	F	T
F	F	T	T	T	T	F	F	T	T
F	F	F	T	T	T	F	F	F	T

21.　　p: The Democrats have a majority.

　　　　q: Smith is appointed.

　　　　r: Student loans are funded.

$p \rightarrow (q \wedge r)$　　　　　　　　　　$\{[p \rightarrow (q \wedge r)] \wedge [q \vee \sim r]\} \rightarrow \sim p$

$q \vee \sim r$　　　　　　The argument is invalid when the Democrats have a majority and

$\sim p$　　　　　　　　Smith is appointed and student loans are funded.

p	q	r	$q \wedge r$	$\sim r$	$p \rightarrow (q \wedge r)$ 1	$q \vee \sim r$ 2	$1 \wedge 2$	$\sim p$	$(1 \wedge 2) \rightarrow \sim p$
T	T	T	T	F	T	T	T	F	F
T	T	F	F	T	F	T	F	F	T
T	F	T	F	F	F	F	F	F	T
T	F	F	F	T	F	T	F	F	T
F	T	T	T	F	T	T	T	T	T
F	T	F	F	T	T	T	T	T	T
F	F	T	F	F	T	F	F	T	T
F	F	F	F	T	T	T	T	T	T

23.　　p: You argue with a police officer.

　　　　q: You get a ticket.

　　　　r: You break the speed limit.

$p \rightarrow q$　　　　　　　　　$\{(p \rightarrow q) \wedge (\sim r \not\rightarrow \sim q)\} \rightarrow (r \rightarrow p)$

$\sim r \not\rightarrow q$

$r \rightarrow p$

The argument is invalid when (1) you do not argue with a police officer, you do get a ticket and you do break the speed limit, *or* (2) you do not argue with a police officer, you do not get a ticket and you do break the speed limit.

p	q	r	~r	~q	1 p→q	2 ~r→~q	1∧2	3 r→p	(1∧2)→3
T	T	T	F	F	T	T	T	T	T
T	T	F	T	F	T	F	F	T	T
T	F	T	F	T	F	T	F	T	T
T	F	F	T	T	F	T	F	T	T
F	T	T	F	F	T	T	T	F	F
F	T	F	T	F	T	F	F	T	T
F	F	T	F	T	T	T	T	F	F
F	F	F	T	T	T	T	T	T	T

25. Rewriting the argument:

If it is a pesticide, then it is harmful to the environment. p: It is a pesticide.

If it is a fertilizer, then it is not a pesticide. q: It is harmful to the environment.

If it is a fertilizer, then it is not harmful to the environment. r: It is a fertilizer.

$p \to q$

$r \not\to p$

$r \not\to\, \sim q$

$$\{(p \to q) \wedge (r \to\, \sim p)\} \to (r \to\, \sim q)$$

The argument is invalid if it is not a pesticide and if it is

harmful to the environment and it is a fertilizer.

p	q	r	~p	~q	1 p→q	2 r→~p	1∧2	3 r→~q	(1∧2)→3
T	T	T	F	F	T	F	F	F	T
T	T	F	F	F	T	T	T	T	T
T	F	T	F	T	F	F	F	T	T
T	F	F	F	T	F	T	F	T	T
F	T	T	T	F	T	T	T	F	F
F	T	F	T	F	T	T	T	T	T
F	F	T	T	T	T	T	T	T	T
F	F	F	T	T	T	T	T	T	T

27.   Rewriting the argument:

If you are a poet, then you are a loner.	p: You are a poet.
<u>If you are a loner, then you are a taxi driver.</u>	q: You are a loner.
If you are a poet, then you are a taxi driver.	r: You are a taxi driver.

$p \rightarrow q$

$\qquad \{(p \rightarrow q) \wedge (q \rightarrow r)\} \rightarrow (p \rightarrow r)$

$\dfrac{q \rightarrow r}{p \rightarrow r}$ 　　　The argument is valid.

			1	2		3	
p	q	r	$p \rightarrow q$	$q \rightarrow r$	$1 \wedge 2$	$p \rightarrow r$	$(1 \wedge 2) \rightarrow 3$
T	T	T	T	T	T	T	T
T	T	F	T	F	F	F	T
T	F	T	F	T	F	T	T
T	F	F	F	T	F	F	T
F	T	T	T	T	T	T	T
F	T	F	T	F	F	T	T
F	F	T	T	T	T	T	T
F	F	F	T	T	T	T	T

29.   Rewriting the argument:

If you are a professor, then you are not a millionaire.	p: You are a professor.
<u>If you are a millionaire, then you are literate.</u>	q: You are a millionaire.
If you are a professor, then you are literate.	r: You are illiterate.

$p \rightarrow \sim q$

$\qquad \{(p \rightarrow q) \wedge (q \rightarrow \sim r)\} \rightarrow (p \rightarrow \sim r)$

$\dfrac{q \rightarrow \sim r}{p \rightarrow \sim r}$ 　　The argument is invalid if you are a professor who is not a millionaire and is illiterate.

				1		2		3	
p	q	r	$\sim q$	$p \rightarrow \sim q$	$\sim r$	$q \rightarrow \sim r$	$1 \wedge 2$	$p \rightarrow \sim r$	$(1 \wedge 2) \rightarrow 3$
T	T	T	F	F	F	F	F	F	T
T	T	F	F	F	T	T	F	T	T
T	F	T	T	T	F	T	T	F	F
T	F	F	T	T	T	T	T	T	T
F	T	T	F	T	F	F	F	T	T
F	T	F	F	T	T	T	T	T	T
F	F	T	T	T	F	T	T	T	T
F	F	F	T	T	T	T	T	T	T

31.   Rewriting the argument:

If you are a lawyer, then you study logic. p: You are a lawyer.

If you study logic, then you are a scholar. q: You study logic

<u>You are not a scholar.</u> r: You are a scholar.

You are not a lawyer.

$p \rightarrow q$ $\{(p \rightarrow q) \wedge (q \rightarrow r) \wedge \sim r\} \rightarrow \sim p$

$q \rightarrow r$ The argument is valid.

$\underline{\sim r}$

$\sim p$

			1	2	3			
p	q	r	$p \rightarrow q$	$q \rightarrow r$	$\sim r$	$1 \wedge 2 \wedge 3$	$\sim p$	$(1 \wedge 2 \wedge 3) \rightarrow \sim p$
T	T	T	T	T	F	F	F	T
T	T	F	T	F	T	F	F	T
T	F	T	F	T	F	F	F	T
T	F	F	F	T	T	F	F	T
F	T	T	T	T	F	F	T	T
F	T	F	T	F	T	F	T	T
F	F	T	T	T	F	F	T	T
F	F	F	T	T	T	T	T	T

33.   Rewriting the argument:

If you are drinking espresso, then you are not sleeping. p: You are drinking espresso.

If you are on a diet, then you are not eating dessert. q: You are sleeping.

<u>If you are not eating dessert then you are drinking.</u> r: You are eating dessert.

If you are sleeping, then you are not on a diet. t: You are on a diet.

$p \rightarrow \sim q$ $\{(p \rightarrow \sim q) \wedge (t \rightarrow \sim r) \wedge (\sim r \rightarrow p)\} \rightarrow (q \rightarrow \sim t)$

$t \rightarrow \sim r$ The argument is valid.

$\underline{\sim r \rightarrow p}$

$q \rightarrow \sim t$

p	q	r	t	$\sim q$	$\sim r$	$\sim t$	1 $p\to\sim q$	2 $t\to\sim r$	3 $\sim r\to p$	$1\wedge2\wedge3$	4 $q\to\sim t$	$(1\wedge2\wedge3)\to4$
T	T	T	T	F	F	F	F	F	T	F	F	T
T	T	T	F	F	F	T	F	T	T	F	T	T
T	T	F	T	F	T	F	F	T	T	F	F	T
T	T	F	F	F	T	T	F	T	T	F	T	T
T	F	T	T	T	F	F	T	F	T	F	T	T
T	F	T	F	T	F	T	T	T	T	T	T	T
T	F	F	T	T	T	F	T	T	T	T	T	T
T	F	F	F	T	T	T	T	T	T	T	T	T
F	T	T	T	F	F	F	T	F	T	F	F	T
F	T	T	F	F	F	T	T	T	T	T	T	T
F	T	F	T	F	T	F	T	T	F	F	F	T
F	T	F	F	F	T	T	T	T	F	F	T	T
F	F	T	T	T	F	F	T	F	T	F	T	T
F	F	T	F	T	F	T	T	T	T	T	T	T
F	F	F	T	T	T	F	T	T	F	F	T	T
F	F	F	F	T	T	T	T	T	F	F	T	T

35. p: The defendant is innocent.

   q: The defendant goes to jail.

$$p\to\sim q \qquad\qquad \{(p\to q)\wedge q\}\to p$$

$$\underline{q\qquad\qquad} \qquad\qquad \text{The argument is valid.}$$

$$\sim p$$

p	q	$\sim q$	$p\to\sim q$	$(p\to q)\wedge q$	$\sim p$	$\{(p\to q)\wedge q\}\to p$
T	T	F	F	F	F	T
T	F	T	T	F	F	T
F	T	F	T	T	T	T
F	F	T	T	F	T	T

37. p: You are in a hurry.

   q: You eat at Lulu's Diner.

   r: You eat good food.

$\sim p \to q$

$p \to \sim r$

$\underline{q\qquad\qquad}$

r

$\{(\sim p \to q) \land (p \to \sim r) \land q\} \to r$

The argument is invalid when (1) you are in a hurry and you eat at Lulu's Diner and you do not eat good food, *or* (2) you are not in a hurry and you eat at Lulu's Diner and you do not eat good food.

p	q	r	$\sim p$	$\sim r$	1 $\sim p \to q$	2 $p \to \sim r$	$1 \land 2 \land q$	$(1 \land 2 \land q) \to r$
T	T	T	F	F	T	F	F	T
T	T	F	F	T	T	T	T	F
T	F	T	F	F	T	F	F	T
T	F	F	F	T	T	T	F	T
F	T	T	T	F	T	T	T	T
F	T	F	T	T	T	T	T	F
F	F	T	T	F	F	T	F	T
F	F	F	T	T	F	T	F	T

39. p: You listen to rock and roll.

   q: You go to heaven.

   r: You are a moral person.

$p \to \sim q$

$\underline{r \to q\qquad\quad}$

$p \to \sim r$

$\{(p \to \sim q) \land (r \to q)\} \to (p \to \sim r)$

The argument is valid.

p	q	r	$\sim q$	1 $p \to \sim q$	2 $r \to q$	$1 \land 2$	$\sim r$	3 $p \to \sim r$	$(1 \land 2) \to 3$
T	T	T	F	F	T	F	F	F	T
T	T	F	F	F	T	F	T	T	T
T	F	T	T	T	F	F	F	F	T
T	F	F	T	T	T	T	T	T	T
F	T	T	F	T	T	T	F	T	T
F	T	F	F	T	T	T	T	T	T
F	F	T	T	T	F	F	F	T	T
F	F	F	T	T	T	T	T	T	T

23

41. p: The water is cold.

q: I go swimming.

r: I have goggles.

$\sim p \to q$

$q \to r$

$\underline{\sim r}$

p

$\{(\sim p \to q) \land (q \to r) \land \sim r\} \to p$

The argument is valid.

p	q	r	$\sim p$	$\sim r$	1 $\sim p \to q$	2 $q \to r$	$1 \land 2 \land \sim r$	$(1 \land 2 \land \sim r) \to p$
T	T	T	F	F	T	T	F	T
T	T	F	F	T	T	F	F	T
T	F	T	F	F	T	T	F	T
T	F	F	F	T	T	T	T	T
F	T	T	T	F	T	T	F	T
F	T	F	T	T	T	F	F	T
F	F	T	T	F	F	T	F	T
F	F	F	T	T	F	T	F	T

43. p: It is medicine.

q: It is nasty.

$\{(p \to q) \land p\} \to q$

$p \to q$

$\underline{p}$

q

The argument is valid.

p	q	$p \to q$	$(p \to q) \land p$	$\{(p \to q) \land p\} \to q$
T	T	T	T	T
T	F	F	F	T
F	T	T	F	T
F	F	T	F	T

45. p: It is intelligible.

   q: It puzzles me.

   r: It is logic.

$p \rightarrow \sim q$

$\dfrac{r \rightarrow q}{r \rightarrow \sim p}$

$\{(p \rightarrow \sim q) \wedge (r \rightarrow q)\} \rightarrow (r \rightarrow \sim p)$

The argument is valid.

p	q	r	$\sim q$	**1** $p \rightarrow \sim q$	**2** $r \rightarrow q$	$1 \wedge 2$	$\sim p$	**3** $r \rightarrow \sim p$	$(1 \wedge 2) \rightarrow 3$
T	T	T	F	F	T	F	F	F	T
T	T	F	F	F	T	F	F	T	T
T	F	T	T	T	F	F	F	F	T
T	F	F	T	T	T	T	F	T	T
F	T	T	F	T	T	T	T	T	T
F	T	F	F	T	T	T	T	T	T
F	F	T	T	T	F	F	T	T	T
F	F	F	T	T	T	T	T	T	T

47. p: A person is a Frenchman.

   q: A person likes plum pudding.

   r: A person is an Englishman.

$p \rightarrow \sim q$

$\dfrac{r \rightarrow q}{r \rightarrow \sim p}$

$\{(p \rightarrow \sim q) \wedge (r \rightarrow q)\} \rightarrow (r \rightarrow \sim p)$

The argument is valid.

p	q	r	$\sim q$	**1** $p \rightarrow \sim q$	**2** $r \rightarrow q$	$1 \wedge 2$	$\sim p$	**3** $r \rightarrow \sim p$	$(1 \wedge 2) \rightarrow 3$
T	T	T	F	F	T	F	F	F	T
T	T	F	F	F	T	F	F	T	T
T	F	T	T	T	F	F	F	F	T
T	F	F	T	T	T	T	F	T	T
F	T	T	F	T	T	T	T	T	T
F	T	F	F	T	T	T	T	T	T
F	F	T	T	T	F	F	T	T	T
F	F	F	T	T	T	T	T	T	T

49. p: An animal is a wasp.

    q: An animal is friendly.

    r: An animal is a puppy.

$p \to \sim q$

$r \to q$

$\overline{}$

$r \to \sim p$

$\{(p \to \sim q) \wedge (r \to q)\} \to (r \to \sim p)$

The argument is valid.

p	q	r	$\sim q$	1 $p \to \sim q$	2 $r \to q$	$1 \wedge 2$	$\sim p$	3 $r \to \sim p$	$(1 \wedge 2) \to 3$
T	T	T	F	F	T	F	F	F	T
T	T	F	F	F	T	F	F	T	T
T	F	T	T	T	F	F	F	F	T
T	F	F	T	T	T	T	F	T	T
F	T	T	F	T	T	T	T	T	T
F	T	F	F	T	T	T	T	T	T
F	F	T	T	T	F	F	T	T	T
F	F	F	T	T	T	T	T	T	T

Chapter 1 Review

1.   a. Inductive   b. Deductive

3.   9. Think of each number in the sequence as a time (1:00, 6:00, etc.). Add five hours to get the next number in the sequence.

5.

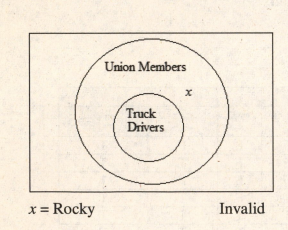

x = Rocky   Invalid

7.

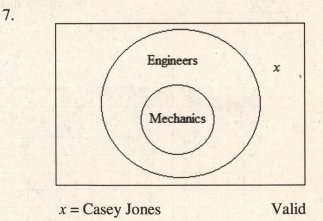

x = Casey Jones   Valid

9.

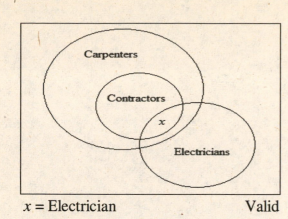

$x =$ Electrician   Valid

11.   a. Statement. It is either true or false.   b. Statement. It is either true or false.

c. Not a statement. It is a question.   d. Not a statement. It is an opinion.

13.   a. His car is new.   b. No building is earthquake proof.

c. Some children do not eat candy.   d. Sometimes I cry in a movie theater.

15.   a. $q \rightarrow p$ b. $\sim r \wedge \sim q \wedge p$ c. $r \rightarrow \sim p$ d. $p \wedge \sim (q \vee r)$

e. $p \rightarrow (\sim r \vee \sim q)$ f. $(r \wedge q) \rightarrow \sim p$

17.   a. If the movie is critically acclaimed or a box office hit, then the movie is available on

videotape.

b. If the movie is critically acclaimed and not a box office hit, then the movie is not available

on videotape.

c. The movie is not critically acclaimed or a box office hit and it is available on videotape.

d. If the movie is not available on video tape, then the movie is not critically acclaimed and it

is not a box office hit.

19. $p \wedge \sim q$ 21. $(p \wedge q) \rightarrow \sim q$

p	q	$\sim q$	$p \wedge \sim q$
T	T	F	F
T	F	T	T
F	T	F	F
F	F	T	F

p	q	$p \wedge q$	$\sim q$	$(p \wedge q) \rightarrow \sim q$
T	T	T	F	F
T	F	F	T	T
F	T	F	F	T
F	F	F	T	T

27

23. $\sim p \rightarrow (q \vee r)$

p	q	r	$\sim p$	$q \vee r$	$\sim p \rightarrow (q \vee r)$
T	T	T	F	T	T
T	T	F	F	T	T
T	F	T	F	T	T
T	F	F	F	F	T
F	T	T	T	T	T
F	T	F	T	T	T
F	F	T	T	T	T
F	F	F	T	F	F

25. $(p \vee r) \rightarrow (q \wedge \sim r)$

p	q	r	$\sim r$	$p \vee r$	$q \wedge \sim r$	$(p \vee r) \rightarrow (q \wedge \sim r)$
T	T	T	F	T	F	F
T	T	F	T	T	T	T
T	F	T	F	T	F	F
T	F	F	T	T	F	F
F	T	T	F	T	F	F
F	T	F	T	F	T	T
F	F	T	F	T	F	F
F	F	F	T	F	F	T

27. p: I get a raise.

    q: I buy a new car.

    $p \rightarrow q$ and $\sim p \rightarrow \sim q$

p	q	$p \rightarrow q$	$\sim p$	$\sim q$	$\sim p \rightarrow \sim q$
T	T	T	F	F	T
T	F	F	F	T	T
F	T	T	T	F	F
F	F	T	T	T	T

The statements are not equivalent.

29. *p*: The raise is unjustified.

    q: Management opposes the raise.

    $\sim p \wedge q$ and $\sim(p \vee \sim q)$

p	*q*	$\sim p$	$\sim p \wedge q$	$\sim q$	$p \vee \sim q$	$\sim(p \vee \sim q)$
T	T	F	F	F	T	F
T	F	F	F	T	T	F
F	T	T	T	F	F	T
F	F	T	F	T	T	F

    The statements are equivalent.

31. Jesse did not have a party or somebody came.

33. I am not the winner and you are not blind.

35. His application is ignored and the selection procedure has not been violated.

37. A person is drinking espresso and the person is sleeping.

39. a. If you are an avid jogger, then you are healthy.

    b. If you are healthy, then you are an avid jogger.

    c. If you are not an avid jogger, then you are not healthy.

    d. If you are not healthy, then you are not an avid jogger.

    e. You are an avid jogger if and only if you are healthy.

41. a. Being lost is sufficient for not having a map.

    b. Not having a map is necessary for being lost.

43. a. *Premise*: The economy improves. *Conclusion*: Unemployment goes down.

    b. If the economy improves, then unemployment goes down.

45. a. *Premise*: It is a computer. *Conclusion*: It is repairable.

    b. If it is a computer, then it is repairable.

47. a. *Premise*: The fourth Thursday in November. *Conclusion*: The U. S. Post Office is closed.

    b. If it is the fourth Thursday in November, then the U. S. Post Office is closed.

49. p: You are allergic to dairy products.

       q: You can eat cheese.

       $p \rightarrow \sim q$ and $\sim q \rightarrow p$

p	q	$\sim q$	$p \rightarrow \sim q$	$\sim q \rightarrow p$
T	T	F	F	T
T	F	T	T	T
F	T	F	T	T
F	F	T	T	F

       The statements are not equivalent.

51.   i and iii are equivalent; ii and iv are equivalent.

53. p: You pay attention.

       q: You learn the new method.

       $\sim p \rightarrow \sim q$ $\left[(\sim p \rightarrow \sim q) \wedge q\right] \rightarrow p$

       $\dfrac{q}{p}$ The argument is valid.

p	q	$\sim p$	$\sim q$	$\sim p \rightarrow \sim q$	$(\sim p \rightarrow \sim q) \wedge q$	$\left[(\sim p \rightarrow \sim q) \wedge q\right] \rightarrow p$
T	T	F	F	T	T	T
T	F	F	T	T	F	T
F	T	T	F	F	F	T
F	F	T	T	T	F	T

55. p: The Republicans have a majority.

       q: Farnsworth is appointed.

       r: No new taxes are imposed.

       $p \rightarrow (q \wedge r)$ $\left\{\left[p \rightarrow (q \wedge r)\right] \wedge \sim r\right\} \rightarrow (\sim p \vee \sim q)$

       $\dfrac{\sim r}{\sim p \vee \sim q}$ The argument is valid.

p	q	r	$q \wedge r$	1 $p \rightarrow (q \wedge r)$	2 $\sim r$	$1 \wedge 2$	$\sim p$	$\sim q$	3 $\sim p \vee \sim q$	$(1 \wedge 2) \rightarrow 3$
T	T	T	T	T	F	F	F	F	F	T
T	T	F	F	F	T	F	F	F	F	T
T	F	T	F	F	F	F	F	T	T	T
T	F	F	F	F	T	F	F	T	T	T
F	T	T	T	T	F	F	T	F	T	T
F	T	F	F	T	T	T	T	F	T	T
F	F	T	F	T	F	F	T	T	T	T
F	F	F	F	T	T	T	T	T	T	T

57. p: You are practicing.

q: You are making mistakes.

r: You receive an award.

$p \rightarrow \sim q$ $\{[p \rightarrow \sim q] \wedge [\sim r \rightarrow q] \wedge \sim r\} \rightarrow \sim p$

$\sim r \rightarrow q$ The argument is valid.

$\sim r$
_____

$\sim p$

p	q	r	$\sim p$	$\sim q$	$\sim r$	1 $p \rightarrow \sim q$	2 $\sim r \rightarrow q$	$1 \wedge 2 \wedge \sim r$	$(1 \wedge 2 \wedge \sim r) \rightarrow \sim p$
T	T	T	F	F	F	F	T	F	T
T	T	F	F	F	T	F	T	F	T
T	F	T	F	T	F	T	T	F	T
T	F	F	F	T	T	T	F	F	T
F	T	T	T	F	F	T	T	F	T
F	T	F	T	F	T	T	T	T	T
F	F	T	T	T	F	T	T	F	T
F	F	F	T	T	T	T	F	F	T

59. p: I will go to the concert.

q: You buy me a ticket.

$p \rightarrow q$ $[(p \rightarrow q) \wedge q] \rightarrow p$

q
_____   The argument is invalid.

p

p	q	$p \to q$	$(p \to q) \wedge q$	$[(p \to q) \wedge q] \to p$
T	T	T	T	T
T	F	F	F	T
F	T	T	T	F
F	F	T	F	T

61. p: Our oil supply is cut off.

q: Our economy collapses.

r: We go to war.

$p \to q$ $\{[p \to q] \wedge [r \to \sim q]\} \to (\sim p \to \sim r)$

$\underline{r \to \sim q}$ The argument is invalid.

$\sim p \to \sim r$

p	q	r	1 $p \to q$	$\sim q$	2 $r \to \sim q$	$1 \wedge 2$	$\sim p$	$\sim r$	3 $\sim p \to \sim r$	$(1 \wedge 2) \to 3$
T	T	T	T	F	F	F	F	F	T	T
T	T	F	T	F	T	T	F	T	T	T
T	F	T	F	T	T	F	F	F	T	T
T	F	F	F	T	T	F	F	T	T	T
F	T	T	T	F	F	F	T	F	F	T
F	T	F	T	F	T	T	T	T	T	T
F	F	T	T	T	T	T	T	F	F	F
F	F	F	T	T	T	T	T	T	T	T

63. p: A person is a professor.

q: A person is educated. $\{[p \to q] \wedge [r \to \sim p]\} \to (r \to \sim q)$

r: An animal is a monkey.   The argument is invalid.

$p \to q$

$\underline{r \to \sim p}$

$r \to \sim q$

32

p	q	r	1 p→q	~p	2 r→~p	1∧2	~q	3 r→~q	(1∧2)→3
T	T	T	T	F	F	F	F	F	T
T	T	F	T	F	T	T	F	T	T
T	F	T	F	F	F	F	T	T	T
T	F	F	F	F	T	F	T	T	T
F	T	T	T	T	T	T	F	F	F
F	T	F	T	T	T	T	F	T	T
F	F	T	T	T	T	T	T	T	T
F	F	F	T	T	T	T	T	T	T

65.   p: You are investing in the stock market.

q: The invested money is to be guaranteed.

r: You retire at an early age.

$q \rightarrow \sim p$

$\sim q \rightarrow \sim r$

$\underline{}$

$r \rightarrow \sim p$

$\{(q \rightarrow \sim p) \wedge (\sim q \rightarrow \sim r)\} \rightarrow (r \rightarrow \sim p)$

The argument is valid.

p	q	r	~p	~q	~r	1 q→~p	2 ~q→~r	3 r→~p	(1∧2)→3
T	T	T	F	F	F	F	T	F	T
T	T	F	F	F	T	F	T	T	T
T	F	T	F	T	F	T	F	F	T
T	F	F	F	T	T	T	T	T	T
F	T	T	T	F	F	T	T	T	T
F	T	F	T	F	T	T	T	T	T
F	F	T	T	T	F	T	F	T	T
F	F	F	T	T	T	T	T	T	T

67.   a.

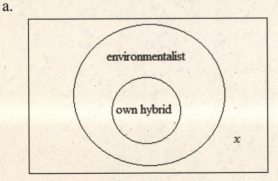

x = Not environmentalist

b. Translate the statements into symbols and construct a truth table.

     p: You own a hybrid car.   *q*: You are an environmentalist.

     $p \rightarrow q$ If you own a hybrid car, then you are an environmentalist.

     <u>$\sim q$</u>   You are not an environmentalist.

     $\sim p$ Therefore, you do not own a hybrid car.

p	q	$p \rightarrow q$	$\sim q$	$(p \rightarrow q) \wedge \sim q$	$\sim p$	$[(p \rightarrow q) \wedge \sim q] \rightarrow \sim p$
T	T	T	F	F	F	T
T	F	F	T	F	F	T
F	T	T	F	F	T	T
F	F	T	T	T	T	T

c. Because the last column contains all T's, the conditional $[(p \rightarrow q) \wedge \sim q] \rightarrow \sim p$ is a tautology; the argument is valid.

2.1 Sets and Set Operations

1. a. "The set of all black automobiles" is well-defined because an automobile is either black or not black.

   b. "The set of all inexpensive automobiles" is not well-defined because we need a level of cost (e.g., under $8,000) specified to determine whether an automobile is expensive or inexpensive.

   c. "The set of all prime numbers" is well-defined as it can be determined whether or not a number is prime or not prime.

   d. "The set of all large numbers" is not well-defined because different people will have different definitions for large.

3. Proper: { }, {Lennon}, {McCartney}; Improper: {Lennon, McCartney}

5. Proper: { }, {yes}, {no}, {undecided}, {yes, no}, {yes, undecided}, {no, undecided}
   Improper: {yes, no, undecided}

7. a. {4, 5}   b. {1, 2, 3, 4, 5, 6, 7, 8}   c. {0, 6, 7, 8, 9}   d. {0, 1, 2, 3, 9}

9. a. { }   b. {0, 1, 2, 3, 4, 5, 6, 7, 8, 9} c. {0, 2, 4, 6, 8}   d. {1, 3, 5, 7, 9}

11. {Friday}

13. {Monday, Tuesday, Wednesday, Thursday}

15. {Friday, Saturday, Sunday}

17.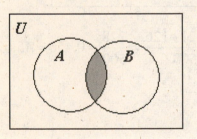

19.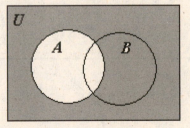

21.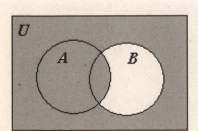

23.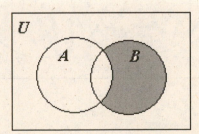

25.

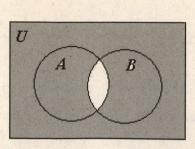

27.   a.   Rearranging the Cardinal Number Formula for the Union of Sets:

$$n(A \cap B) = n(A) + n(B) - n(A \cup B)$$
$$n(A \cap B) = 37 + 84 - 100$$
$$n(A \cap B) = 21$$

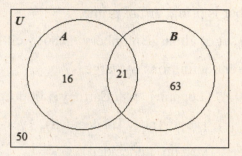

   b.   Using the formula from part a:

$$n(A \cap B) = n(A) + n(B) - n(A \cup B)$$
$$n(A \cap B) = 37 + 84 - 121$$
$$n(A \cap B) = 0$$

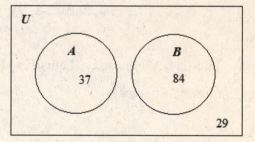

29.   a.

b. $\dfrac{81 + 21 + 126}{500} = \dfrac{228}{500} = 0.456 = 45.6\%$

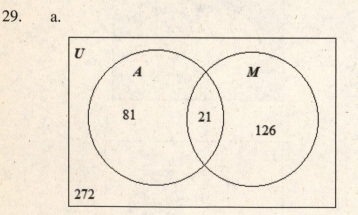

36

31.   a. Rearranging the Cardinal Number Formula for the Union of Sets:

$$n(child \cap career) = n(child) + n(career) - n(child \cup career)$$
$$n(child \cap career) = 285 + 316 - (700 - 196)$$
$$n(child \cap career) = 97$$

b. $\dfrac{97}{700} \approx 0.13857 = 13.857\%$

For questions 33 – 35, the following facts are useful:

$n(U) = 50$

$n(A) = 4$, since A = {Alabama, Alaska, Arizona, Arkansas}

$n(I) = 4$, since I = {Idaho, Illinois, Indiana, Iowa}

$n(M) = 8$, since M = {Maine, Maryland, Massachusetts, Michigan, Minnesota, Mississippi, Missouri, Montana}

$n(N) = 8$, since N = {Nebraska, Nevada, New Hampshire, New Jersey, New Mexico, New York, North Carolina, North Dakota}

$n(O) = 3$, since O = {Ohio, Oklahoma, Oregon}

33. $n(M') = n(U) - n(M) = 50 - 8 = 42$

35.   Rearranging the Cardinal Number Formula for the Union of Sets:

$$n(I' \cap O') = n(I') + n(O') - n(I' \cup O')$$

But, $n(I') = n(U) - n(I) = 50 - 4 = 46$ and $n(O') = n(U) - n(O) = 50 - 3 = 47$.

Thus, $n(I' \cap O') = 46 + 47 - 50 = 43$

For questions 37 – 39, the following facts are useful:

$n(U) = 12$

$n(J) = 3$, since J = {January, June, July}

$n(Y) = 4$, since Y = {January, February, May, July}

$n(V) = 3$, since V = {April, August, October}

$n(R) = 4$, since R = {September, October, November, December}

37. $n(R') = n(U) - n(R) = 12 - 4 = 8$

39. $n(J \cup Y) = n(J) + n(Y) - n(J \cap Y) = 3 + 4 - 2 = 5$

41. $n(S \cup A) = n(S) + n(A) - n(S \cap A) = 13 + 4 - 1 = 16$

Note: $n(S \cap A) = 1$, since there is only one ace which is also a spade.

43. $n(F \cup B) = n(F) + n(B) - n(F \cap B) = 12 + 26 - 6 = 32$

Note: $n(F \cap B) = 6$, since there are six face cards that are also black.

45. $n(F \cap B) = 6$, since there are 6 black face cards.

47. $n(A \cup E) = n(A) + n(E) - n(A \cap E) = 4 + 4 - 0 = 8$

Note: $n(A \cap E) = 0$, since there are no aces which are also eights.

49. $n(A \cap E) = 0$, since there are no aces which are also eights.

51.   a. {1, 2, 3}   b. {1, 2, 3, 4, 5, 6}   c. $E \subseteq F$ d. $E \subseteq F$

53.   a. { }, {a} 2 subsets   b. { }, {a}, {b}, {a, b} 4 subsets

c. { }, {a}, {b}, {c}, {a, b}, {a, c}, {b, c}, {a, b, c} 8 subsets

d. { }, {a}, {b}, {c}, {d}, {a, b}, {a, c}, {a, d}, {b, c}, {b, d}, {c, d}, {a, b, c}, {a, b, d},

{a, c, d}, {b, c, d}, {a, b, c, d} 16 subsets

e. Yes! The number of subsets of $A = 2^{n(A)}$

f. Since $n(A) = 6$, the number of subsets is $2^6 = 64$.

65.   a. doesn't conform   M & O are in the same group

b. doesn't conform   J & P are in the same group

c. doesn't conform   N is in John's group, but P is not in Juneko's group

d. conforms   e. doesn't conform   M & O are in the same group

67.   c. L, M, P   Juneko's group must contain both M & P and Juneko's group can't contain O or J.

69.   e. If K is in John's group, then M must be in Juneko's group. But, M & O cannot be in the same group, so O cannot be in Juneko's group. O could be in John's group, but it could also be the poster that is not used. Nothing is known about J, L, or N which means answer (e) is the only one that can be true.

2.2 Applications of Venn Diagrams

1.

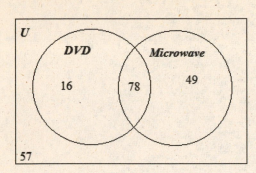

a. $16 + 78 + 49 = 143$

b. 16

c. 49

d. 57

3.

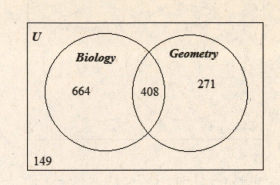

a. 408

b. $664 + 408 + 271 = 1,343$

c. 664

d. 149

5.

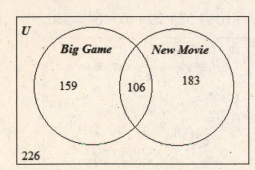

a. 106

b. 448

c. 265

d. 159

7.

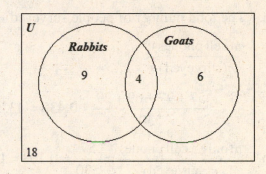

a. $\dfrac{19}{37} \approx 0.51351 = 51.351\%$

b. $\dfrac{9}{37} \approx 0.24324 = 24.324\%$

c. $\dfrac{6}{37} \approx 0.162162 = 16.216\%$

9.

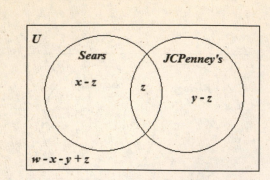

$$w - (x - z) - z - (y - z) = w - x - y + z$$

a. $(x - z) + z + (y - z) = x + y - z$

b. $x - z$

c. $y - z$

d. $w - x - y + z$

11. a. The total number of people surveyed is $63 + 77 + 69 + 104 + 57 + 29 + 67 + 64 = 530$.

$$\frac{n(\text{cell phones})}{n(\text{surveyed})}$$

$$= \frac{57 + 77 + 69 + 29}{530} \approx 0.438 = 43.8\%$$

b. $\dfrac{n(\text{only a cell phone})}{n(\text{surveyed})} = \dfrac{57}{530} \approx 0.108 = 10.8\%$

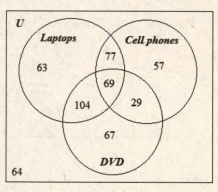

13. a. The total number of college students is $58 + 95 + 117 + 111 + 130 + 266 + 240 + 513 = 1{,}530$.

$$\frac{n(\text{saw at least one band})}{n(\text{college students})} = \frac{1017}{1530} \approx 0.664706 = 66.471\%$$

b. $\dfrac{n(\text{saw exactly one band})}{n(\text{college students})} = \dfrac{58 + 130 + 240}{1530}$

$$= \frac{428}{1530} \approx 0.27974 = 27.974\%$$

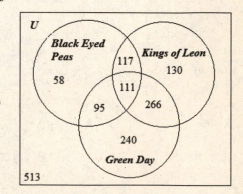

40

15.

$x + y + 9 + 24 = 77$

$x + z + 9 + 18 = 65$

$x + y + z + 9 + 24 + 18 = 101$

$y + z + a + 9 = 27$

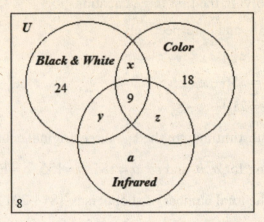

Rewriting the first three equations:

$x + y = 44$

$x + z = 38$

$x + y + z = 50$

Subtracting the first equation from the third: $z = 6$

Substituting into the second equation: $x = 32$

Substituting into the third equation: $y = 12$

Substituting into the fourth original equation: $a = 0$

a. The total number of members is $24 + 32 + 9 + 12 + 18 + 6 + 0 + 8 = 109$.

$$\frac{n(\text{only Infrared})}{n(\text{members})} = \frac{0}{109} = 0\,\%$$

b. $\dfrac{n(\text{at least two types})}{n(\text{members})}$

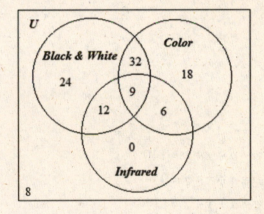

$$= \frac{32 + 12 + 6 + 9}{109} \approx 0.541 = 54.1\%$$

17.

$a + x + y + 157 = 621$

$b + x + z + 157 = 513$

$c + y + z + 157 = 367$

$b + c + x + y + z + 157 = 723$

$a + c + x + y + z + 157 = 749$

$a + b + x + y + z + 157 = 776$

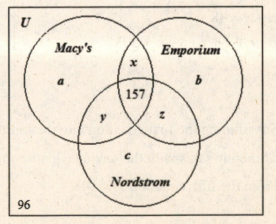

Solving the first three equations for

a, b, and c, respectively.

$a = 464 - x - y$

$b = 356 - x - z$

$c = 210 - y - z$

Substituting into the last three original equations: $z = 0$, $y = 82$, $x = 201$

Solving for a, b, and c: $a = 181$, $b = 155$, $c = 128$

a. The total number of shoppers is $181 + 201 + 157 + 82 + 155 + 0 + 128 + 96 = 1,000$.

$$\frac{n(\text{shopped at more than one store})}{n(\text{shoppers})} = \frac{201 + 82 + 0 + 157}{1,000} = 0.440 = 44.0\%$$

b. $\dfrac{n(\text{shopped exclusively at Nordstorm})}{n(\text{shoppers})}$

$= \dfrac{128}{1,000} \approx 0.128 = 12.8\%$

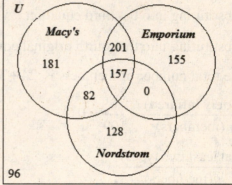

19.

$a + d + x + y = 140$

$b + d + x + z = 95$

$c + d + y + z = 134$

$a + b + d + x + y + z = 235$

$d + y = 48$

$b + c + d + x + y + z = 208$

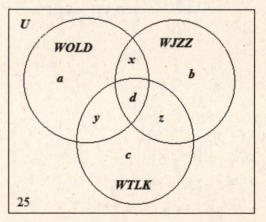

Substituting the first equation into the fourth: $b + z = 95$

Combining this with the second equation: $d + x = 0$, which means that $d = x = 0$

From the fifth equation: $y = 48$

Substituting these values into the other original equations:

$$a = 92, \ b = 74, \ c = 65, \ z = 21$$

a. The total number of listeners is $92 + 74 + 65 + 0 + 0 + 48 + 21 + 25 = 325$.

$$\frac{n(\text{listeners who only listen to WTLK})}{n(\text{listeners})} = \frac{65}{325}$$

$$= 0.2 = 20\%$$

b. $\dfrac{n(\text{listeners who do not listen to WTLK})}{n(\text{listeners})}$

$$= \frac{25 + 92 + 74}{325} \approx 0.588 = 58.8\%$$

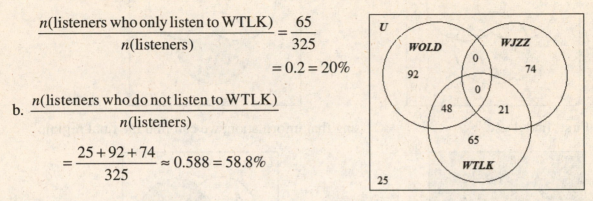

21. Starting with those who own all four pets, work backward to fill in the appropriate numbers in each region:

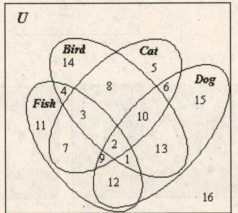

The number of owners who have no fish, no birds, no cats and no dogs is 16.

23. Using De Morgan's Law: $(A' \cup B)' = (A')' \cap B' = A \cap B'$

Since $B' = \{0, 3, 4, 5, 6\}$, $(A' \cup B)' = \{0, 4, 5\}$

25. Using De Morgan's Law: $(A \cap B')' = A' \cup (B')' = A' \cup B$

Since $A' = \{1, 3, 6, 7, 8\}$, $(A \cap B')' = \{1, 2, 3, 6, 7, 8, 9\}$

27.

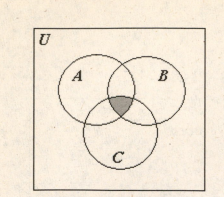

29.

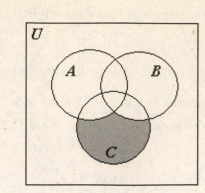

31.   First, find $A \cup C'$:   Using that information, we can find the final region:

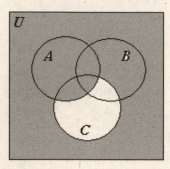

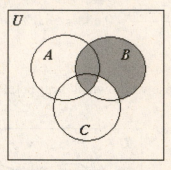

33. $A \cap B$ $(A \cap B)'$ A'

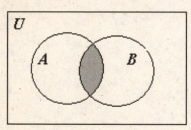

B' $A' \cup B'$

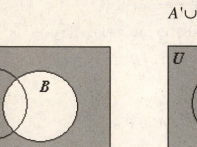

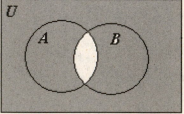

35.   a. Use the figure to find the sum of blood types with Rh positive: $38 + 34 + 9 + 3 = 84$

   84% of all people in the U. S. have blood that is Rh positive.

   b. Use the same method as in part a, or use the result from part a: $100 - 84 = 16$

   16% of all people in the U. S. have blood that is Rh negative.

37. a. Use the figure, concentrating on blood type A: $\dfrac{\text{number with A}+}{\text{number with A}} = \dfrac{34}{40} = 0.85$

85% of all people in the U. S. who have type A blood have Rh positive.

   b. Use the same method as in part a, or use the result from part a: $100 - 85 = 15$

15% of all people in the U. S. who have type A blood have Rh negative.

39. a. Use the figure, concentrating on blood type AB: $\dfrac{\text{number with AB}+}{\text{number with AB}} = \dfrac{3}{4} = 0.75$

75% of all people in the U. S. who have type AB blood have Rh positive.

   b. Use the same method as in part a, or use the result from part a: $100 - 75 = 25$

25% of all people in the U. S. who have type AB blood have Rh negative.

41. Type B blood type cannot be mixed with any blood containing the A molecule (type A or type AB). Therefore, a person with type B blood can only receive a transfusion of type B or type O blood.

43. The transfusion cannot contain a blood molecule (A or B) that the receiver does not already have. Consequently, a person with type O blood may only receive a transfusion of type O blood.

47. e. The answer can't be (c) or (d) because it can't contain Delores. It can't be (a) since Ed can only attend if Grant is there. The answer also can't be (b) since Frank can only attend if Angela and Carmen are there.

49. b. The answer can't be (c) or (e) since Carmen can't attend. It can't be (d) since Delores can't attend. The answer also can't be (a) since Frank can attend only those meetings that both Angela and Carmen attend.

51. b. Grant, Ed, Carmen, Angela and Frank can't go.

2.3 Introduction to Combinatorics

1. a. $2 \cdot 2 \cdot 2 = 8$ b.

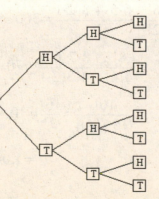   {HHH,HHT,HTH,HTT,THH,THT,TTH,TTT}

45

3.   a. $2 \cdot 3 \cdot 2 = 12$

      b. {Mega-BW-BN, Mega-BW-NW, Mega-WW-BN,

          Mega-WW-NW, Mega-GW-BN, Mega-GW-NW,

          BB-BW-BN, BB-BW-NW, BB-WW-BN,

          BB-WW-NW, BB-GW-BN, BB-GW-NW}

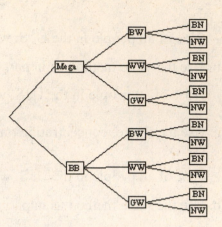

5. $3 \cdot 4 \cdot 2 = 24$ 7. $4 \cdot 3 \cdot 6 \cdot 10 = 720$ 9. $14 \cdot 7 \cdot 9 \cdot 3 = 2,646$

11. $3 \cdot 4 \cdot 6 \cdot 3 = 216$ 13. $10 \cdot 10 \cdot 10 \cdot 10 \cdot 10 \cdot 10 \cdot 10 \cdot 10 \cdot 10 = 10^9 = 1,000,000,000$

15. $10 \cdot 10 \cdot 10 \cdot 10 \cdot 10 \cdot 10 \cdot 10 \cdot 10 \cdot 10 \cdot 10 = 10^{10} = 10,000,000,000$

17.   a. $8 \cdot 2 \cdot 8 = 128$ b. $8 \cdot 10 \cdot 10 = 800$ c. The phone company needed more.

19.   a. $26 \cdot 10 \cdot 9 \cdot 8 \cdot 7 \cdot 6 \cdot 5 \cdot 4 \cdot 3 \cdot 25 = 1,179,360,000 = 1.17936 \times 10^9$

      b. $26 \cdot 10 \cdot 10 \cdot 10 \cdot 10 \cdot 10 \cdot 10 \cdot 10 \cdot 10 \cdot 26 = 67,600,000,000 = 6.76 \times 10^{10}$

      c. $21 \cdot 10 \cdot 10 \cdot 10 \cdot 10 \cdot 10 \cdot 10 \cdot 10 \cdot 10 \cdot 20 = 42,000,000,000 = 4.2 \times 10^{10}$

21. $9 \cdot 10 \cdot 10 \cdot 10 \cdot 5 \cdot 4 \cdot 3 = 540,000$

23. $4! = 4 \cdot 3 \cdot 2 \cdot 1 = 24$

25. $10! = 10 \cdot 9 \cdot 8 \cdot 7 \cdot 6 \cdot 5 \cdot 4 \cdot 3 \cdot 2 \cdot 1 = 3,628,800$

27. $20! = 2,432,902,008,176,640,000 = 2.432902008 \times 10^{18}$

29. $6!4! = 6 \cdot 5 \cdot 4 \cdot 3 \cdot 2 \cdot 1 \cdot 4 \cdot 3 \cdot 2 \cdot 1 = 17,280$

31.   a. $\dfrac{6!}{4!} = \dfrac{6 \cdot 5 \cdot 4 \cdot 3 \cdot 2 \cdot 1}{4 \cdot 3 \cdot 2 \cdot 1} = 30$ b. $\dfrac{6!}{2!} = \dfrac{6 \cdot 5 \cdot 4 \cdot 3 \cdot 2 \cdot 1}{2 \cdot 1} = 360$

33. $\dfrac{8!}{5!3!} = \dfrac{8 \cdot 7 \cdot 6}{3 \cdot 2 \cdot 1} = 56$ 35. $\dfrac{8!}{4!4!} = \dfrac{8 \cdot 7 \cdot 6 \cdot 5}{4 \cdot 3 \cdot 2 \cdot 1} = 70$

37. $\dfrac{82!}{80!2!} = \dfrac{82 \cdot 81}{2 \cdot 1} = 3,321$ 39. $\dfrac{16!}{(16-14)!} = \dfrac{16!}{2!} = 10,461,394,944,000$

$$= 1.046139494 \times 10^{13}$$

41. $\dfrac{5!}{(5-5)!}=\dfrac{5!}{0!}=120$ 43. $\dfrac{7!}{(7-3)!3!}=\dfrac{7!}{4!3!}=\dfrac{7\cdot6\cdot5}{3\cdot2\cdot1}=35$

45. $\dfrac{5!}{(5-5)!5!}=\dfrac{5!}{5!}=1$

51. b. If space #6 was empty, then the gray and white cars would have to park next to each other.

53. d. The white car can't park in #5. If it did (and pink is in #3), the gray car would have to park in #1 or #2. But, then there aren't two adjacent spots for the black and yellow cars.

2.4 Permutations and Combinations

1. a. $_7P_3=\dfrac{7!}{(7-3)!}=\dfrac{7!}{4!}=7\cdot6\cdot5=210$ b. $_7C_3=\dfrac{7!}{(7-3)!3!}=\dfrac{7!}{4!3!}=\dfrac{7\cdot6\cdot5}{3\cdot2\cdot1}=35$

3. a. $_5P_5=\dfrac{5!}{(5-5)!}=\dfrac{5!}{0!}=5!=120$ b. $_5C_5=\dfrac{5!}{(5-5)!5!}=\dfrac{5!}{0!5!}=1$

5. a. $_{14}P_1=\dfrac{14!}{(14-1)!}=\dfrac{14!}{13!}=14$ b. $_{14}C_1=\dfrac{14!}{(14-1)!1!}=\dfrac{14!}{13!1!}=14$

7. a. $_{100}P_3=\dfrac{100!}{(100-3)!}=\dfrac{100!}{97!}=100\cdot99\cdot98=970,200$

    b. $_{100}C_3=\dfrac{100!}{(100-3)!3!}=\dfrac{100!}{97!3!}=\dfrac{100\cdot99\cdot98}{3\cdot2\cdot1}=161,700$

9. a. $_xP_{x-1}=\dfrac{x!}{\left[x-(x-1)\right]!}=\dfrac{x!}{1!}=x!$ b. $_xC_{x-1}=\dfrac{x!}{\left[x-(x-1)\right]!(x-1)!}=\dfrac{x!}{1!(x-1)!}=x$

11. a. $_xP_2=\dfrac{x!}{(x-2)!}=x\cdot(x-1)=x^2-x$ b. $_xC_2=\dfrac{x!}{(x-2)!2!}=\dfrac{x\cdot(x-1)}{2}=\dfrac{x^2-x}{2}$

13. a. $_3P_2=\dfrac{3!}{(3-2)!}=\dfrac{3!}{1!}=3!=6$ b. $\{a,b\},\{a,c\},\{b,c\},\{b,a\},\{c,a\},\{c,b\}$

15. a. $_4C_2=\dfrac{4!}{(4-2)!2!}=\dfrac{4!}{2!2!}=\dfrac{4\cdot3}{2\cdot1}=6$ b. $\{a,b\},\{a,c\},\{a,d\},\{b,c\},\{b,d\},\{c,d\}$

17. a. Since order does matter, we use permutations: b. Only 1 ordering is possible.

     $_{11}P_{11}=\dfrac{11!}{(11-11)!}=\dfrac{11!}{0!}=11!=39,916,800$

47

19.   Since order does matter, we use permutations: $_4P_4 = \dfrac{4!}{(4-4)!} = \dfrac{4!}{0!} = 4! = 24$

21.   Since order does not matter, we use combinations: $_{13}C_2 = \dfrac{13!}{(13-2)!2!} = \dfrac{13!}{11!2!} = \dfrac{13 \cdot 12}{2 \cdot 1} = 78$

23.   Since order does matter, we use permutations: $_{13}P_3 = \dfrac{13!}{(13-3)!} = \dfrac{13!}{10!} = 13 \cdot 12 \cdot 11 = 1,716$

25.   a. Using the Fundamental Principle of Counting:

   (number of ways to choose 2 women) × (number of ways to choose 2 men)

   $_9C_2 \cdot {_6C_2} = \dfrac{9!}{(9-2)!2!} \cdot \dfrac{6!}{(6-2)!2!} = \dfrac{9!}{7!2!} \cdot \dfrac{6!}{4!2!} = \dfrac{9 \cdot 8}{2 \cdot 1} \cdot \dfrac{6 \cdot 5}{2 \cdot 1} = 540$

   b. $_{15}C_4 = \dfrac{15!}{(15-4)!4!} = \dfrac{15!}{11!4!} = \dfrac{15 \cdot 14 \cdot 13 \cdot 12}{4 \cdot 3 \cdot 2 \cdot 1} = 1,365$

   c. We find the number of ways to choose 3 women and add it to the number of ways to

   choose 4 women.

   (number of ways to choose 3 women) × (number of ways to choose 1 man)

   $_9C_3 \cdot {_6C_1} = \dfrac{9!}{(9-3)!3!} \cdot \dfrac{6!}{(6-1)!1!} = \dfrac{9!}{6!3!} \cdot \dfrac{6!}{5!1!} = \dfrac{9 \cdot 8 \cdot 7}{3 \cdot 2 \cdot 1} \cdot 6 = 504$

   (number of ways to choose 4 women) × (number of ways to choose no men)

   $_9C_4 \cdot {_6C_0} = \dfrac{9!}{(9-4)!4!} \cdot \dfrac{6!}{(6-0)!0!} = \dfrac{9!}{5!4!} \cdot \dfrac{6!}{6!0!} = \dfrac{9 \cdot 8 \cdot 7 \cdot 6}{4 \cdot 3 \cdot 2 \cdot 1} \cdot 1 = 126$

   Thus, the total number of number of committees possible: $504 + 126 = 630$

27.   Since order is not important, we use combinations:

   $_{52}C_5 = \dfrac{52!}{(52-5)!5!} = \dfrac{52!}{47!5!} = \dfrac{52 \cdot 51 \cdot 50 \cdot 49 \cdot 48}{5 \cdot 4 \cdot 3 \cdot 2 \cdot 1} = 2,598,960$

29.   a. Using the Fundamental Principle of Counting:

   (number of ways to choose 3 aces) × (number of ways to choose 2 other cards)

   $_4C_3 \cdot {_{48}C_2} = \dfrac{4!}{(4-3)!3!} \cdot \dfrac{48!}{(48-2)!2!} = \dfrac{4!}{1!3!} \cdot \dfrac{48!}{46!2!} = 4 \cdot \dfrac{48 \cdot 47}{2} = 4,512$

   b. Since there are 13 different ranks: $13 \cdot 4,512 = 58,656$

31. Using the Fundamental Principle of Counting:

    (number of ways to choose 2 ranks) × (number of ways to choose a pair of the first rank) ×

    (number of ways to choose a pair of the second rank) × (number of ways to choose one other

    card)

    $$_{13}C_2 \cdot _4C_2 \cdot _4C_2 \cdot _{44}C_1 = \frac{13!}{(13-2)!2!} \cdot \frac{4!}{(4-2)!2!} \cdot \frac{4!}{(4-2)!2!} \cdot \frac{44!}{(44-1)!1!}$$

    $$= \frac{13!}{11!2!} \cdot \frac{4!}{2!2!} \cdot \frac{4!}{2!2!} \cdot \frac{44!}{43!1!} = \frac{13 \cdot 12}{2 \cdot 1} \cdot \frac{4 \cdot 3}{2 \cdot 1} \cdot \frac{4 \cdot 3}{2 \cdot 1} \cdot 44 = 123,552$$

33. Since order is not important: $_{53}C_6 = \frac{53!}{(53-6)!6!} = \frac{53!}{47!6!} = \frac{53 \cdot 52 \cdot 51 \cdot 50 \cdot 49 \cdot 48}{6 \cdot 5 \cdot 4 \cdot 3 \cdot 2 \cdot 1} = 22,957,480$

35. Since order is not important: $_{36}C_5 = \frac{36!}{(36-5)!5!} = \frac{36!}{31!5!} = \frac{36 \cdot 35 \cdot 34 \cdot 33 \cdot 32}{5 \cdot 4 \cdot 3 \cdot 2 \cdot 1} = 376.992$

37. It is easier to win a 5/36 lottery since there are fewer possible tickets.

39. The first five rows of triangle are:

    ```
              1
            1   1
          1   2   1
        1   3   3   1
      1   4   6   4   1
    ```

    a. 1   b. 2   c. 4   d. 8   e. 16   f. Yes. The sum of each row is twice the previous sum, or,

    sum of entries in n^{th} row is 2^{n-1}.

    g. We predict $2 \cdot 16 = 32$.   h. The sum is $1+5+10+10+5+1 = 32$. Our prediction

    i. 2^{n-1} was correct.

41. Using the answers to problem 40:

    a. fifth row   b. $(n+1)^{st}$ row   c. No.   d. Yes.

    e. $_nC_r$ is the $(r+1)^{st}$ number in the $(n+1)^{st}$ row.

43. The word "ALASKA" has $n = 6$ letters; only A is repeated $x = 3$ times: $\frac{n!}{x!} = \frac{6!}{3!} = 120$

45. The word "ILLINOIS" has $n = 8$ letters; I is repeated $x = 3$ times, and L is repeated $y = 2$ times:

    $\frac{n!}{x!y!} = \frac{8!}{3!2!} = 3,360$

47.   The word "INDIANA" has $n = 7$ letters; I is repeated $x = 2$ times, N is repeated $y = 2$ times, and

A is repeated $z = 2$ times: $\dfrac{n!}{x!\,y!\,z!} = \dfrac{7!}{2!2!2!} = 630$

49.   The word "TALLAHASSEE" has $n = 11$ letters; A is repeated $w = 3$ times, L is repeated

$x = 2$ times, S is repeated $y = 2$ times, and E is repeated $z = 2$ times:

$$\dfrac{n!}{w!\,x!\,y!\,z!} = \dfrac{11!}{3!2!2!2!} = 831,600$$

51.   a. The word "PIER" has $n = 4$ letters and none are repeated: $n! = 4! = 24$

b. The word "PEER" has $n = 4$ letters and E is repeated $x = 2$ times: $\dfrac{n!}{x!} = \dfrac{4!}{2!} = 12$

53.   a. The word "STEAL" has $n = 5$ letters and none are repeated: $n! = 5! = 120$

b. The word "STEEL" has $n = 5$ letters and E is repeated $x = 2$ times: $\dfrac{n!}{x!} = \dfrac{5!}{2!} = 60$

57.   c. Each team must play the other 5 teams.

59.   c. B plays E first and C third, so it can't be E or C. A plays F second, so it can't be A or F.

Therefore, B must play D second.

61.   e. Team B must lose at least one game (the one with Team D!).

2.5 Infinite Sets

1. $n(S) = 4,\ n(C) = 4$ They are equivalent. Match each state to its capital.

3. $n(R) = 3,\ n(G) = 4$ They are not equivalent.

5. $n(C) = 22,\ n(D) = 22$ They are equivalent.   Match each multiple of 3 to the corresponding

multiple of 4; $3n \leftrightarrow 4n$.

7. $n(G) = 250,\ n(H) = 251$ They are not equivalent.

9. $n(A) = 62,\ n(B) = 62$ They are equivalent.   For n from 1 to 62, match the term described by

$2n - 1$ in A to the term described by $2(n + 62) - 1$ in B; $2n - 1 \leftrightarrow 2n + 123$.

11.   a. For n starting at 1, match the term described by n in N to the term described by $2n - 1$ in O;

$n \leftrightarrow 2n - 1$

b. Solving for n:

$$2n-1 = 1835$$
$$2n = 1836$$
$$n = 918$$

c. Solving for n:

$$2n-1 = x$$
$$2n = x+1$$
$$n = \tfrac{1}{2}(x+1)$$

$$n = \frac{x+1}{2}$$

d. $782 \to 2(782)-1 = 1{,}564-1 = 1{,}563$

e. $n \to 2n-1$

13. a. For n starting at 1, match the term described by n in N to the term described by $3n$ in T;

$n \leftrightarrow 3n$

b. $3n = 936$

$n = 312$

c. $3n = x$

$n = \tfrac{1}{3}x$

d. $n = 936 \to 3n = 2{,}808$

e. $n \to 3n$

15. a. $345 \to 1-345 = -344$ b. $248 \to 248$ c. $1-n = -754$ d. $n(A) = \aleph_0$

$-n = -755$

$n = 755$

17. Match any real number $x \in [0,1]$ to $3x \in [0,3]$.

19. From the center of the circle, match the corresponding points that lie on the radial line.

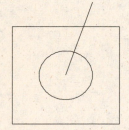

21. From the inside circle, match the corresponding points that lie on the same radial line.

23. First, the semicircle is equivalent to $[0,1]$. Draw a vertical line to match points.

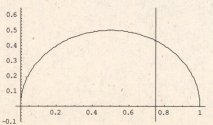

To match the semicircle and line, draw a line passing through $\left(\tfrac{1}{2},0\right)$ (not horizontal).

51

Chapter 2 Review

1.   a. Well defined.   b. Not well defined.   c. Not well defined.   d. Well defined.

3.   a. $A \cup B = \{$Maria, Nobuku, Leroy, Mickey, Kelly, Rachel, Deanna$\}$

       b. $A \cap B = \{$Leroy, Mickey$\}$

5.   a. $n(A \cap B) = n(A) + n(B) - n(A \cup B) = 32 + 26 - 40 = 18$

       b.

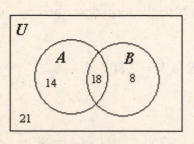

7.

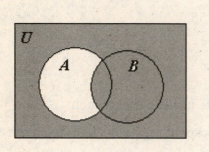

9.

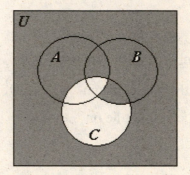

11.

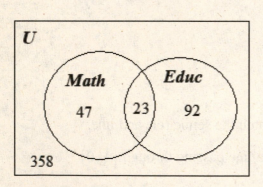

      a. $\dfrac{47 + 23 + 92}{520} = 0.31154 = 31.154\%$

      b. $\dfrac{47}{520} = 0.09038 = 9.038\%$

      c. $\dfrac{92}{520} = 0.17692 = 17.692\%$

13.

$$x + y + 95 + 87 \quad = \quad 305$$
$$x + z + 87 + 192 \quad = \quad 393$$
$$x + y + z + 95 + 87 + 192 \quad = \quad 510$$
$$y + z + a + 87 \quad = \quad 163$$

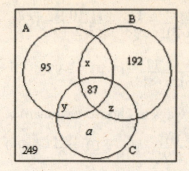

Rewriting the first 3 equations:

$$x + y = 123 \rightarrow y = 123 - x$$

$$x + z = 114 \rightarrow z = 114 - x$$

$$x + y + z = 136$$

Substituting: $x + (123 - x) + (114 - x) = 136$

$$237 - x = 136$$

$$x = 101, \ y = 22, \ z = 13$$

Substituting into 4th equation

$$22 + 13 + a + 87 = 163$$

$$a = 41$$

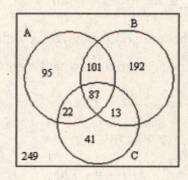

Percentage more than 1: $\dfrac{101 + 22 + 13 + 87}{95 + 101 + 87 + 22 + 192 + 13 + 41 + 249} = \dfrac{223}{800} = 0.27875 = 27.875\%$

15.   a. 45 have type O; 40 have type A: $n(O \cup A) = n(O) + n(A) = 45 + 40 + 85$

b. None have type O and type A: $n(O \cap A) = 0$

c. Using De Morgan's Law: $(O \cup A)' = O' \cap A'$

$$n(O' \cap A') = 4 + 11 = 15$$

17.   a.   (number choices for art) × (number of choices for dinner)

× (numbers of choices for dancing): $2 \cdot 3 \cdot 2 = 12$

b.   {MMA-S-LC, MMA-S-L, MMA-J-LC,

MMA-J-L, MMA-C-LC, MMA-C-L,

NPG-S-C, NPG-S-L, NPG-J-LC,

NPG-J-L, NPG-C-LC, NPG-C-L}

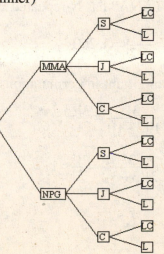

53

19. $9 \cdot 10 \cdot 10 \cdot 10 \cdot 10 \cdot 4 \cdot 3 = 1,080,000$

21. a. $_{11}C_3 = \dfrac{11!}{3!(11 \cdot 3)!} = \dfrac{11!}{3!8!} = \dfrac{11 \cdot 10 \cdot 9}{3 \cdot 2 \cdot 1} = 165$

    b. $_{11}P_3 = \dfrac{11!}{(11-3)!} = \dfrac{11!}{8!} = 11 \cdot 10 \cdot 9 = 990$

23. a. (number of ways to choose 1 woman) $\times$ (number of ways to choose 2 men)

    $_{10}C_1 \cdot {}_{12}C_2 = \dfrac{10!}{1!9!} \cdot \dfrac{12!}{2!10!} = 10 \cdot \dfrac{12 \cdot 11}{2} = 660$

    b. Total people $= 10 + 12 = 22$

    $_{22}C_3 = \dfrac{22!}{3!19!} = \dfrac{22 \cdot 21 \cdot 20}{3 \cdot 2 \cdot 1} = 1,540$

    c. We know there are 660 committees with 2 men.

    The number of committees with 3 men:

    $_{12}C_3 = \dfrac{12!}{3!9!} = \dfrac{12 \cdot 11 \cdot 10}{3 \cdot 2} = 220$

    Total ways: $660 + 220 = 880$

25. $_{10}P_3 = \dfrac{10!}{(10-3)!} = \dfrac{10!}{7!} = 10 \cdot 9 \cdot 8 = 720$

27. $_{54}C_7 = \dfrac{54!}{7!(54-7)!} = \dfrac{54!}{7!47!} = \dfrac{54 \cdot 53 \cdot 52 \cdot 51 \cdot 50 \cdot 49 \cdot 48}{7 \cdot 6 \cdot 5 \cdot 4 \cdot 3 \cdot 2 \cdot 1} = 177,100,560$

29. Each word has $n = 7$ letters.

    a. "FLORIDA" has no repeated letters: $n! = 7! = 5,040$

    b. "ARIZONA" has A repeated $x = 2$ times: $\dfrac{n!}{x!} = \dfrac{7!}{2!} = 2,520$

    c. "MONTANA" has N repeated $x = 2$ times and A repeated $y = 2$ times:

    $\dfrac{n!}{x!y!} = \dfrac{7!}{2!2!} = 1,260$

31. Order is important in permutations, whereas it is not important in combinations.

33. a. $_3C_1 = \dfrac{3!}{1!(3-1)!} = \dfrac{3!}{2!} = 3$ b. $_3C_2 = \dfrac{3!}{2!(3-2)!} = \dfrac{3!}{2!} = 3$

    c. $_3C_3 = \dfrac{3!}{3!(3-3)!} = \dfrac{3!}{3!} = 1$ d. $_3C_0 = \dfrac{3!}{0!(3-0)!} = \dfrac{3!}{3!} = 1$

    e. 8   f. $2^{n(S)} = 2^3 = 8$

35. They are equivalent. For n from 1 to 450, match the term described by $2n$ in D to the term described by $2n+1$ in C: $n(C) = 450, n(D) = 450;$ $2n+1 \leftrightarrow 2n$

37. a. Match each number in N to its square in S: $n \leftrightarrow n^2$ b. $n^2 = 841 \rightarrow n = 29$

    c. $n^2 = x \rightarrow n = \sqrt{x}$ d. $(144)^2 = 20,736$ e. $n \rightarrow n^2$

39. Match any real number $x \in [0,1]$ to $\pi x \in [0, \pi]$.

45. e.   47.   b.

3.1 History of Probability

1.   Individual solutions will vary depending on what happens when the single die is rolled. The following table records the results of a trial of this experiment.

Result (Number of 6's rolled in four rolls of die)	Number of times in 10 trials that result occurred
0	3
1	4
2	3
3	0
4	0

Betting that you could roll at least one six, you would have won 7 times ($70) and lost three times ($30). Thus, you would have won $40(70 − 30 = 40).

3.   a. Answers will vary.

     b. Answers will vary.

5.   a. Answers will vary.

     b. Answers will vary.

7.   a. Answers will vary.

     b. Answers will vary.

9.   a. win $35 \cdot 10 = \$350$

     b. lose $10

11.   a. win $17 \cdot 5 = \$85$

     b. win $17 \cdot 5 = \$85$

     c. lose $5

13.   a. lose $20

     b. win $8 \cdot 20 = \$160$

     c. lose $20

15.   a. lose $10

     b. win $5 \cdot 10 = \$50$

17.   a. win $2 \cdot 25 = \$50$

     b. lose $25

19.   a. lose $50

     b. win $1 \cdot 50 = \$50$

21.   a. lose $\$20 + \$25 = \$45$

     b. win $35 \cdot 25 = \$875$, lose $20 for a total of $855

     c. win $35 \cdot 20 = \$700$, lose $25 for a total of $675

23.   a. win $17 \cdot 30 = \$510$, lose $15 for a total of $495   b. win $17 \cdot 30 + 15 = \$525$

     c. lose $30 + 15 = \$45$ d. win $15, lose $30 for a loss of $15

25. $35x = 100$

     $x \approx 2.86$

     You would have to bet $3.

27. $2x = 1000$

     $x = \$500$

29. a. 13 b. $\frac{1}{4}$ 31. a. 12 b. $\frac{12}{52} = \frac{3}{13}$ 33. a. 4 b. $\frac{4}{52} = \frac{1}{13}$

3.2 Basic Terms of Probability

1. Choose 1 jellybean and look at its color.

3. $p(\text{black}) = \dfrac{n(\text{black})}{n(S)} = \dfrac{12}{35} \approx 0.34 = 34\%$

5. $p(\text{red or yellow}) = \dfrac{n(\text{red or yellow})}{n(S)} = \dfrac{8+10}{35} = \dfrac{18}{35} \approx 0.51 = 51\%$

7. $p(\text{not yellow}) = \dfrac{n(\text{not yellow})}{n(S)} = \dfrac{25}{35} = \dfrac{5}{7} \approx 0.71 = 71\%$

9. $p(\text{white}) = \dfrac{n(\text{white})}{n(S)} = \dfrac{0}{35} = 0 = 0\%$

11. $o(\text{black}) = n(\text{black}) : n(\text{not black}) = 12 : 23$

13. $o(\text{red or yellow}) = n(\text{red or yellow}) : n(\text{not red or yellow}) = 18 : 17$

15. Draw or selecting a card from a deck

17. a. $\frac{26}{52} = \frac{1}{2}$ b. $26 : 26$ or $1 : 1$ 19. a. $\frac{4}{52} = \frac{1}{13}$ b. $4 : 48$ or $1 : 12$

21. a. $\frac{1}{52}$ b. $1 : 51$ 23. a. $\frac{12}{52} = \frac{3}{13}$ b. $12 : 40$ or $3 : 10$

25. a. $\frac{40}{52} = \frac{10}{13}$ b. $40 : 12$ or $10 : 3$ 27. a. $\frac{12}{52} = \frac{3}{13}$ b. $12 : 40$ or $3 : 10$

29. a. $\frac{1}{38}$ b. $1 : 37$ 31. a. $\frac{3}{38}$ b. $3 : 35$

33. a. $\frac{5}{38}$ b. $5 : 33$ 35. a. $\frac{12}{38} = \frac{6}{19}$ b. $12 : 26$ or $6 : 13$

37. a. $\frac{18}{38} = \frac{9}{19}$ b. $18 : 20$ or $9 : 10$

39. a. $\dfrac{n(\text{female})}{n(S)} = \dfrac{154,135}{149,925+154,135} = \dfrac{154,135}{304,060} = 50.7\%$ b. $\dfrac{n(\text{male 20 - 44})}{n(S)} = \dfrac{53,060}{304,060} = 17.5\%$

41. Total area: $\pi r^2 = \pi(5)^2 = 25\pi$ Red area: $\pi r_R^2 = \pi(1)^2 = \pi$

Yellow area: $\pi r_Y^2 - \pi r_R^2 = \pi(3^2 - 1^2) = 8\pi$ Green area: $\pi r_G^2 - 8\pi - \pi = \pi(5^2 - 8 - 1) = 16\pi$

a. $p(\text{red}) = \dfrac{\text{red area}}{\text{total area}} = \dfrac{\pi}{25\pi} = \dfrac{1}{25} = 4\%$ b. $p(\text{yellow}) = \dfrac{\text{yellow area}}{\text{total area}} = \dfrac{8\pi}{25\pi} = \dfrac{8}{25} = 32\%$

c. $p(\text{green}) = \dfrac{\text{green area}}{\text{total area}} = \dfrac{16\pi}{25\pi} = \dfrac{16}{25} = 64\%$

43. $p(\text{breaks in first half hour}) = \dfrac{\text{number of minutes in first half hour}}{\text{total number of minutes}} = \dfrac{30}{145} = \dfrac{6}{29}$

45. $1:4$ 47. $\frac{3}{5}$ 49. $a:(b-a)$

51.   a. $\dfrac{18}{38} = \dfrac{9}{19}$ b. $18:(38-18) = 18:20$ or $9:10$

53.   a. House odds. sportsbook.com determines their own odds based on their own set of rules.

        b. New York Yankees: $p(\text{win}) = \dfrac{5}{5+2} = \dfrac{5}{7}$ New York Mets: $p(\text{win}) = \dfrac{7}{7+1} = \dfrac{7}{8}$

        c. The New York Mets are more likely to win because $\dfrac{7}{8} > \dfrac{5}{7}$

55.   a. Intentional self-harm has the highest odds (1: 9,085), so it is the most likely cause of death in

          one year.

        b. Once again, intentional self-harm has the highest odds (1: 117), so it is the most likely cause

          of death in a lifetime.

57.   a. The odds are 1: 20,331. Therefore, the probability is $\dfrac{1}{1+20,331} = \dfrac{1}{20,332}$.

        b. The odds are 1: 502,554. Therefore, the probability is $\dfrac{1}{1+502,554} = \dfrac{1}{502,555}$.

59.   a. The car has the highest odds (1: 20,331), so it has the highest probability of $\dfrac{1}{1+20,331} = \dfrac{1}{20,332}$.

        b. The airplane has the lowest odds (1: 502,554), so it has the lowest probability of

          $\dfrac{1}{1+502,554} = \dfrac{1}{502,555}$.

61.   a. $S = \{bb, gb, bg, gg\}$ b. $E = \{gb, bg\}$ c. $F = \{gb, bg, gg\}$ d. $G = \{gg\}$

        e. $\dfrac{2}{4} = \dfrac{1}{2}$ f. $\dfrac{3}{4}$ g. $\dfrac{1}{4}$ h. $2:2$ or $1:1$ i. $3:1$ j. $1:3$

63.   a. $S = \{ggg, ggb, gbg, bgg, gbb, bgb, bbg, bbb\}$ b. $E = \{ggb, gbg, bgg\}$

        c. $F = \{ggb, gbg, bgg, ggg\}$ d. $G = \{ggg\}$

        e. $\dfrac{3}{8}$ f. $\dfrac{4}{8} = \dfrac{1}{2}$ g. $\dfrac{1}{8}$ h. $3:5$ i. $4:4$ or $1:1$ j. $1:7$

65. $S = \{bb, bg, gb, gg\}$ a. $\dfrac{1}{4}$ b. $\dfrac{2}{4} = \dfrac{1}{2}$ c. $\dfrac{1}{4}$

        d. same sex: $\dfrac{2}{4}$, different sex: $\dfrac{2}{4}$. They are equally likely.

67. $\{ggg, ggb, gbg, bgg, gbb, bgb, bbg, bbb\}$

        Same sex: $\dfrac{2}{8} = \dfrac{1}{4}$, Different sex: $\dfrac{6}{8} = \dfrac{3}{4}$ Different sex is more likely. $\dfrac{3}{4} > \dfrac{1}{4}$

69. a. $S = \begin{Bmatrix} (1,1), (1,2), (1,3), (1,4), (1,5), (1,6), \\ (2,1), (2,2), (2,3), (2,4), (2,5), (2,6), \\ (3,1), (3,2), (3,3), (3,4), (3,5), (3,6), \\ (4,1), (4,2), (4,3), (4,4), (4,5), (4,6), \\ (5,1), (5,2), (5,3), (5,4), (5,5), (5,6), \\ (6,1), (6,2), (6,3), (6,4), (6,5), (6,6) \end{Bmatrix}$

b. $E = \{(1,6), (2,5), (3,4), (4,3), (5,2), (6,1)\}$ c. $F = \{(5,6), (6,5)\}$

d. $G = \{(1,1), (2,2), (3,3), (4,4), (5,5), (6,6)\}$ e. $\dfrac{6}{36} = \dfrac{1}{6}$ f. $\dfrac{2}{36} = \dfrac{1}{18}$

g. $\dfrac{6}{36} = \dfrac{1}{6}$ h. 6: 30 or 1: 5   i. 2: 34 or 1 : 17   j. 6: 30 or 1 : 5

71. RR – red, WW – white, RW and WR - pink

	R	W
R	RR	WR
W	RW	WW

a. $\dfrac{1}{4}$ b. $\dfrac{1}{4}$ c. $\dfrac{2}{4} = \dfrac{1}{2}$

73.

	N	s
N	NN	sN
s	Ns	ss

a. $\dfrac{1}{4}$ b. $\dfrac{2}{4} = \dfrac{1}{2}$ c. $\dfrac{1}{4}$

75.

	N	N
N	NN	NN
t	Nt	Nt

a. $\dfrac{0}{4} = 0$ b. $\dfrac{2}{4} = \dfrac{1}{2}$ c. $\dfrac{4}{4} = 1$

77.

	N	h
N	NN	hN
N	NN	hN

a. $\dfrac{2}{4} = \dfrac{1}{2}$ b. $\dfrac{0}{4} = 0$ c. $\dfrac{2}{4} = \dfrac{1}{2}$

3.3 Basic Rules of Probability

1. E and F are not mutually exclusive. There are many women doctors.

3. E and F are mutually exclusive. One cannot be both single and married.

5. E and F are not mutually exclusive. A brown haired person may have some gray.

7. E and F are mutually exclusive. One can't wear both boots and sandals (excluding one on each foot!)

9. E and F are mutually exclusive. Four is not odd.

11.   a. $p(J \cap R) = \frac{2}{52} = \frac{1}{26}$

      b. $p(J \cup R) = p(J) + p(R) - p(J \cap R) = \frac{4}{52} + \frac{26}{52} - \frac{2}{52} = \frac{28}{52} = \frac{7}{13}$

      c. $p(\text{not a red Jack}) = 1 - p(J \cap R) = 1 - \frac{1}{26} = \frac{25}{26}$

13.   a. $p(T \cap S) = \frac{1}{52}$ b. $p(T \cup S) = p(T) + (S) - p(T \cap S) = \frac{4}{52} + \frac{13}{52} - \frac{1}{52} = \frac{16}{52} = \frac{4}{13}$

      c. $p(\text{not a 10 of spades}) = 1 - p(T \cap S) = \frac{51}{52}$

15.   a. $p(\text{under 4}) = p(2 \cup 3) = p(2) + p(3) = \frac{4}{52} + \frac{4}{52} = \frac{8}{52} = \frac{2}{13}$

      b. $p(\text{above 9}) = p(10 \cup J \cup Q \cup K \cup A) = p(10) + p(J) + P(Q) + p(K) + p(A) = \frac{20}{52} = \frac{5}{13}$

      c. $p(\text{under 4 and above 9}) = 0$ d. $p(\text{under 4 or over 9}) = \frac{8}{52} + \frac{20}{52} = \frac{28}{52} = \frac{7}{13}$

17.   a. $p(\text{above 5}) = p(6 \cup 7 \cup 8 \cup 9 \cup 10 \cup J \cup Q \cup K \cup A) = \frac{36}{52} = \frac{9}{13}$

      b. $p(\text{below 10}) = p(2 \cup 3 \cup 4 \cup 5 \cup 6 \cup 7 \cup 8 \cup 9) = \frac{32}{52} = \frac{8}{13}$

      c. $p(\text{above 5 and below 10}) = p(6 \cup 7 \cup 8 \cup 9) = \frac{16}{52} = \frac{4}{13}$

      d. $p(\text{above 5 or below 10}) = 1$

19. $p(\text{not a Queen}) = 1 - p(Q) = 1 - \frac{4}{52} = \frac{48}{52} = \frac{12}{13}$

21. $p(\text{not a face card}) = 1 - p(\text{face card}) = 1 - p(J \cup Q \cup K) = 1 - \frac{12}{52} = \frac{40}{52} = \frac{10}{13}$

23. $p(\text{above 3}) = 1 - p(\leq 3) = 1 - p(2 \cup 3) = 1 - \frac{8}{52} = \frac{44}{52} = \frac{11}{13}$

25. $p(\text{below a Jack}) = 1 - p(\geq J) = 1 - p(J \cup Q \cup K \cup A) = 1 - \frac{16}{52} = \frac{36}{52} = \frac{9}{13}$

27. $p(E) = \frac{5}{14}$, $p(E') = 1 - p(E) = 1 - \frac{5}{14} = \frac{9}{14}$, $o(E') = 9:5$

29. $o(E) = 2:5$, $o(E') = 5:2$

31. $p(E) = \frac{a}{a+b}$; $p(E') = 1 - \frac{a}{a+b} = \frac{a+b}{a+b} - \frac{a}{a+b} = \frac{a+b-a}{a+b} = \frac{b}{a+b}$; $o(E') = b:a$

33. $p(\text{not a King}) = 1 - p(K) = 1 - \frac{4}{52} = \frac{48}{52} = \frac{12}{13}$. The odds are $12:1$.

35. $p(\text{not a face card}) = 1 - p(\text{face card}) = 1 - p(J \cup Q \cup K) = 1 - \frac{12}{52} = \frac{40}{52} = \frac{10}{13}$. The odds are $10:3$.

37. $p(\text{above a 4}) = 1 - p(\leq 4) = 1 - p(2 \cup 3 \cup 4) = 1 - \frac{12}{52} = \frac{40}{52} = \frac{10}{13}$. The odds are $10:3$.

39. a. $\frac{133 + 151}{700} = \frac{71}{175}$ b. $1 - \frac{71}{175} = \frac{104}{175}$

41. a. $\frac{151}{700}$ b. $\frac{133 + 151 + 201}{700} = \frac{97}{140}$

43. a. $\frac{208 + 112}{1,000} = \frac{8}{25}$ b. $1 - \frac{8}{25} = \frac{17}{25}$

45. a. $\frac{183 + 177}{1,000} = \frac{9}{25}$ b. $1 - \frac{9}{25} = \frac{16}{25}$

47. a. $E = \{(1,6), (2,5), (3,4), (4,3), (5,2), (6,1)\}$; $p(E) = \frac{6}{36} = \frac{1}{6}$

    b. $E = \{(3,6), (4,5), (5,4), (6,3)\}$; $p(E) = \frac{4}{36} = \frac{1}{9}$ c. $E = \{(5,6), (6,5)\}$; $p(E) = \frac{2}{36} = \frac{1}{18}$

49. a. $E = \{(1,6), (2,5), (3,4), (4,3), (5,2), (6,1), (5,6), (6,5)\}$; $p(E) = \frac{8}{36} = \frac{2}{9}$

    b. $E = \{(1,6), (2,5), (3,4), (4,3), (5,2), (6,1), (5,6), (6,5), (1,1), (2,2), (3,3), (4,4), (5,5), (6,6)\}$;

    $p(E) = \frac{14}{36} = \frac{7}{18}$

51. a. $E = \{(3,6), (4,5), (5,4), (6,3), (5,6), (6,5)\}$; $p(E) = \frac{6}{36} = \frac{1}{6}$

    b. $E' = \{(1,1), (1,3), (2,2), (3,1), (1,5), (2,4), (3,3), (4,2), (5,1)\}$;

    $p(E') = \frac{9}{36} = \frac{1}{4}$ $p(E) = 1 - p(E') = 1 - \frac{9}{36} = \frac{27}{36} = \frac{3}{4}$

53. a. $E = \{(1,1), (2,2), (3,3), (4,4), (5,5), (6,6)\}$; $p(E) = \frac{6}{36} = \frac{1}{6}$

    b. $E = \{(1,1), (1,3), (2,2), (3,1), (1,5), (2,4), (3,3), (4,2), (5,1), (2,6), (3,5), (4,4), (5,3), (6,2),$

    $(4,6), (5,5), (6,4), (6,6)\}$; $p(E) = \frac{18}{36} = \frac{1}{2}$

55. Let $A = \{\text{drives } 20,000 \leq \text{ miles/yr}\}$ and $B = \{\text{has an accident}\}$

    a. $p(A \cup B) = p(A) + p(B) - p(A \cap B) = 0.50 + 0.45 - 0.30 = 0.65$

    b. $p(A \cap B) = 0.30$ c. $1 - p(A \cup B) = 1 - 0.65 = 0.35$

57. Relative frequencies; data was collected from client's records and the probabilities were calculated from this data.

59. Let O = {supported Obama} and M = {is a male}.

a. $p(O \cup M) = p(O) + p(M) - p(O \cap M) = 0.527 + 0.47 - 0.23 = 0.767 = 76.7\%$

b. $p(O \cap M) = 23\%$ c. $1 - p(O \cup M) = 1 - 0.767 = 0.233 = 23.3\%$

61. Relative frequencies; they were calculated from a poll.

63. Let S = {interested in Snackolas} and C = {interested in Chippers}

a. $p(S \cup C) = p(S) + p(C) - p(S \cap C) = \frac{15}{25} + \frac{12}{25} - \frac{10}{25} = \frac{17}{25}$

b. $p(C) - p(S \cap C) = \frac{12}{25} - \frac{10}{25} = \frac{2}{25}$

65. a. p(interested in Wi-Fi and not using cable modem)

$= p$(interested in Wi-Fi) $- p$(interested in switching)

$= \frac{726}{1,451} - \frac{514}{1,451} = \frac{212}{1,451}$

b. p(interested in Wi-Fi and is currently using cable modem)

$= p$(interested in switching from cable modem to Wi-Fi)

$= \frac{514}{1,451}$

67. a. $n(C \cup L) = n(C) + n(L) - n(C \cap L)$

696 total, 572 cell, 612 land

$696 = 572 + 612 - n(C \cap L)$

$n(C \cap L) = 488$ both cell/land line

$p(\text{cell only}) = \frac{n(\text{cell}) - n(C \cap L)}{n(\text{total})} = \frac{572 - 488}{696} = \frac{84}{696} = \frac{7}{58}$

b. $p(\text{land only}) = \frac{n(\text{land}) - n(C \cap L)}{n(\text{total})} = \frac{612 - 488}{696} = \frac{124}{696} = \frac{31}{174}$

69. a. $\frac{3}{4}$ b. $\frac{2}{4} = \frac{1}{2}$ c. $\frac{4}{4} = 1$

	N	t
N	NN	tN
t	Nt	tt

	N	t
N	NN	tN
N	NN	tN

	N	N
t	Nt	Nt
t	Nt	Nt

71.   Use Exercise 69 tables.

     a. $1 - \dfrac{3}{4} = \dfrac{1}{4}$ b. $1 - \dfrac{2}{4} = \dfrac{1}{2}$ c. $1 - \dfrac{4}{4} = 0$

73.   Let P = passing photography and E = passing Economics.

     $p(P) = 0.75$

     $p(E) = 1 - 0.65 = 0.35$

     $p(P \cup E) = 0.85$

     $p(P \cap E) = p(P) + p(E) - p(P \cup E)$

                 $= 0.75 + 0.35 - 0.85 = 0.25$

     a. 0.35   b. 0.25   c. $1 - (0.5 + 0.25 + 0.1) = 0.15$ d. $0.5 + 0.1 = 0.6$

75.   a. $0.15 + 0.10 - 0.05 = 0.20$

     b. $1 - 0.05 = 0.95$

     c. $1 - 0.20 = 0.80$

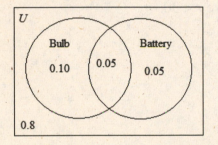

77.   The Union/Intersection Rule: $p(E \cup F) = p(E) + p(F) - p(E \cap F)$

     Known: $n(E \cup F) = n(E) + n(F) - n(E \cap F)$

     Dividing by $n(S)$:

$$\frac{n(E \cup F)}{n(S)} = \frac{n(E)}{n(S)} + \frac{n(F)}{n(S)} - \frac{n(E \cap F)}{n(S)}$$

         $p(E \cup F) = p(E) + p(F) - p(E \cap F)$

79.   The Complement Rule: $p(E) + p(E') = 1$

     By Rule 4: $p(E \cup E') = p(E) + p(E') - p(E \cap E')$

     Note: $E \cup E' = S$, so $p(E \cup E') = 1$, by Rule 2.

        $E \cap E' = \varnothing$, so $p(E \cap E') = 0$, by Rule 1.

     Thus, $1 = p(E) + p(E')$.

83. a. $\dfrac{18}{33} = \dfrac{6}{11}$ b. $\dfrac{6}{11}$ 85. a. $\dfrac{6}{15} \cdot \dfrac{10}{21} = \dfrac{4}{21}$ b. $\dfrac{4}{21}$

87. a. $\dfrac{6}{15} + \dfrac{10}{21} = \dfrac{42}{105} + \dfrac{50}{105} = \dfrac{92}{105}$ b. $\dfrac{92}{105}$

89. a. $\dfrac{7}{6} - \dfrac{5}{7} + \dfrac{9}{14} = \dfrac{49}{42} - \dfrac{30}{42} + \dfrac{27}{42} = \dfrac{46}{42} = \dfrac{23}{21}$ b. $\dfrac{23}{21}$

91. Don't press $" \rightarrow$ Frac"

3.4 Combinatorics and Probability

1. $n(S) = 365^{30}$

$$n(E') = {}_{365}P_{30} = \frac{365!}{(365-30)!} = 365 \cdot 364 \cdots 336 = 2.1710302 \times 10^{76}$$

$$p(E) = 1 - p(E') = 1 - \frac{n(E')}{n(S)} = 1 - 0.29368 \approx 0.7063, \text{ over } 70\%$$

3. From Exercise 1, $r = 30$ and $p = 0.706$. Therefore, here $r < 30$.

Try: $r = 20$, $1 - \dfrac{{}_{365}P_{20}}{365^{20}} = 0.411$, too small

$r = 25$, $1 - \dfrac{{}_{365}P_{25}}{365^{25}} = 0.569$, too large

$r = 22$, $1 - \dfrac{{}_{365}P_{22}}{365^{22}} = 0.476$, too small

$r = 23$, $1 - \dfrac{{}_{365}P_{23}}{365^{23}} = 0.507$, you need 23 people in the group.

5. a. $n(S) = {}_{49}C_6 = \dfrac{49!}{6!(49-6)!} = \dfrac{49!}{6!43!} = \dfrac{49 \cdot 48 \cdot 47 \cdot 46 \cdot 45 \cdot 44}{6 \cdot 5 \cdot 4 \cdot 3 \cdot 2 \cdot 1} = 13{,}983{,}816$

$p = \dfrac{1}{13{,}983{,}816}$

b. $n(S) = {}_{53}C_6 = \dfrac{53!}{6!(53-6)!} = \dfrac{53!}{6!47!} = \dfrac{53 \cdot 52 \cdot 51 \cdot 50 \cdot 49 \cdot 48}{6 \cdot 5 \cdot 4 \cdot 3 \cdot 2 \cdot 1} = 22{,}957{,}480$

$p = \dfrac{1}{22{,}957{,}480}$

c. $n(S) = {}_{51}C_6 = \dfrac{51!}{6!(51-6)!} = \dfrac{51!}{6!45!} = \dfrac{51 \cdot 50 \cdot 49 \cdot 48 \cdot 47 \cdot 46}{6 \cdot 5 \cdot 4 \cdot 3 \cdot 2 \cdot 1} = 18{,}009{,}460$

$p = \dfrac{1}{18{,}009{,}460}$

d. $\dfrac{1}{13,983,816} - \dfrac{1}{22,957,480} = 2.795 \times 10^{-8}$; $\dfrac{2.795 \times 10^{-8}}{\dfrac{1}{22,957,480}} \approx 0.64 = 64\%$ more likely

e. Answers will vary.

7. 5/39: $n(S) = {}_{39}C_5 = \dfrac{39!}{5!(39-5)!} = \dfrac{39!}{5!34!} = \dfrac{39 \cdot 38 \cdot 37 \cdot 36 \cdot 35}{5 \cdot 4 \cdot 3 \cdot 2 \cdot 1} = 575,757$

   a. $p(\text{first prize}) = \dfrac{1}{575,757}$

   b. $n(E) = $ (number of ways to choose 4 winners) $\times$ (number of ways to choose 1 loser)

   $n(E) = {}_5C_4 \cdot {}_{34}C_1 = \dfrac{5!}{4!(5-4)!} \cdot \dfrac{34!}{1!33!} = 170$

   $p(\text{second prize}) = \dfrac{170}{575,757} \approx 0.00029 \approx 0.0003$ or $\dfrac{3}{10,000}$

9. 5/35:

   a. $n(S) = {}_{35}C_5 = \dfrac{35!}{5!30!} = \dfrac{35 \cdot 34 \cdot 33 \cdot 32 \cdot 31}{5 \cdot 4 \cdot 3 \cdot 2 \cdot 1} = 324,632$

   $p(\text{first prize}) = \dfrac{1}{324,632}$

   b. $n(E) = $ (number of ways to choose 4 winners) $\times$ (number of ways to choose 1 loser)

   $n(E) = {}_5C_4 \cdot {}_{30}C_1 = \dfrac{5!}{4!1!} \cdot \dfrac{30!}{1!29!} = 150$

   $p(\text{second prize}) = \dfrac{150}{324,632} \approx 0.000462 \approx 0.0005 = \dfrac{5}{10,000} = \dfrac{1}{2,000}$

11. $n(S) = {}_{56}C_5 \cdot {}_{46}C_1 = \dfrac{56!}{5!51!} \cdot \dfrac{46!}{1!45!} = \dfrac{56 \cdot 55 \cdot 54 \cdot 53 \cdot 52}{5 \cdot 4 \cdot 3 \cdot 2 \cdot 1} \cdot 46 = 175,711,536$

   $p(\text{first prize}) = \dfrac{1}{175,711,536}$

13. $n(S) = {}_{31}C_5 \cdot {}_{16}C_1 = \dfrac{31!}{5!26!} \cdot \dfrac{16!}{1!15!} = \dfrac{31 \cdot 30 \cdot 29 \cdot 28 \cdot 27}{5 \cdot 4 \cdot 3 \cdot 2 \cdot 1} \cdot 16 = 2,718,576$

   $p(\text{first prize}) = \dfrac{1}{2,718,576}$

15. 4/26 is better than 4/77 (it is easier to choose 4 out of 26 than 4 out of 77!)

    Therefore, we can cross off many of the lotteries.

    Also, 5/35 is better than 6/35. This allows us to cross more off.

    In the end, we have to compare: 4/26, 5/30, 6/25

    $$_{26}C_4 = \frac{26!}{4!22!} = \frac{26 \cdot 25 \cdot 24 \cdot 23}{4 \cdot 3 \cdot 2 \cdot 1} = 14,950$$

    $$_{30}C_5 = \frac{30!}{5!25!} = \frac{30 \cdot 29 \cdot 28 \cdot 27 \cdot 26}{5 \cdot 4 \cdot 3 \cdot 2 \cdot 1} = 142,506$$

    $$_{25}C_6 = \frac{25!}{6!19!} = \frac{25 \cdot 24 \cdot 23 \cdot 22 \cdot 21 \cdot 20}{6 \cdot 5 \cdot 4 \cdot 3 \cdot 2 \cdot 1} = 177,100$$

    4/26 easiest

    We use a similar strategy for finding the hardest to win: 4/77 is harder than 4/26; 6/35 is harder than 5/35. In end, we have to compare: 4/77, 5/50, 6/54

    $$_{77}C_4 = \frac{77!}{4!73!} = \frac{77 \cdot 76 \cdot 75 \cdot 74}{4 \cdot 3 \cdot 2 \cdot 1} = 1,353,275$$

    $$_{50}C_5 = \frac{50!}{5!45!} = \frac{50 \cdot 49 \cdot 48 \cdot 47 \cdot 46}{5 \cdot 4 \cdot 3 \cdot 2 \cdot 1} = 2,118,760$$

    $$_{54}C_6 = \frac{54!}{6!48!} = \frac{54 \cdot 53 \cdot 52 \cdot 51 \cdot 50 \cdot 49}{6 \cdot 5 \cdot 4 \cdot 3 \cdot 2 \cdot 1} = 25,827,165$$

    6/54 hardest

17. $$n(S) = {}_{80}C_8 = \frac{80!}{8!72!} = \frac{80 \cdot 79 \cdot 78 \cdot 77 \cdot 76 \cdot 75 \cdot 74 \cdot 73}{8 \cdot 7 \cdot 6 \cdot 5 \cdot 4 \cdot 3 \cdot 2} = 28,987,537,150$$

    $$n(E) = {}_{20}C_8 \cdot {}_{60}C_0 = \frac{20!}{8!12!} = \frac{20 \cdot 19 \cdot 18 \cdot 17 \cdot 16 \cdot 15 \cdot 14 \cdot 13}{8 \cdot 7 \cdot 6 \cdot 5 \cdot 4 \cdot 3 \cdot 2} = 125,970$$

    $$p_8 = \frac{125,970}{28,987,537,150} = 0.000004$$

    $$n(E) = {}_{20}C_7 \cdot {}_{60}C_1 = \frac{20!}{7!13!} \cdot \frac{60!}{1!59!} = \frac{20 \cdot 19 \cdot 18 \cdot 17 \cdot 16 \cdot 15 \cdot 14}{7 \cdot 6 \cdot 5 \cdot 4 \cdot 3 \cdot 2} \cdot 60 = 4,651,200$$

    $$p_7 = \frac{4,651,200}{28,987,537,150} = 0.000160$$

    $$n(E) = {}_{20}C_6 \cdot {}_{60}C_2 = \frac{20!}{6!14!} \cdot \frac{60!}{2!58!} = \frac{20 \cdot 19 \cdot 18 \cdot 17 \cdot 16 \cdot 15}{6 \cdot 5 \cdot 4 \cdot 3 \cdot 2} \cdot \frac{60 \cdot 59}{2} = 68,605,200$$

$$p_6 = \frac{68,605,200}{28,987,537,150} = 0.002367$$

$$n(E) = {}_{20}C_5 \cdot {}_{60}C_3 = \frac{20!}{5!15!} \cdot \frac{60!}{3!57!} = \frac{20 \cdot 19 \cdot 18 \cdot 17 \cdot 16}{5 \cdot 4 \cdot 3 \cdot 2 \cdot 1} \cdot \frac{60 \cdot 59 \cdot 58}{3 \cdot 2 \cdot 1} = 530,546,880$$

$$p_5 = \frac{530,546,880}{28,987,537,150} = 0.018303$$

$$n(E) = {}_{20}C_4 \cdot {}_{60}C_4 = \frac{20!}{4!16!} \cdot \frac{60!}{4!56!} = \frac{20 \cdot 19 \cdot 18 \cdot 17}{4 \cdot 3 \cdot 2} \cdot \frac{60 \cdot 59 \cdot 58 \cdot 57}{4 \cdot 3 \cdot 2} = 2,362,591,575$$

$$p_4 = \frac{2,362,591,575}{28,987,537,150} = 0.081504$$

less than 4: $28,987,537,150 - 125,970 - 4,651,200 - 68,605,200 - 530,546,880 -$

$$2,362,591,575 = 26,021,016,325$$

$$p(\text{less than } 4) = \frac{26,021,016,325}{28,987,537,150} = 0.897662$$

Outcome	Probability
8 winning spots	0.0000043457
7 winning spots	0.0001604552
6 winning spots	0.0023667137
5 winning spots	0.0183025856
4 winning spots	0.0815037015
Fewer than 4 winning spots	0.8976621984

19. a. $10 \cdot 10 \cdot 10 = 1,000$ b. There are 1,000 straight play numbers.

c. $p(\text{straight play}) = \dfrac{1}{1,000}$

21. a. Total number of hands ${}_{52}C_5 = \dfrac{52!}{5!47!} = \dfrac{52 \cdot 51 \cdot 51 \cdot 49 \cdot 48}{5 \cdot 4 \cdot 3 \cdot 2} = 2,598,960$

$n(E) = {}_{13}C_5 = \dfrac{13!}{5!8!} = \dfrac{13 \cdot 12 \cdot 11 \cdot 10 \cdot 9}{5 \cdot 4 \cdot 3 \cdot 2} = 1,287 \qquad p = \dfrac{1,287}{2,598,960} = 0.000495$

b. The number of hands all spades $= 1,287$.

The number of hands of all one suit $= 4 \cdot (1287) = 5,148 \qquad p = \dfrac{5,148}{2,598,960} = 0.001981$

c. Total number of straight flushes $= 4(10) = 40 \qquad p = \dfrac{40}{2,598,960} = 0.000015$

d. Total number of non-straight-flush flushes $= 5,148 - 40 = 5,108$

$$p = \frac{5,108}{2,598,960} = 0.0019654015$$

For 23 – 29, total ways to choose 3: $\quad _{12}C_3 = \frac{12!}{3!9!} = \frac{12 \cdot 11 \cdot 10}{3 \cdot 2} = 220$

23. $\quad _5C_3 = \frac{5!}{3!2!} = \frac{5 \cdot 4}{2} = 10$

$$p = \frac{10}{220} = 0.045454545 \approx 0.05$$

25. $\quad _5C_1 \cdot _7C_2 = \frac{5!}{1!4!} \cdot \frac{7!}{2!5!} = 5 \cdot \frac{7 \cdot 6}{2} = 105$

$$p = \frac{105}{220} = 0.477272727 \approx 0.48$$

27. Number at most one $= 35 + 105 = 140 \qquad p = \frac{140}{220} = 0.63636363 \approx 0.64$

29. Number at least two $= 70 + 10 = 80 \qquad p = \frac{80}{220} = 0.36363636 \approx 0.36$

31. Total number of ways to choose 2 people: $\quad _{200}C_2 = \frac{200!}{2!198!} = \frac{200 \cdot 199}{2} = 19,900$

   a. 2 women: $\quad _{60}C_2 = \frac{60!}{2!58!} = \frac{60 \cdot 59}{2} = 1,770 \qquad p = \frac{1,770}{19,900} = 0.0889447236 \approx 0.09$

   b. 1 woman, 1 man: $\quad _{60}C_1 \cdot _{140}C_1 = \frac{60!}{1!59!} \cdot \frac{140!}{1!139!} = 60 \cdot 140 = 8,400$

$$p = \frac{8,400}{19,900} = 0.4221105528 \approx 0.42$$

   c. 2 men: $\quad _{140}C_2 = \frac{140!}{2!138!} = \frac{140 \cdot 139}{2} = 9,730 \qquad p = \frac{9,730}{19,900} = 0.4889447236 \approx 0.49$

33. $\quad _6C_6 = \frac{6!}{6!(6-6)!} = \frac{6!}{6!0!} = 1$

3.5 Expected Value

1. a. $EV = \frac{2}{38}(17) + \frac{36}{38}(-1) = -\$0.0526315789$

3. a. $EV = \frac{4}{38}(8) + \frac{34}{38}(-1) = -\$0.0526315789$

5. a. $EV = \frac{6}{38}(5) + \frac{32}{38}(-1) = -\$0.0526315789$

7. a. $EV = \frac{18}{38}(1) + \frac{20}{38}(-1) = -\$0.0526315789$

9. a. $EV = \frac{18}{38}(1) + \frac{20}{38}(-1) = -\$0.0526315789$

11. Income $= -50(24) + 7,000(.053) + 4,000(.053) + 4,000(.053) + 3,000(.053)$

$$+ 7,000(.053) + 8,000(.053) = \$549.00$$

13. $EV = 0(0.15) + 1(0.35) + 2(0.25) + 3(0.15) + 4(0.05) + 5(0.05) = 1.75$ books

15. $EV = 8.50(0.2) + 9.00(0.15) + 9.50(0.25) + 10.00(0.2) + 12.50(0.15) + 15.00(0.05) = \$10.05$

17. $EV = \frac{2}{6}(50) + \frac{1}{6}(-20) + \frac{3}{6}(-30) = -\frac{10}{6} = -\$1.67$; Since $-\$1.67 < 0$, don't play.

19. $EV = 10 \cdot \left[\frac{2}{6}(50) + \frac{1}{6}(20) + \frac{3}{6}(-30) \right] = 10 \cdot 5 = \$50.00$; No, since the expected value of 10 games,

$50, is less than $100, don't play. Accept $100.

21. $\dfrac{35 + 37 \cdot (-1)}{38} = \dfrac{35 \cdot 1 + 37(-1)}{38} = \dfrac{35 \cdot 1}{38} + \dfrac{37(-1)}{38} = 35 \cdot \dfrac{1}{38} + (-1)\dfrac{37}{38} = -\dfrac{2}{38} \approx -0.053$

23. The bank's savings account has a value of +4.1%.

To find the fund's expected value:

Outcome	Probability	Value
Up 10%	1/3	+10%
Down 19%	1/3	−19%
Up 14%	1/3	+14%

EV = (+10%)(1/3) + (−19%)(1/3) + (+14%)(1/3) = +1.6666…%

Decision theory indicates that the bank's savings account is the better investment.

25. Let p be the probability that the investment succeeds.

Outcome	Probability	Value
Investment succeeds	p	$+50\% = +0.5$
Investment fails	$1-p$	$-60\% = -0.6$

$EV = (+0.5)(p) + (-0.6)(1-p) = 1.1p - 0.6$

The speculative investment is the better choice if $1.1p - 0.6 > 0.045$, that is,

if $p > 645/1100 = 0.586... \approx 0.59$.

27. a. $EV = \frac{1}{5}(1) + \frac{4}{5}\left(-\frac{1}{4}\right) = 0$ b. $EV = \left(\frac{1}{4}\right)(1) + \frac{3}{4}\left(-\frac{1}{4}\right) = \frac{1}{16}$

    c. $EV = \frac{1}{2}(1) + \frac{1}{2}\left(-\frac{1}{4}\right) = \frac{3}{8}$

    d. If you can eliminate one or more answers, guessing is a winning strategy.

29. $EV = 0.081504(0) + 0.018303(4) + 0.002367(99)$

$+0.000160(1,479) + 0.000004(18,999) + 0.897662(-1) = -0.277481 \approx -0.28$

$-\$0.28$; You should expect to lose about 28 cents for every dollar you bet, if you play a long time.

31. $n(S) = {}_{26}C_4 = \dfrac{26!}{4!22!} = \dfrac{26 \cdot 25 \cdot 24 \cdot 23}{4 \cdot 3 \cdot 2 \cdot 1} = 14,950$

    a. $p(\text{first prize}) = \dfrac{1}{14,950}$

    $p(\text{second prize}) = \dfrac{{}_4C_3 \cdot {}_{22}C_1}{{}_{26}C_4} = \dfrac{88}{14,950} \approx 0.00589 \approx 0.006 = \dfrac{6}{1,000} = \dfrac{3}{500}$

    $p(\text{third prize}) = \dfrac{{}_4C_2 \cdot {}_{22}C_2}{{}_{26}C_4} = \dfrac{1,386}{14,950} \approx 0.0927 \approx 0.09 = \dfrac{9}{100}$

    b. $p(\text{losing}) = 1 - \left(p(\text{first}) + p(\text{second}) + p(\text{third}) \right)$

    $= 1 - \left(\dfrac{1}{14,950} + \dfrac{88}{14,950} + \dfrac{1,386}{14,950} \right) = 1 - \dfrac{1,475}{14,950} = \dfrac{13,475}{14,950} \approx 0.901 \approx \dfrac{9}{10}$

    c. $EV = \dfrac{1}{14,950}(9,999) + \dfrac{88}{14,950}(24) + \dfrac{1,386}{14,950}(1) + \dfrac{13,475}{14,950}(-1) = 0.00147 \approx \$0.001$

33. a. $p(\text{losing}) = 1 - (p(\text{first}) + p(\text{second}) + p(\text{third}))$

$$= 1 - \left(\frac{1}{1,000} + \frac{6}{1,000} + \frac{3}{1,000}\right) = 1 - \frac{10}{1,000} = 0.99$$

  b. $EV = \dfrac{1}{1,000}(499) + \dfrac{6}{1,000}(79) + \dfrac{3}{1,000}(159) + 0.99(-1) = \$0.46$

35. Let p = insurance premium.

$$EV(\text{for insurance company}) = (0.01)(p - 120,000) + (0.99)(p) = 0$$

$$0.01p - 1200 + 0.99p = 0$$

$$p = \$1200 \text{ to break even.}$$

To make a profit, the price should be more than \$1,200.

37. Back: $EV = (0.3)(60) + (0.7)(-6) = \$13.80$ million

Front: $EV = (0.4)(40) + (0.6)(-6) = \$12.40$ million   They should drill in the back yard.

39. a. The value of each prize is actually \$15 less since you are paying for each ticket.

$$EV = 21565\left(\frac{1}{1000}\right) + 925\left(\frac{1}{1000}\right) + 485\left(\frac{2}{1000}\right) + 85\left(\frac{2}{1000}\right) + 165\left(\frac{20}{1000}\right) - 15\left(\frac{974}{1000}\right)$$

  $= \$12.32$ You should buy a ticket.

  b. $$EV = 21565\left(\frac{1}{2000}\right) + 925\left(\frac{1}{2000}\right) + 485\left(\frac{2}{2000}\right) + 85\left(\frac{2}{2000}\right) + 165\left(\frac{20}{2000}\right) - 15\left(\frac{1974}{2000}\right)$$

  $= -\$1.34$ You should not buy a ticket.

  c. $$EV = 21565\left(\frac{1}{3000}\right) + 925\left(\frac{1}{3000}\right) + 485\left(\frac{2}{3000}\right) + 85\left(\frac{2}{3000}\right) + 165\left(\frac{20}{3000}\right) - 15\left(\frac{2974}{3000}\right)$$

  $= -\$5.89$ You should not buy a ticket.

41. Values of prizes are \$5 too high because you pay for ticket.

  a. There are 25 winners and 975 losers.

$$EV = 2320\left(\frac{1}{1000}\right) + 420\left(\frac{2}{1000}\right) + 315\left(\frac{3}{1000}\right) + 18\left(\frac{4}{1000}\right) + 15\left(\frac{15}{1000}\right) - 5\left(\frac{975}{1000}\right) = -\$0.473$$

  You should not buy a ticket.

  b. There are 25 winners and 1,975 losers.

$$EV = 2320\left(\frac{1}{2000}\right) + 420\left(\frac{2}{2000}\right) + 315\left(\frac{3}{2000}\right) + 18\left(\frac{4}{2000}\right) + 15\left(\frac{15}{2000}\right) - 5\left(\frac{1975}{2000}\right) = -\$2.74$$

  You should not buy a ticket.

c. There are 25 winners and 2,975 losers.

$$EV = 2320\left(\frac{1}{3000}\right) + 420\left(\frac{2}{3000}\right) + 315\left(\frac{3}{3000}\right) + 18\left(\frac{4}{3000}\right) + 15\left(\frac{15}{3000}\right) - 5\left(\frac{2975}{3000}\right) = -\$3.49$$

You should not buy a ticket.

43. $EV = \left(\frac{5,000,000}{7,000,000}\right)(15,500,000) + \left(\frac{2,000,000}{7,000,000}\right)(-11,500,000) \approx \$7.786$ million or $\$7.8$ million

45. a. $n(S) = {}_{49}C_6 = \dfrac{49!}{6!43!} = \dfrac{49 \cdot 48 \cdot 47 \cdot 46 \cdot 45 \cdot 44}{6 \cdot 5 \cdot 4 \cdot 3 \cdot 2} = 13,983,816$

b. $13,983,816

c. $13,983,816$ minutes $= 9,710.9833$ days

Since we have 100 people buying, we only need 97.109833 days.

47. You can bet and lose 6 times. If you won the 5th bet, you would win

$32. You would have lost $16 + 8 + 4 + 2 + 1 = \$31$. The net winnings is $1.

Bet	Amount Bet	Amount left after loss
1	$1	$99
2	$2	$97
3	$4	$93
4	$8	$85
5	$16	$69
6	$32	$37
7	$64	----

3.6 Conditional Probability

1. a. $p(H \mid Q)$; Conditional since the sample space is limited to well-qualified candidates.

b. $p(H \cap Q)$; Not conditional.

3. a. $p(S \mid D)$; Conditional since the sample space is limited to users that are dropped a lot.

b. $p(S \cap D)$; Not conditional.

c. $p(D \mid S)$; Conditional since the sample space is limited to users that switch carriers.

5. a. Start at the condition, A: $p(B \mid A) = \dfrac{3}{4}$

b. Start at the beginning of the tree: $p(B \cap A) = p(B \mid A) \cdot p(A) = \dfrac{3}{4} \cdot \dfrac{2}{3} = \dfrac{1}{2}$

7. a. Start at the condition, A': $p(B|A') = \dfrac{3}{8}$

    b. Start at the beginning of the tree: $p(B \cap A') = p(B|A') \cdot p(A') = \dfrac{3}{8} \cdot \dfrac{1}{3} = \dfrac{1}{8}$

9. a. $p(N) = \dfrac{140}{600} \approx 0.2333$ About 23.3% of those surveyed said no.

    b. $p(W) = \dfrac{320}{600} \approx 0.5333$ About 53.3% of those surveyed were women.

    c. $p(N|W) = \dfrac{45}{320} \approx 0.1406$ About 14.1% of the women said no.

    d. $p(W|N) = \dfrac{45}{140} \approx 0.3214$ About 32.1% of those who said no were women.

    e. $p(N \cap W) = \dfrac{45}{600} \approx 0.08$ Exactly 8% of the respondents said no and were women.

    f. $p(W \cap N) = \dfrac{45}{600} \approx 0.08$ Exactly 8% of the respondents were women and said no.

11. a. p(dies of pedestrian transportation accident | dies of a transportation accident in one year)

    $= \dfrac{6,122}{48,817} \approx 0.13$. In one year, 13% of those who die of a transportation accident die of a pedestrian transportation accident.

    b. p(dies of pedestrian transportation accident | dies of a transportation accident in lifetime)

    $= \dfrac{80}{628} \approx 0.13$. In a lifetime, 13% of those who die of a transportation accident die of a pedestrian transportation accident.

    c. p(dies of pedestrian transportation accident | dies of non-transportation accident in

    lifetime) $= 0$. In a lifetime, none of those who die of a non-transportation accident die of a pedestrian transportation accident.

13. a. p(dies from earthquake | dies from a non-transportation accident in lifetime) =

    $\dfrac{56}{103,005} \approx 0.00054$. In a lifetime, approximately 0.05% of those who die of a non-transportation accident die from an earthquake.

    b. p(dies from earthquake | dies from a non-transportation accident in one year) =

    $\dfrac{4,275}{8,013,705} \approx 0.00053$. In one year, approximately 0.05% of those who die of a non-transportation accident die from an earthquake.

c. $p(\text{dies from earthquake} \mid \text{dies from external cause in one year}) = \dfrac{\frac{1}{8,013,705}}{\frac{1}{4,275} + \frac{1}{6,122}} \approx 0.00031$.

In one year, approximately 0.03% of those who die from an external cause die from an earthquake.

15. a. $\frac{13}{52} = \frac{1}{4}$ b. $\frac{12}{51} = \frac{4}{17}$

c. $\frac{13}{52} \cdot \frac{12}{51} = \frac{1}{17}$

d.

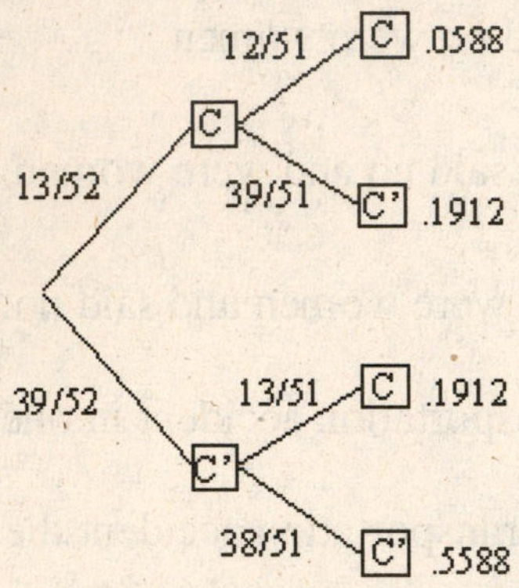

17. a. $\frac{13}{52} = \frac{1}{4}$ b. $\frac{13}{51}$

c. $\frac{13}{52} \cdot \frac{13}{51} = \frac{13}{204}$

d.

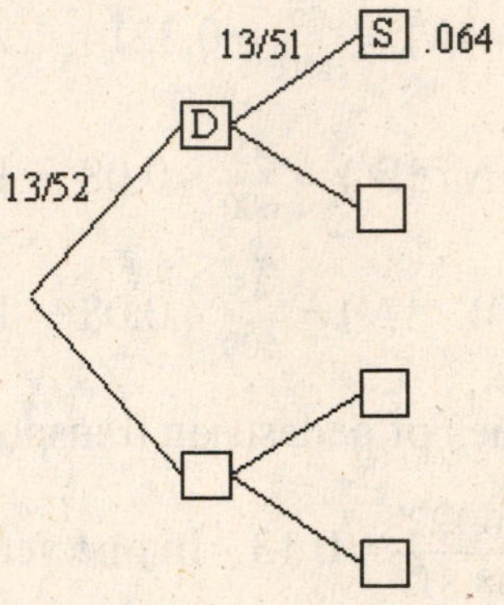

19. a. $p(B|A)$ b. $p(A')$ c. $p(C|A')$

21. a. $\frac{1}{6}$ b. $\frac{1}{3}$ c. $\frac{0}{3} = 0$ d. 1

22. a. $\frac{1}{6}$ b. $\frac{0}{3} = 0$ c. $\frac{1}{3}$ d. 1

In exercises 23 – 25, the following chart is useful:

$$S = \left\{ \begin{array}{l} (1,1),(1,2),(1,3),(1,4),(1,5),(1,6), \\ (2,1),(2,2),(2,3),(2,4),(2,5),(2,6), \\ (3,1),(3,2),(3,3),(3,4),(3,5),(3,6), \\ (4,1),(4,2),(4,3),(4,4),(4,5),(4,6), \\ (5,1),(5,2),(5,3),(5,4),(5,5),(5,6), \\ (6,1),(6,2),(6,3),(6,4),(6,5),(6,6) \end{array} \right\}$$

23. a. $E = \{(1,5), (2,4), (3,3), (4,2), (5,1)\}$; $p = \frac{5}{36}$

b. $p = \frac{5}{18}$ c. $p = 0$ d. $p = 1$

25. a. $p = \frac{3}{36} = \frac{1}{12}$ b. $p = \frac{3}{10}$ c. $p = 1$

27. E_2 is most likely; E_3 is least likely

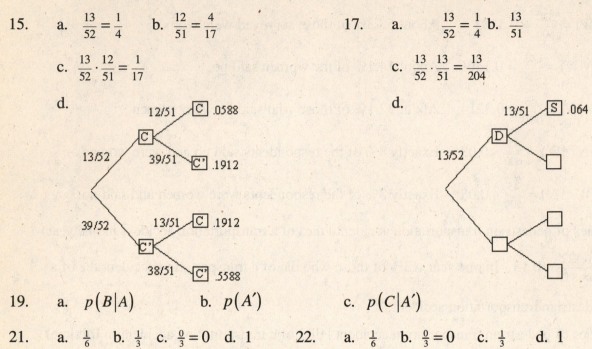

29.   236 made a purchase; 464 did not make a purchase

        $125 + 148 = 273$ were happy; $111 + 316 = 427$ were unhappy

$$p(\text{purchase}|\text{happy}) = \frac{n(\text{purchase and happy})}{n(\text{happy})} = \frac{125}{273} \approx 0.46$$

        About 46% of those who were happy with the service made a purchase.

31. $\dfrac{13}{52} \cdot \dfrac{12}{51} \cdot \dfrac{11}{50} \cdot \dfrac{10}{49} \cdot \dfrac{9}{48} \approx 0.0005$

33. $p(\text{last four are spades}|\text{first is spade})$

$$= \frac{p(\text{last four are spades and first is spade})}{p(\text{first is spade})} = \frac{\frac{13}{52} \cdot \frac{12}{51} \cdot \frac{11}{50} \cdot \frac{10}{49} \cdot \frac{9}{48}}{\frac{13}{52}} = \frac{12}{51} \cdot \frac{11}{50} \cdot \frac{10}{49} \cdot \frac{9}{48} \approx 0.0020$$

The following tree diagram is useful for Exercise 35:

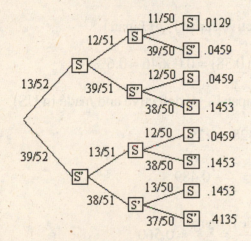

35. $0.0459 + 0.0459 + 0.0459 \approx 0.14$

The following tree diagram is useful for Exercise 37:

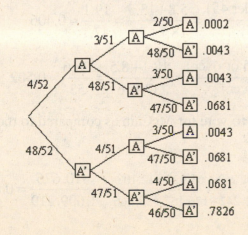

37. $0.0681 + 0.0681 + 0.0681 \approx 0.20$

The following tree diagram is useful for Exercise 39:

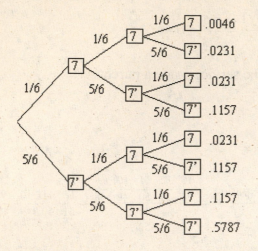

39. $0.0231 + 0.0231 + 0.0231 \approx 0.07$

41. $p\left(\text{defective and made in Japan}\right) = p\left(\text{defective}\middle|\text{Japan}\right) \cdot p\left(\text{Japan}\right)$

$$= (0.017)(0.38) = 0.00646 \approx 0.6\%$$

43. $p\left(\text{defective}\right) = p\left(\text{defective and made in Japan}\right) + p\left(\text{defective and made in US}\right)$

$$= 0.00646 + 0.00682 = 0.01328 \approx 1.3\%$$

45.   a. $p\left(\text{Obama}\middle|\text{male}\right) = \dfrac{n\left(\text{Obama} \cap \text{male}\right)}{n\left(\text{male}\right)} = \dfrac{23}{47} \approx 0.489$

       b. $p\left(\text{Obama}\middle|\text{female}\right) = \dfrac{n\left(\text{Obama} \cap \text{female}\right)}{n\left(\text{female}\right)} = \dfrac{29.7}{53} = 0.560$

       c.   For those voting for Obama, a higher percentage was women.

47.   a. $p\left(\text{McCain}\middle|\text{under 45}\right) = \dfrac{n\left(\text{McCain} \cap \text{under 45}\right)}{n\left(\text{under 45}\right)} = \dfrac{5.8 + 13.3}{18.1 + 29} = \dfrac{19.1}{47.1} \approx 0.406$

       b. $p\left(\text{McCain}\middle|\text{45 or over}\right) = \dfrac{n\left(\text{McCain} \cap \text{45 or over}\right)}{n\left(\text{45 or over}\right)} = \dfrac{18.1 + 8.5}{37 + 16} = \dfrac{26.6}{53} \approx 0.502$

       c.   Those 45 years or over were more likely to vote for McCain as compared to those under 45 years.

49.   a. $p\left(\text{male}\right) = \dfrac{n\left(\text{male}\right)}{n\left(\text{total}\right)} = \dfrac{487,695 + 175,704 + 71,242 + 63,927 + 12,108}{487,695 + 255,859 + 71,242 + 176,157 + 18,266} = \dfrac{810,676}{1,009,219} = 0.803$

$$p(\text{female}) = \frac{n(\text{female})}{n(\text{total})} = \frac{80,155 + 112,230 + 6,158}{1,009,219} = \frac{198,543}{1,009,219} = 0.197$$

b. $p(\text{injection drug use}|\text{male}) = \dfrac{n(\text{injection drug use} \cap \text{male})}{n(\text{male})} = \dfrac{175,704}{810,676} = 0.217$

$p(\text{injection drug use}|\text{female}) = \dfrac{n(\text{injection drug use} \cap \text{female})}{n(\text{female})} = \dfrac{80,155}{198,543} = 0.404$

c. $p(\text{injection drug use} \cap \text{male}) = p(\text{injection drug use} | \text{male}) \cdot p(\text{male})$

$$= 0.217 \cdot 0.803 = 0.174$$

$p(\text{injection drug use} \cap \text{female}) = p(\text{injection drug use} | \text{female}) \cdot p(\text{female})$

$$= 0.404 \cdot 0.197 = 0.080$$

d. $p(\text{heterosexual contact}|\text{male}) = \dfrac{n(\text{heterosexual} \cap \text{male})}{n(\text{male})} = \dfrac{63,927}{810,676} = 0.079$

$p(\text{heterosexual contact}|\text{female}) = \dfrac{n(\text{heterosexual} \cap \text{female})}{n(\text{female})} = \dfrac{112,230}{198,543} = 0.565$

e. $p(\text{heterosexual contact} \cap \text{male}) = p(\text{heterosexual contact} | \text{male}) \cdot p(\text{male})$

$$= 0.079 \cdot 0.803 = 0.063$$

$p(\text{heterosexual contact} \cap \text{female}) = p(\text{heterosexual contact} | \text{female}) \cdot p(\text{female})$

$$= 0.565 \cdot 0.197 = 0.111$$

f. Part (b) is the percentage of males that were exposed by injection drug use. Part (c) is the percentage of the total that are both male and exposed by injection drug use. Similarly for parts (d) and (e).

51. a. $p(\text{obese} | \text{man}) = 32\%$ (given)

b. $p(\text{obese} \cap \text{man}) = p(\text{obese} | \text{man}) \cdot p(\text{man}) = 0.32 \cdot \dfrac{148 \text{ million}}{148 \text{ million} + 152 \text{ million}} \approx 15.8\%$

c. $p(\text{obese} | \text{woman}) = 35\%$ (given)

d. $p(\text{obese} \cap \text{woman}) = p(\text{obese} | \text{woman}) \cdot p(\text{woman})$

$$= 0.35 \cdot \dfrac{152 \text{ million}}{148 \text{ million} + 152 \text{ million}} \approx 17.7\%$$

e. $p(\text{obese} \mid \text{male or female}) = \dfrac{n(\text{obese} \cap \text{male or female})}{n(\text{male or female})}$

$$= \dfrac{(0.32 \cdot 148) + (0.35 \cdot 152)}{148 + 152} \approx 33.5\%$$

f. Part (c) is the percentage of adult women who are obese. Part (d) is the percentage of adults who are both female and obese.

53.　a.　$p = \dfrac{n(\text{death penalty imposed} \cap \text{victim white and defendant white})}{n(\text{victim white and defendant white})} = \dfrac{19}{19 + 132} \approx 12.6\%$

　　b.　$p = \dfrac{n(\text{death penalty imposed} \cap \text{victim white and defendant black})}{n(\text{victim white and defendant black})} = \dfrac{11}{11 + 152} \approx 6.7\%$

　　c.　$p = \dfrac{n(\text{death penalty imposed} \cap \text{victim black and defendant white})}{n(\text{victim black and defendant white})} = \dfrac{0}{0 + 9} = 0\%$

　　d.　$p = \dfrac{n(\text{death penalty imposed} \cap \text{victim black and defendant black})}{n(\text{victim black and defendant black})} = \dfrac{6}{6 + 97} \approx 5.8\%$

55.　Both parents are carriers:

	N	s
N	NN	sN
s	Ns	ss

Since we know he does not have sickle-cell anemia, he cannot be ss: $p(\text{Ns} \mid \text{Ns or NN}) = \dfrac{2}{3}$

The following tree diagram is useful for Exercise 57:

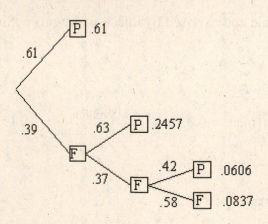

57.　$0.61 + 0.2457 + 0.0606 = 0.9163 \approx 92\%$

59.　$1 - 0.61 = 0.39 = 39\%$

61.   The number of 10, J, Q, K cards = 16.

$$p\left(\text{Ace then 10, J, Q, or K}\right)=\tfrac{4}{52}\cdot\tfrac{16}{51}=\tfrac{16}{663}$$

$$p\left(\text{10, J, Q or K then A}\right)=\tfrac{16}{52}\cdot\tfrac{4}{51}=\tfrac{16}{663}$$

$$\tfrac{16}{663}+\tfrac{16}{663}=\tfrac{32}{663}\approx 0.04826546\approx 0.05$$

63.   a. $p\left(N'|W\right)=\dfrac{n\left(N'\cap W\right)}{n\left(W\right)}=\dfrac{256+19}{320}\approx 0.86=86\%$

   b. $p\left(N|W'\right)=\dfrac{n\left(N\cap W'\right)}{n\left(W'\right)}=\dfrac{95}{280}\approx 0.34=34\%$

   c. $p\left(N'|W'\right)=\dfrac{n\left(N'\cap W'\right)}{n\left(W'\right)}=\dfrac{162+23}{280}\approx 0.66=66\%$

   d. $N'|W$; $p\left(N|W\right)=\tfrac{45}{320}=0.14=1-0.86$

65.   The complement is $A'\big|B$.

67. $p\left(N|W\right)=\dfrac{p\left(N\cap W\right)}{p\left(W\right)}=\dfrac{\frac{45}{600}}{\frac{320}{600}}=\dfrac{45}{320}\approx 0.14$

3.7 Independence; Trees in Genetics

1.   a. E and F are dependent. Knowing F affects E's probability.

   b. E and F are not mutually exclusive. There are many women doctors.

3.   a. E and F are dependent. Knowing F affects E's probability.

   b. E and F are mutually exclusive. One cannot be both single and married.

5.   a. E and F are independent. Knowing F does not affect E's probability.

   b. E and F are not mutually exclusive. A brown-haired person may have some gray.

7.   a. E and F are dependent. Knowing F affects E's probability.

   b. E and F are mutually exclusive. One can't wear both shoes and sandals (excluding one on each foot!)

9.   a. $p\left(E\right)=\tfrac{1}{6}$ $p\left(F\right)=\tfrac{3}{6}$ $p\left(E|F\right)=0$

       E and F are dependent.

   b. E and F are mutually exclusive.

c. Knowing that you got an odd number

changes the probability of getting a 4.

You cannot get both a 4 and an odd number.

11. E = responding yes, F = being a woman

$$p(E) = \frac{418}{600}, \quad p(F) = \frac{320}{600} \qquad p(E|F) = \frac{n(E \cap F)}{n(F)} = \frac{256}{320}$$

E and F are not independent.

13. a. $p(5) = \frac{1}{6}$ b. $p(5|\text{even}) = \frac{p(5 \cap \text{even})}{p(\text{even})} = \frac{0}{0.5} = 0$

c. No. If you roll a 5, the probability of rolling an even number is zero.

d. Yes. 5 is not an even numbers.

15. a. $p(\text{jack}) = \frac{4}{52} = \frac{1}{13}$ b. $p(\text{jack} \mid \text{red}) = \frac{p(\text{jack} \cap \text{red})}{p(\text{red})} = \frac{\frac{2}{52}}{\frac{26}{52}} = \frac{2}{26} = \frac{1}{13}$

c. Yes. Being dealt a red card doesn't change the probability of being dealt a jack.

d. No. There are red jacks.

17. E = happy with service, F = making a purchase

$$p(E) = \frac{125 + 148}{700} = \frac{273}{700} = 0.39, \quad p(F) = \frac{125 + 111}{700} = \frac{236}{700} \approx 0.337$$

$$p(F \mid E) = \frac{n(F \cap E)}{n(E)} = \frac{125}{273} = 0.457... \approx 0.46 > p(F)$$

The events are dependent. Being happy with the service increases the probability of a purchase.

19. E = defective, F = Japanese made

$p(E) = 0.01328$, $p(F) = 0.38$ (from Exercises 41 – 44 in Section 3.6)

$p(E|F) = 0.017$ E and F are dependent. Knowing that the chip is made in Japan increases

the probability that it is defective.

21. a. $p(\text{vegetarian}) = \frac{189 + 36}{365} \approx 0.616$; $p(\text{vegetarian} \mid \text{healthy}) = \frac{189}{281} \approx 0.673$

Since $0.616 \neq 0.673$, they are not independent.

b. $p(\text{vegetarian} \cap \text{healthy}) = p(\text{vegetarian} \mid \text{healthy}) \cdot p(\text{healthy}) = \frac{189}{281} \cdot \frac{281}{365} \approx 0.518$

Since $0.518 \neq 0$, they are not mutually exclusive.

23. a. $p(A) = \dfrac{518+89}{122+518+89+223} \approx 0.638$; $\quad p(A|B) = \dfrac{89}{89+223} \approx 0.285$

    Since $0.638 \neq 0.285$, they are not independent.

    b. No; $p(A \cap B) \neq 0$; the circles overlap with 89 in both.

25. $p(\text{HAL}) = 0.60$, $p(\text{quit}) = 0.04$, $p(\text{HAL} \cap \text{quit}) = 0.03$

    $$p(\text{quit}|\text{HAL}) = \frac{p(\text{quit} \cap \text{HAL})}{p(HAL)} = \frac{0.03}{0.6} = 0.05$$

    They are dependent. HAL users were more likely to quit.

27. $p(\text{SmellSoGood}) = 0.40$, $p(\text{quit}) = 0.10$, $p(\text{quit} \cap \text{SmellSoGood}) = 0.04$

    $$p(\text{quit}|\text{SmellSoGood}) = \frac{p(\text{quit} \cap \text{SmellSoGood})}{\text{p}(\text{SmellSoGood})} = \frac{0.04}{0.40} = 0.1$$

    They are independent. Smell So Good users quit at the same rate as all deodorant users.

29. a. $p = (0.01)(0.01)(0.01) = 0.000001$

    b. $(0.01)^5 \leq 0.000000001$; so 4 backup systems.

31. $p(\text{ill}|-) = \dfrac{p(\text{ill} \cap -)}{p(-)} = \dfrac{0.005}{0.005+0.49} = \dfrac{0.005}{0.495} \approx 0.010101 \approx 1.0\%$

33. $p(\text{ill}) = \dfrac{1,000,000}{287,000,000} = 0.0034843206$

    $p(\text{healthy}) = 0.9965156794$

    $p(-|\text{healthy}) = 0.996$, so $p(+|\text{healthy}) = 0.004$

    $p(+|\text{ill}) = 0.999$, so $p(-|\text{ill}) = 0.001$

    $p(- \cap \text{healthy}) = p(-|\text{healthy}) \cdot p(\text{healthy}) = (0.996)(0.9965156794) = 0.9925296167$

    $p(+ \cap \text{healthy}) = p(+|\text{healthy}) \cdot p(\text{healthy}) = (0.004)(0.9965156794) = 0.0039860627$

    $p(+ \cap \text{ill}) = p(+|\text{ill}) \cdot p(\text{ill}) = (0.999)(0.0034843206) = 0.0034808363$

    $p(- \cap \text{ill}) = p(-|\text{ill}) \cdot p(\text{ill}) = (0.001)(0.0034843206) = 0.0000034843206$

a. $p(\text{ill}|+) = \dfrac{p(\text{ill} \cap +)}{p(+)} = \dfrac{0.00348}{0.00348 + 0.00399} \approx 0.47$

b. $p(\text{healthy}|-) = \dfrac{p(\text{healthy} \cap -)}{p(-)} = \dfrac{0.99253}{0.99253 + 0.000003} \approx 0.999996$

c. $p(\text{healthy}|+) = \dfrac{p(\text{healthy} \cap +)}{p(+)} = \dfrac{0.00399}{0.00399 + 0.00348} \approx 0.53$

d. $p(\text{ill}|-) = \dfrac{p(\text{ill} \cap -)}{p(-)} = \dfrac{0.000003}{0.992533 + 0.000003} \approx 0.000004$

e. The results from (a) and (c) would be informative because a positive test doesn't necessarily mean you are ill.

f. (c) is a false positive because the test was positive, but he was healthy. (d) is a false negative because the test was negative, but he was ill.

37. a. Three Punnett squares are used.

Type A Both Parents are Carriers		
	N	c
N	NN	cN
c	Nc	cc

Type B Only One Parent is a Carrier		
	N	c
N	NN	cN
N	NN	cN

Type C Neither Parent is a Carrier		
	N	N
N	NN	NN
N	NN	NN

The grandfather comes from Type A but does not have cystic fibrosis. Therefore, his probability of being a carrier is 2/3 and of not being a carrier is 1/3. (cc possibility is eliminated.)

The grandfather's children (parents of the cousins) each have a probability of being a carrier of ½ if the grandfather is a carrier (Type B) and a probability of 0 if the grandfather is not a carrier (Type C).

Likewise, the cousins each have a probability of being a carrier of ½ if the parent is a carrier (Type B) and a probability of 0 if the parent is not a carrier (Type C).

Construct a tree diagram to find the probability of a cousin being a carrier.

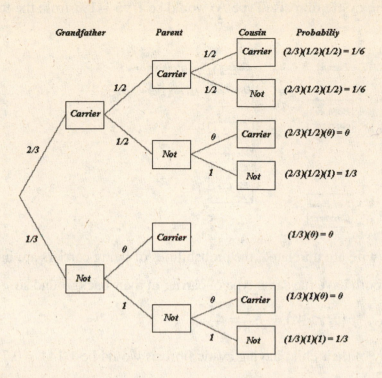

Since the husband (H) and wife (W) are related, the event that they are both carriers $(H \cap W)$ is dependent. Thus, $p(H \cap W) = p(H \mid W) \cdot p(W)$. From the tree above, the probability that a cousin, the wife for example, is a carrier is 1/6. If the wife is a carrier, the grandfather must have been a carrier.

The new tree for the husband follows.

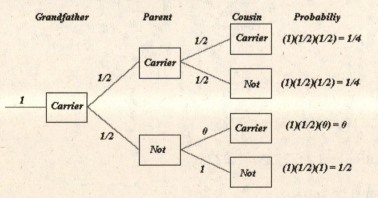

The probability the husband is a carrier given that the wife is a carrier is ¼.

$$p(H \cap W) = p(H \mid W) \cdot p(W) = \frac{1}{4} \cdot \frac{1}{6} = \frac{1}{24}$$

Therefore, the probability of both cousins being a carrier would be 1/24. And the probability of their child having cystic fibrosis (Type A) would be $1/96 \approx 1\%$ from the following tree diagram.

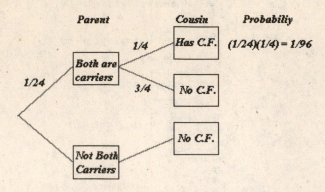

b. Since husband and wife are unrelated, the probabilities of being carriers are independent. The husband and wife both have the same type of carrier in their background so

$$p(H) = p(W) = \frac{1}{6}; \quad p(H \cap W) = \frac{1}{6} \cdot \frac{1}{6} = \frac{1}{36}$$

And the probability of their child having cystic fibrosis would be $1/144 \approx 0.7\%$ as seen in the following tree diagram.

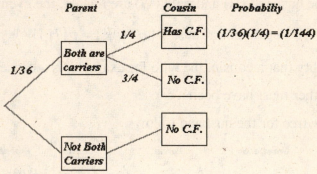

39. $p\left(+|\text{carrier}\right) = 0.85$

$p\left(\text{both test positive}\right) = \left(0.85\right)\left(0.85\right) = 0.7225$

$p\left(\text{don't both test positive}\right) = 1 - 0.7225 = 0.2775$

41.   Mrs. Jones' parents must be carriers.

	N	a
N	NN	aN
a	Na	aa

Mrs. Jones has $\frac{1}{3}$ chance of not being a carrier, and $\frac{2}{3}$ chance of being a carrier.

If she is not a carrier, the child cannot be albino.

	a	a
N	aN	aN
N	aN	aN

If she is a carrier, the child has $\frac{1}{2}$ chance of being albino.

	a	a
N	aN	aN
a	aa	aa

$$p(\text{child is albino}) = \frac{2}{3} \cdot \frac{1}{2} = \frac{1}{3}$$

43. If the first child is albino, Mrs. Jones must be a carrier. Thus, the child has $\frac{1}{2}$ chance of being albino.

45.

	M^{Bd}	M^{Bw}
M^{Bk}	$M^{Bd}M^{Bk}$	$M^{Bw}M^{Bk}$
M^{Bk}	$M^{Bd}M^{Bk}$	$M^{Bw}M^{Bk}$

$p(M^{Bd}M^{Bk}) = \frac{1}{2}$, $p(M^{Bw}M^{Bk}) = \frac{1}{2}$

	R^{+}	R^{+}
R^{-}	$R^{+}R^{-}$	$R^{+}R^{-}$
R^{-}	$R^{+}R^{-}$	$R^{+}R^{-}$

$p(R^{+}R^{-}) = 1$

$\frac{1}{2}$ chance of chestnut hair, $\frac{1}{2}$ chance of shiny dark brown

47.

	M^{Bw}	M^{Bk}
M^{Bd}	$M^{Bw}M^{Bd}$	$M^{Bk}M^{Bd}$
M^{Bd}	$M^{Bw}M^{Bd}$	$M^{Bk}M^{Bd}$

$p(M^{Bd}M^{Bw}) = \frac{1}{2}$, $p(M^{Bd}M^{Bk}) = \frac{1}{2}$

	R^{+}	R^{-}
R^{+}	$R^{+}R^{+}$	$R^{-}R^{+}$
R^{-}	$R^{+}R^{-}$	$R^{-}R^{-}$

$p(R^{+}R^{+}) = \frac{1}{4}$, $p(R^{+}R^{-}) = \frac{1}{2}$, $p(R^{-}R^{-}) = \frac{1}{4}$

The child has a $\frac{1}{8}$ chance of the following hair colors: dark red, light brown, auburn or medium brown. The child has a $\frac{1}{4}$ chance of the following: reddish brown, chestnut.

49.

	M^{Bw}	M^{Bk}
M^{Bd}	$M^{Bw}M^{Bd}$	$M^{Bk}M^{Bd}$
M^{Bk}	$M^{Bw}M^{Bk}$	$M^{Bk}M^{Bk}$

	R^+	R^-
R^+	R^+R^+	R^-R^+
R^-	R^+R^-	R^-R^-

$p\left(M^{Bd}M^{Bw}\right)=\frac{1}{4}$, $p\left(M^{Bd}M^{Bk}\right)=\frac{1}{4}$,

$p\left(M^{Bw}M^{Bk}\right)=\frac{1}{4}$, $p\left(M^{Bk}M^{Bk}\right)=\frac{1}{4}$

$p\left(R^+R^+\right)=\frac{1}{4}$, $p\left(R^+R^-\right)=\frac{1}{2}$,

$p\left(R^-R^-\right)=\frac{1}{4}$

The child has a $\frac{1}{8}$ chance of the following hair colors: reddish brown, chestnut, shiny dark brown, shiny black. The child has a $\frac{1}{16}$ chance of the following hair colors: dark red, light brown, auburn, medium brown, glossy dark brown, dark brown, glossy black, black.

51.
a. $\frac{1}{6}$ b. $\frac{5}{6}$ c. $\frac{5}{6}\cdot\frac{5}{6}\cdot\frac{5}{6}\cdot\frac{5}{6}=\frac{625}{1,296}\approx 0.48$

d. $1-\frac{625}{1,296}=\frac{671}{1,296}\approx 0.52$ e. $EV=(0.518)(1)+(0.482)(-1)=0.036\approx \$0.04$

53. He won because the expected value was positive. He lost because the expected value was negative. $\$0.04>-\$0.017$

55. $p\left(\text{ANT}\,|\,\text{have symptoms}\right)=0.25$; $p\left(\text{not ANT}\,|\,\text{have symptoms}\right)=0.75$

$p\left(+\,|\,\text{ill}\right)=0.88$, $p\left(-\,|\,\text{ill}\right)=0.12$ $p\left(-\,|\,\text{healthy}\right)=0.92$, $p\left(+\,|\,\text{healthy}\right)=0.08$

$p\left(\text{side effects}\right)=0.02$, $p\left(\text{die}\,|\,\text{untreated}\right)=0.90$, $p\left(\text{live}\,|\,\text{untreated}\right)=0.10$

a. $p\left(\text{die}\right)=(0.25)(0.12)(0.9)=0.027$ b. $p\left(\text{die}\right)=(0.25)(0.9)=0.225$

$p\left(\text{live}\right)=1-0.027=0.973=97.3\%$ $p\left(\text{live}\right)=1-0.225=0.775$

c. $p\left(\text{good health returned}\right)=1-0.02=0.98$ d. Yes.

Chapter 3 Review

1. Experiment is pick one card from a deck of 52 cards. Sample space S = {possible outcomes} = {jack of hears, ace of spades, ...}; n(S) = 52.

3. $p=\frac{13}{52}=\frac{1}{4}$, odds 1:3 If you deal a card many times, you should expect to be dealt a club approximately ¼ the time, and you should expect to be dealt a club approximately one time for every 3 times you are dealt something else.

5. $p(Q \cup \text{club}) = p(Q) + p(\text{club}) - p(Q \cap \text{club}) = \frac{4}{52} + \frac{13}{52} - \frac{1}{52} = \frac{16}{52} = \frac{4}{13}$

odds $4:9$ If you deal a card many times, you should expect to be dealt a queen or a club approximately 4/13 the time, and you should expect to be dealt a queen of clubs approximately four times for every 9 times you are dealt something else.

7.   a.   Flip three coins and observe the results.

       b. $S = \{\text{HHH,HHT,HTH,HTT,TTT,TTH,THT,THH}\}$

9. $F = \{\text{HTT,TTH,THT,TTT}\}$

11. $p(F) = \frac{4}{8} = \frac{1}{2}$, odds $1:1$.   13. $p = \frac{6}{36} = \frac{1}{6}$.   15. $p = \frac{6+2+6}{36} = \frac{14}{36} = \frac{7}{18}$.

17. $p\left(3 \text{ or } 5 \text{ or } 7 \text{ or } 9 \text{ or } 10 \text{ or } 11 \text{ or } 12\right) = \frac{2+4+6+4+3+2+1}{36} = \frac{22}{36} = \frac{11}{18}$.

19. $p = \frac{13}{52} \cdot \frac{12}{51} \cdot \frac{11}{50} = \frac{1,716}{132,600} = \frac{11}{850} \approx 0.013$

21. $p = \frac{11}{850} + \frac{117}{850} = \frac{128}{850} = \frac{64}{425} \approx 0.151$

23. $p(\text{2nd Heart} \mid \text{1st Heart}) = \dfrac{p(\text{2nd Heart} \cap \text{1st Heart})}{p(\text{1st Heart})} = \dfrac{\frac{13}{52} \cdot \frac{12}{51}}{\frac{13}{52}} = \frac{12}{51}$

25. $p = \frac{1}{6} \cdot \frac{1}{6} \cdot \frac{1}{6} = \frac{1}{216}$

27. $p = \frac{1}{216} + \frac{15}{216} = \frac{16}{216} = \frac{2}{27}$

29. $p(\text{2nd 7} \mid \text{first 7}) = \dfrac{p(\text{2nd 7} \cap \text{first 7})}{p(\text{first 7})} = \dfrac{\frac{1}{6} \cdot \frac{1}{6}}{\frac{1}{6}} = \frac{1}{6}$

31. $\frac{2}{4} = \frac{1}{2}$ 33. $\frac{1}{4}$ 35. $\frac{1}{4}$

	L	s
s	Ls	ss
s	Ls	ss

	N	c
N	NN	cN
c	Nc	cc

37. $\frac{2}{4} = \frac{1}{2}$

	S	s
S	SS	Ss
s	Ss	ss

39.　$\dfrac{2}{4} = \dfrac{1}{2}$　　　　41.　$\dfrac{2}{4} = \dfrac{1}{2}$

	H	h
h	Hh	hh
h	Hh	hh

43.　$p(2 \text{ tens and 3 jacks}) = \dfrac{_4C_2 \cdot _4C_3}{_{52}C_5} = \dfrac{6 \cdot 4}{2,598,960} \approx 9.23 \times 10^{-6}$

45.　a.

Winning Spot	Probability
9	$_{20}C_9 \cdot _{60}C_0 / _{80}C_9 = 0.000000724$
8	$_{20}C_8 \cdot _{60}C_1 / _{80}C_9 = 0.000032592$
7	$_{20}C_7 \cdot _{60}C_2 / _{80}C_9 = 0.000591678$
6	$_{20}C_6 \cdot _{60}C_3 / _{80}C_9 = 0.005719558$
5	$_{20}C_5 \cdot _{60}C_4 / _{80}C_9 = 0.032601481$
4 or less	0.961053966

　　b.　$EV = -1(0.961053966) + 0(0.032601481) + 49(0.005719558) + 389(0.000591678)$

　　　$+ 5999(0.000032592) + 24,999(0.000000724) = -\$0.24$; You would expect to lose $0.24.

47.　$EV = \dfrac{3(6.5) + 3(7) + 4(8.5) + 4(10) + 4(13.5) + 2(25)}{3 + 3 + 4 + 4 + 4 + 2} = \dfrac{218.5}{20} \approx \$10.93$

49.　$p(\text{O'Neill} | \text{rural}) = \dfrac{131}{131 + 181 + 16} = \dfrac{131}{328} \approx 0.40$;　$p(\text{Bell} | \text{rural}) = \dfrac{181}{131 + 181 + 16} = \dfrac{181}{328} \approx 0.55$

51.　$p(\text{urban} | \text{Bell}) = \dfrac{n(\text{urban} \cap \text{Bell})}{n(\text{Bell})} = \dfrac{184}{184 + 181} = \dfrac{184}{365} \approx 0.50$

　　$p(\text{rural} | \text{Bell}) = \dfrac{n(\text{rural} \cap \text{Bell})}{n(\text{Bell})} = \dfrac{181}{365} \approx 0.50$

53.　The urban residents prefer O'Neill.　The rural residents prefer Bell.

55.　Since the governors election depends only on who gets the most votes, O'Neill is ahead with

　　49.6% to Bell's $\dfrac{184 + 181}{800} \approx 45.6\%$.

57.　Independent.　Not mutually exclusive.

59.　Dependent.　Not mutually exclusive.

61.　$p(\text{returned}) = p(\text{return} \mid \text{GG}) \cdot p(\text{GG}) + p(\text{return} \mid \text{HH}) \cdot p(\text{HH})$

　　　　$= 0.20 \cdot 0.40 + 0.10 \cdot 0.60 = 0.14 = 14\%$

63. $p(\text{returned} \cap \text{GG}) = p(\text{returned} \mid \text{GG}) \cdot p(\text{GG}) = 0.20 \cdot 0.40 = 0.08 = 8\%$

65. $p(\text{HH} \mid \text{returned}) = \dfrac{p(\text{HH} \cap \text{returned})}{p(\text{returned})} = \dfrac{0.06}{0.14} \approx 0.43 = 43\%$

67. $p(\text{defective} \cap \text{AK}) = p(\text{defective} \mid \text{AK}) \cdot p(\text{AK}) = 0.02 \cdot 0.39 = 0.0078 = 0.78\%$

69. $p(\text{defective}) = p(\text{defective} \mid \text{AK}) \cdot p(\text{AK}) + p(\text{defective} \mid \text{NV}) \cdot p(\text{NV})$

$$= 0.02 \cdot 0.39 + 0.017 \cdot 0.61 = 0.01817 = 1.817\%$$

71. $p(\text{AK} \mid \text{defective}) = \dfrac{p(\text{AK} \cap \text{defective})}{p(\text{defective})} = \dfrac{0.0078}{0.01817} \approx 0.43 = 43\%$

73.   No, since $p(\text{AK}) = 0.39 \neq p(\text{AK} \mid \text{defective}) = 0.43$, they are not independent. Being made in

Arkansas increases the probability of being defective.

4.1 Population, Sample, and Data

1. a.

Number of Visits	Frequency
1	9
2	8
3	2
4	5
5	6
	30

b.

Number of Visits	Frequency	Relative Frequency	Angle (in degrees)
1	9	0.3	108
2	8	0.2667	96
3	2	0.0667	24
4	5	0.1667	60
5	6	0.2	72
	30		

c.

Library Habits - Number of Student Visits

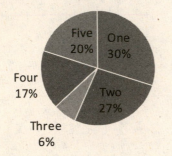

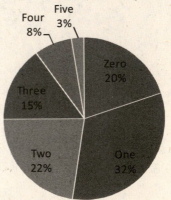

3. a.

Number of Children	Frequency
0	8
1	13
2	9
3	6
4	3
5	1
	40

Families in Manistee, Michigan

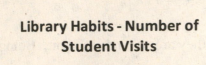

b.

Number of Children	Frequency	Relative Frequency	Angle (in degrees)
0	8	0.2	72
1	13	0.325	117
2	9	0.225	81
3	6	0.15	54
4	3	0.075	27
5	1	0.025	9
	40		

c.

Families in Manistee, Michigan

5.   a.   Range $= 80 - 51 = 29$. So each interval should be $\frac{29}{6} \approx 5$.

x = Speed	Frequency	Relative Frequency
$51 \leq x < 56$	2	0.05
$56 \leq x < 61$	4	0.1
$61 \leq x < 66$	7	0.175
$66 \leq x < 71$	10	0.25
$71 \leq x < 76$	9	0.225
$76 \leq x < 81$	8	0.2
	40	

b.

Speed of Cars on Interstate 40

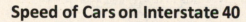

7.

x = Time in Minutes	Number of Freshmen
$0 \leq x < 10$	32
$10 \leq x < 20$	47
$20 \leq x < 30$	36
$30 \leq x < 40$	22
$40 \leq x < 50$	13
$50 \leq x < 60$	10
	160

Time Waiting in Line

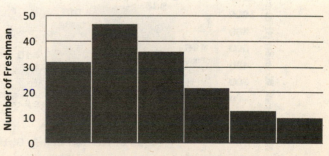

9.

x = Hourly Wage	Number of Employees	Relative Frequency
$\$8.00 \le x < 9.50$	21	0.138
$\$9.50 \le x < 11.00$	35	0.230
$\$11.00 \le x < 12.50$	42	0.276
$\$12.50 \le x < 14.00$	27	0.178
$\$14.00 \le x < 15.50$	18	0.118
$\$15.50 \le x < 17.00$	9	0.059
	152	

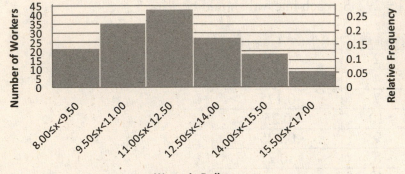

11.

x = Age	Number of Women	Relative Frequency
$15 \le x < 20$	486,000	0.125
$20 \le x < 25$	948,000	0.244
$25 \le x < 30$	1,075,000	0.276
$30 \le x < 35$	891,000	0.229
$35 \le x < 40$	410,000	0.105
$40 \le x < 45$	77,000	0.020
$45 \le x < 50$	4,000	0.001
	3,891,000	

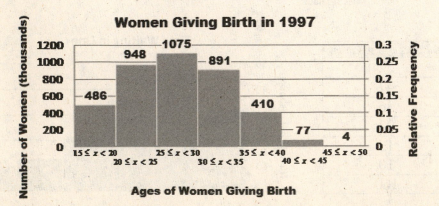

13.

x = Age of Males	Number of Students	Relative Frequency	Range	Relative Frequency Density
$14 \leq x < 18$	94,000	0.0144	4	0.0036
$18 \leq x < 20$	1,551,000	0.2372	2	0.1186
$20 \leq x < 22$	1,420,000	0.2172	2	0.1086
$22 \leq x < 25$	1,091,000	0.1668	3	0.0556
$25 \leq x < 30$	865,000	0.1323	5	0.0265
$30 \leq x < 35$	521,000	0.0797	5	0.0159
$35 \leq x < 60$	997,000	0.1525	25	0.0061
	6,539,000			

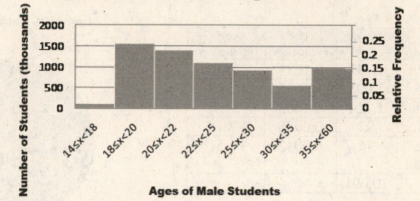

Ages of Male Students in Higher Education

15. a. $\dfrac{1+4+52}{200} = \dfrac{57}{200} = 0.285$ or 28.5%   b. $\dfrac{16+29+19}{200} = \dfrac{64}{200} = 0.32$ or 32%

c. Not possible

d. $\dfrac{52+48+31+16+29+19}{200} = \dfrac{195}{200} = 0.975$ or 97.5%

e. $\dfrac{52+48}{200} = \dfrac{100}{200} = 0.5$ or 50%   f. $\dfrac{1+4+52+29+19}{200} = \dfrac{105}{200} = 0.525$ or 52.5%

17.

Reason	Number of Patients (thousands)	Relative Frequency	Angle (in degrees)
Stomach pain	8,057	0.3185	114.7
Chest pain	6,392	0.2527	91.0
Fever	4,485	0.1773	63.8
Headache	3,354	0.1326	47.7
Shortness of breath	3,007	0.1189	42.8
	25,295		

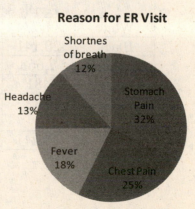

Reason for ER Visit

19.   a.

Race	Male	Relative Frequency	Angle (in degrees)
White	10,027	0.3393	122.2
Black	13,048	0.4416	159.0
Hispanic	5,949	0.2013	72.5
Asian	389	0.0132	4.7
Native American	137	0.0046	1.7
	29,550		

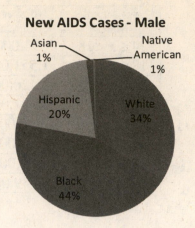

b.

Race	Female	Relative Frequency	Angle (in degrees)
White	1,747	0.1634	58.8
Black	7,093	0.6635	238.8
Hispanic	1,714	0.1603	57.7
Asian	92	0.0086	3.1
Native American	45	0.0042	1.5
	10,691		

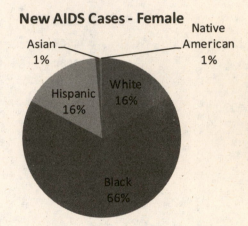

c.   Females have a higher percentage of blacks with new AIDS cases than males do. Males have a higher percentage of whites with new AIDS cases than females do. Both Asian and Native American males and females are less than 1%.

d.

Race	Male and Female	Relative Frequency	Angle (in degrees)
White	11,774	0.2926	105.3
Black	20,141	0.5005	180.2
Hispanic	7,663	0.1904	68.6
Asian	481	0.0120	4.3
Native American	182	0.0045	1.6
	40,241		

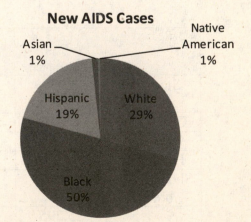

21.   a.

Specialty	Male	Relative Frequency	Angle (in degrees)
Family Practice	54,022	0.1953	70.3
General Surgery	32,329	0.1169	42.1
Internal Medicine	104,688	0.3785	136.3
Obstetrics/Gynecology	24,801	0.0897	32.3
Pediatrics	33,515	0.1212	43.6
Psychiatry	27,213	0.0984	35.4
	276,568		

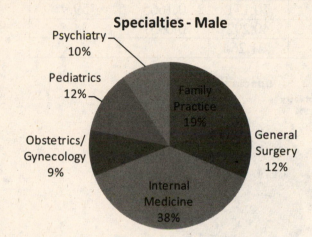

b.

Specialty	Female	Relative Frequency	Angle (in degrees)
Family Practice	26,305	0.1818	65.4
General Surgery	5,173	0.0358	12.9
Internal Medicine	46,245	0.3196	115.1
Obstetrics/Gynecology	17,258	0.1193	42.9
Pediatrics	36,636	0.2532	91.1
Psychiatry	13,079	0.0904	32.5
	144,696		

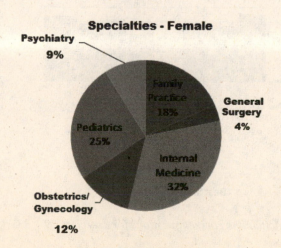

c. A higher percentage of males than females are general surgeons. A higher percentage of females than males are pediatricians.

d.

Specialty	Male and Female	Relative Frequency	Angle (in degrees)
Family Practice	80,327	0.1907	68.6
General Surgery	37,502	0.0890	32.0
Internal Medicine	150,933	0.3583	129.0
Obstetrics/Gynecology	42,059	0.0998	35.9
Pediatrics	70,151	0.1665	59.9
Psychiatry	40,292	0.0957	34.4
	421,264		

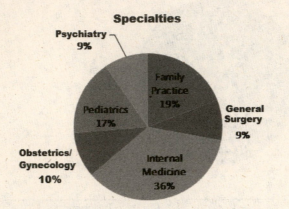

Specialties

*29. The questions should be:

a. Create a histogram to represent the data for the air quality index.

b. What do the results say about these towns?

a.

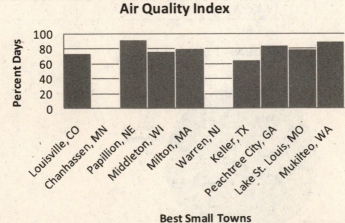

Air Quality Index

* The first printing of the text did not include the correct questions for exercises 29 – 33.

96

b. The air quality index (percent days AQI ranked good) is well over 50%, specifically from 65% to 92%.

*31. *The questions should be:*

a. *Create a histogram to represent the data for the property crime levels.*

b. *What do the results say about these towns?*

a.

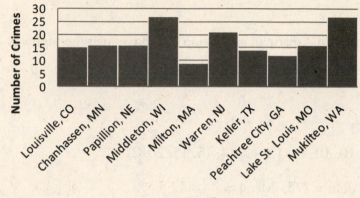

Property Crime Incidents Per 1,000

b. These small towns, considered the best places to live, have a very low rate of property crime incidents (less than 3%).

*33. *The questions should be:*

a. *For each of the two manufacturing regions, create a histogram for highway driving mileage. Use the same categories for the two regions.*

b. *What conclusions can you draw?*

a. For comparison, two sets of histograms have been constructed.

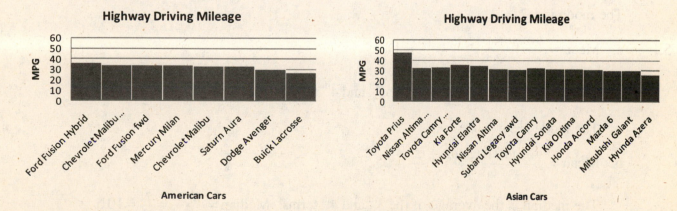

* The first printing of the text did not include the correct questions for exercises 29 – 33.

97

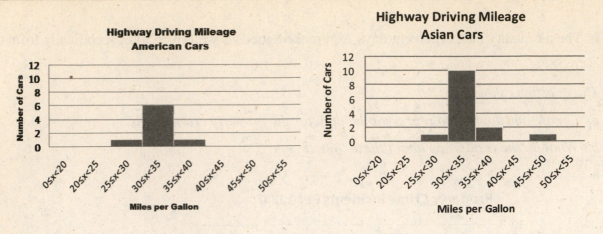

b. In general, the majority of both American and Asian cars have average highway driving mileage between 30 and 35 miles per gallon.

4.2 Measures of Central Tendency

1. Sorting the data: 8, 9, 9, 9, 10, 10, 11, 12, 12, 14, 15, 15, 20, 21

    The sum of all 14 pieces of data = 175. Mean = $\frac{175}{14} = 12.5$

    The median is the average of the 7th and 8th data points: Median = $\frac{11+12}{2} = \frac{23}{2} = 11.5$

    The mode is 9.

3. Sorting the data:

    0.7, 0.7, 0.8, 0.9, 1.0, 1.2, 1.2, 1.3, 1.5, 1.5, 1.7, 1.7, 1.7, 1.8, 1.9, 2.0, 2.2, 2.3

    The sum of all 18 pieces of data = 26.1. Mean = $\frac{26.1}{18} = 1.45$

    The median is the average of the 9th and 10th data points: Median = $\frac{1.5+1.5}{2} = \frac{3}{2} = 1.5$

    The mode is 1.7.

5. a. Mean = $\frac{9+9+10+11+12+15}{6} = \frac{66}{6} = 11$

    The median is the average of the 3rd and 4th terms: Median = $\frac{10+11}{2} = \frac{21}{2} = 10.5$

    The mode is 9.

    b. Mean = $\frac{9+9+10+11+12+102}{6} = \frac{153}{6} = 25.5$

    The median is the average of the 3rd and 4th terms: Median = $\frac{10+11}{2} = \frac{21}{2} = 10.5$

    The mode is 9.

c. The mean is affected by the change. The median and mode stay the same.

7. a. Mean = $\dfrac{2+4+6+8+10+12}{6} = \dfrac{42}{6} = 7$

   The median is the average of the 3rd and 4th terms: Median = $\dfrac{6+8}{2} = \dfrac{14}{2} = 7$; There is no mode.

   b. Mean = $\dfrac{102+104+106+108+110+112}{6} = \dfrac{642}{6} = 107$

   The median is the average of the 3rd and 4th terms: Median = $\dfrac{106+108}{2} = \dfrac{114}{2} = 107$

   There is no mode.

   c. The data in part (b) is 100 more than the data in part (a).

   d. The mean and median are 100 more. Neither data set had a mode.

9. Sorting the data: 3:06, 3:12, 3:12, 3:15, 4:02, 4:03

   After converting ounces to pounds, the sum is 23.125.

   Mean = $\dfrac{23.125}{6} = 3.854 = 3\,\text{lb}, 16(0.854)$ ounces or $3:14$.

   The median is the average of the 3rd and 4th terms: Median = $\dfrac{3:12+3:15}{2} = 3:13.5$

   The mode is 3:12.

11. Sorting the data: 1, 2, 3, 3, 5, 7, 7, 8, 9, 9, 9, 10, 13, 13, 14, 15, 15, 15, 17, 22

    The sum of all 20 pieces of data = 197. Mean = $\dfrac{197}{20} = 9.85$

    The median is the average of the 10th and 11th terms: Median = $\dfrac{9+9}{2} = 9$

    The mode is 9 and 15.

13. Sorting the data: 5, 16, 19, 24, 25, 25, 26, 28, 33, 33, 34, 34, 37, 37, 40, 42, 45, 45, 46, 46, 49, 73

    The sum of all 22 pieces of data = 762. Mean = $\dfrac{762}{22} \approx 34.64$

    The median is the average of the 11th and 12th terms: Median = 34

    Mode: 25, 33, 34, 37, 45, 46 (all have frequency = 2)

15. The total number of scores is $3 + 10 + 9 + 8 + 10 + 2 = 42$

    The sum of all scores is $10\cdot3 + 9\cdot10 + 8\cdot9 + 7\cdot8 + 6\cdot10 + 5\cdot2 = 318$. Mean = $\dfrac{318}{42} \approx 7.6$

    The median is the average of the 21st and 22nd scores. Median = $\dfrac{8+8}{2} = \dfrac{16}{2} = 8$

    The mode is 9 and 6.

17.

x = weight (in ounces)	Number of Boxes = f	Midpoint = m	$f \cdot m$
$15.3 \le x < 15.6$	13	15.45	200.85
$15.6 \le x < 15.9$	24	15.75	378
$15.9 \le x < 16.2$	84	16.05	1348.2
$16.2 \le x < 16.5$	19	16.35	310.65
$16.5 \le x \le 16.8$	10	16.65	166.5
	150		2,404.2

Mean $= \dfrac{2,404.2}{150} = 16.028 \, \text{oz}$

19. Let x = score on 5th exam.

   a. $\dfrac{71+69+85+83+x}{5} = 70$
       b. $\dfrac{308+x}{5} = 80$
       c. $\dfrac{308+x}{5} = 90$

$$\begin{aligned} \frac{308+x}{5} &= 70 \\ 308+x &= 350 \\ x &= 42 \end{aligned}$$
     $\begin{aligned} 308+x &= 400 \\ x &= 92 \end{aligned}$
     $\begin{aligned} 308+x &= 450 \\ x &= 142 \end{aligned}$

Impossible, it would require a score of 142.

21. The total salary of all men is $12 \cdot 58,000 = \$696,000$.

The total salary of all women is $8 \cdot 42,000 = \$336,000$.

Total salaries: $\$1,032,000$ Average $= \dfrac{1,032,000}{12+8} = \dfrac{1,032,000}{20} = \$51,600$

23. distance = rate $\cdot$ time

The Chicago to Milwaukee trip took $\dfrac{90}{60} = 1.5$ hours. The return trip took $\dfrac{90}{45} = 2$ hours.

Mean speed $= \dfrac{90+90}{1.5+2} = \dfrac{180}{3.5} \approx 51.4 \, \text{mph}$

25. The total sum of all ages is $25 \cdot 23.4 = 585$

Let x = age of 26th student.

$$\begin{aligned} \frac{585+x}{26} &= 24 \\ 585+x &= 624 \\ x &= 39 \text{ years old} \end{aligned}$$

27. The total salaries: $8 \cdot 40,000 = 320,000$

   a. new total = 326,000   mean $= \dfrac{326,000}{8} = \$40,750$

b. The median does not change: $42,000.

29.

Department	Education	Health and Human Services	Environmental Protection Agency	Homeland Security
Mean (thousands)	$\dfrac{44,497}{4,201} \approx \$10.592$	$\dfrac{576,747}{62,502} \approx \$9.228$	$\dfrac{195,320}{18,119} \approx \$10.780$	$\dfrac{1,251,754}{159,447} \approx \$7.851$
Department	Navy	Air Force	Army	
Mean (thousands)	$\dfrac{712,172}{175,722} \approx \$4.053$	$\dfrac{635,901}{157,182} \approx \$4.046$	$\dfrac{678,044}{245,599} \approx \$2.761$	

a. Environmental Protection Agency; $10,780

b. Department of the Army; $2,761

c. $\text{mean} = \dfrac{44,497 + 576,747 + 195,320}{4,201 + 62,502 + 18,119} = \dfrac{816,564}{84,822} \approx \$9,627$

d. $\text{mean} = \dfrac{1,251,754 + 712,172 + 635,901 + 678,044}{159,447 + 175,722 + 157,182 + 245,599} = \dfrac{3,277,871}{737,950} \approx \$4,442$

31. a.

Midpoint = x	Frequency = f	$f \cdot x$
2.5	19,176	47,940
7.5	20,550	154,125
12.5	20,528	256,600
20	39,184	783,680
30	39,892	1,196,760
40	44,149	1,765,960
50	37,678	1,883,900
60	24,275	1,456,500
75	30,752	2,306,400
	276,184	9,851,865

$\text{Mean} = \dfrac{9,851,865}{276,184} \approx 35.7 \text{ years old}$

b.

Midpoint = x	Frequency = f	$f \cdot x$
2.5	19,176	47,940
7.5	20,550	154,125
12.5	20,528	256,600
20	39,184	783,680
30	39,892	1,196,760
40	44,149	1,765,960
50	37,678	1,883,900
60	24,275	1,456,500
75	30,752	2,306,400
92.5	4,240	392,200
	280,424	10,244,065

Mean $= \dfrac{10,244,065}{280,424} \approx 36.5$ years old

33.

Midpoint = m	Frequency = f	$f \cdot m$
16	78,000	1,248,000
19	1,907,000	36,233,000
21	1,597,000	33,537,000
23.5	1,305,000	30,667,500
27.5	1,002,000	27,555,000
32.5	664,000	21,580,000
47.5	1,888,000	89,680,000
	8,441,000	240,500,500

a. The sum of all values of $f \cdot m$ for females under 35 is 150,820,500.

The sum of all values of f for females under 35 is 6,553,000.

Mean $= \dfrac{150,820,500}{6,553,000} \approx 23.0$ years old

b. Mean $= \dfrac{240,500,500}{8,441,000} \approx 28.5$ years old

4.3 Measures of Dispersion

1. a. Mean $\overline{x} = \dfrac{3+8+5+3+10+13}{6} = \dfrac{42}{6} = 7$

$$\text{Variance} \quad s^2 = \frac{(3-7)^2 + (8-7)^2 + (5-7)^2 + (3-7)^2 + (10-7)^2 + (13-7)^2}{6-1}$$

$$= \frac{16+1+4+16+9+36}{5} = \frac{82}{5} = 16.4$$

Standard deviation $\quad s \approx 4.0$

b.

x	x^2
3	9
8	64
5	25
3	9
10	100
13	169
$\Sigma x = 42$	$\Sigma x^2 = 376$

$$\text{Variance} \quad s^2 = \frac{1}{5}\left[376 - \frac{42^2}{6}\right] = \frac{1}{5}[376 - 294] = \frac{82}{5} = 16.4$$

Standard deviation $\quad s \approx 4.0$

3. a. Mean $= 10$ $\qquad$ Variance $\quad s^2 = \frac{1}{5}\left[600 - \frac{3600}{6}\right] = 0$ $\qquad$ b. $s = 0$

5. a. Mean $= \frac{12+16+20+24+28+32}{6} = \frac{132}{6} = 22$

$$\text{Variance} \quad s^2 = \frac{(12-22)^2 + (16-22)^2 + (20-22)^2 + (24-22)^2 + (28-22)^2 + (32-22)^2}{5}$$

$$= \frac{100+36+4+4+36+100}{5} = \frac{280}{5} = 56$$

Standard deviation $\quad s \approx 7.5$

b. Mean $= \frac{600+800+1,000+1,200+1,400+1,600}{6} = \frac{6,600}{6} = 1,100$

$$\text{Variance} \quad s^2 = \frac{500^2 + 300^2 + 100^2 + 100^2 + 300^2 + 500^2}{5}$$

$$= \frac{250,000+90,000+10,000+10,000+90,000+250,000}{5} = \frac{700,000}{5} = 140,000$$

Standard deviation $\quad s \approx 374.2$

c. The data in (b) are 50 times the data in (a).

d. The mean and standard deviation are 50 times larger in (b).

7. a. Joey's mean $= \frac{144+171+220+158+147}{5} = \frac{840}{5} = 168$

Dee Dee's mean $= \frac{182+165+187+142+159}{5} = \frac{835}{5} = 167$ $\qquad$ Joey's mean is higher.

103

b. Joey:

$$s^2 = \frac{(144-168)^2 + (171-168)^2 + (220-168)^2 + (158-168)^2 + (147-168)^2}{4} = \frac{3830}{4} = 957.5$$

$s \approx 30.9$

Dee Dee:

$$s^2 = \frac{(182-167)^2 + (165-167)^2 + (187-167)^2 + (142-167)^2 + (159-167)^2}{4} = \frac{1,318}{4} = 329.5$$

$s \approx 18.2$

Dee Dee's standard deviation is smaller.

c. Dee Dee is more consistent than Joey because her standard deviation is lower.

9. Convert weights to a common unit, say ounces.

$n = 6$

$$s^2 = \frac{1}{5}\left[32,494 - \frac{(440)^2}{6}\right]$$

$s^2 = 45.4\overline{6}$

$s \approx 6.74 \text{ ounces} = \frac{6.74}{16} \approx 0.421 \text{ lb}$

$= 0{:}07$

Weight (lb:oz)	Weight (oz) x	x^2
4:12	76	5,776
5:03	83	6,889
4:06	70	4,900
3:15	63	3,969
4:12	76	5,776
4:08	72	5,184
	$\sum x = 440$	$\sum x^2 = 32,494$

11.

$x =$ Home Runs	x^2	$x =$ Home Runs	x^2
16	256	40	1,600
25	625	37	1,369
24	576	34	1,156
19	361	49	2,401
33	1,089	73	5,329
25	625	46	2,116
34	1,156	45	2,025
46	2,116	45	2,025
37	1,369	5	25
33	1,089	26	676
42	1,764	28	784
		$\sum x = 762$	$\sum x^2 = 30,532$

$n = 22 \qquad s^2 = \frac{1}{21}\left[30,532 - \frac{(762)^2}{22}\right] \approx 197.0996 \qquad s \approx 14.039$

13. a. $\bar{x} = \dfrac{68+50+70+67+72+78+69+68+66+67}{10} = 67.5$

    $n = 10$

    $s^2 = \dfrac{1}{9}\left[46,011 - \dfrac{(675)^2}{10}\right] \approx 49.8333$

    $s \approx 7.1$

    b. One standard deviation

    $[67.5-7.1, 67.5+7.1] = [60.4, 74.6]$

    $\dfrac{8}{10} = 0.8 = 80\%$

    c. Two standard deviations

    $\left[67.5 - 2(7.1), 67.5 + 2(7.1)\right]$

    $[53.3, 81.7]$

    $\dfrac{9}{10} = 0.9 = 90\%$

x	x^2
68	4,624
50	2,500
70	4,900
67	4,489
72	5,184
78	6,084
69	4,761
68	4,624
66	4,356
67	4,489
$\Sigma x = 675$	$\Sigma x^2 = 46,011$

15. a. Mean $= \bar{x} = \dfrac{38.1}{12} = 3.175$

    $n = 12$

    $s^2 = \dfrac{1}{11}\left[156.89 - \dfrac{(38.1)^2}{12}\right] \approx 3.26568$

    $s \approx 1.807$

    b. One standard deviation

    $= [3.175 - 1.807, \ 3.175 + 1.807]$

    $= [1.368, \ 4.982]$

    $\dfrac{7}{12} \approx 0.5833 = 58\%$

x	x^2
5.4	29.16
4.0	16
3.8	14.44
2.5	6.25
1.8	3.24
1.6	2.56
0.9	0.81
1.2	1.44
1.9	3.61
3.3	10.89
5.7	32.49
6.0	36
$\Sigma x = 38.1$	$\Sigma x^2 = 156.89$

    c. Two standard deviations

    $= \left[3.175 - 2(1.807), \ 3.175 + 2(1.807)\right] = [-0.439, \ 6.789]$

    $\dfrac{12}{12} = 100\%$

17.   a.

x = Score	f = Number of Students	$f \cdot x$	$f \cdot x^2$
10	5	50	500
9	10	90	810
8	6	48	384
7	8	56	392
6	3	18	108
5	2	10	50
	34	$\Sigma(f \cdot x) = 272$	$\Sigma(f \cdot x^2) = 2244$

$$\text{Mean} = \frac{272}{34} = 8 \qquad s^2 = \frac{1}{33}\left[2244 - \frac{(272)^2}{34}\right] = \frac{68}{33} = 2.06 \qquad s \approx 1.4$$

b. One standard deviation $[8-1.4,\ 8+1.4] = [6.6,\ 9.4]$

$$\frac{8+6+10}{34} = \frac{24}{34} \approx 0.71 = 71\%$$

c. Two standard deviations $\left[8-2(1.4),\ 8+2(1.4)\right] = [5.2,\ 10.8]$

$$\frac{3+8+6+10+5}{34} = \frac{32}{34} \approx 0.94 = 94\%$$

d. Three standard deviations $\left[8-3(1.4),\ 8+3(1.4)\right] = [3.8,\ 12.2]$

$$\frac{2+3+8+6+10+5}{34} = \frac{34}{34} = 1 = 100\%$$

19.

x = midpoint	f = Number of Boxes	$f \cdot x$	$f \cdot x^2$
15.45	13	200.85	3,103.1325
15.75	24	378	5,953.5
16.05	84	1,348.2	21,638.61
16.35	19	310.65	5,079.1275
16.65	10	166.5	2,772.225
	150	$\Sigma(f \cdot x) = 2,404.2$	$\Sigma(f \cdot x^2) = 38,546.595$

$$s^2 = \frac{1}{149}\left[38,546.595 - \frac{(2404.2)^2}{150}\right] \approx 0.081 \qquad s \approx 0.285$$

21.

x = Midpoint	f = Number of Women	$f \cdot x$	$f \cdot x^2$
17.5	549,000	9,607,500	168,131,250
22.5	872,000	19,620,000	441,450,000
27.5	897,000	24,667,500	678,356,250
32.5	859,000	27,917,500	907,318,750
37.5	452,000	16,950,000	635,625,000
42.5	137,000	5,822,500	247,456,250
	3,766,000	$\Sigma(f \cdot x) = 104,585,000$	$\Sigma(f \cdot x^2) = 3,078,337,500$

$$s^2 = \frac{1}{3,765,999}\left[3,078,337,500 - \frac{(104,585,000)^2}{3,766,000}\right] \approx 46.1826 \qquad s \approx 6.80$$

29.   a. $\bar{x} = \dfrac{640}{8} = 80$

b. Sample standard deviation would be more appropriate because this does not include all the towns identified as best places to live.

c. $s^2 = \dfrac{1}{7}\left[51,720 - \dfrac{(640)^2}{8}\right] \approx 74.286 \Rightarrow s \approx 8.62$

d. The air quality index averages 80.0% with 62.5% of all indices falling within one standard deviation of the mean air quality index.

31.   a. $\bar{x} = \dfrac{173}{10} = 17.3$

b. Sample standard deviation would be more appropriate because this does not include all the towns identified as best places to live.

c. $s^2 = \dfrac{1}{9}\left[3313 - \dfrac{(173)^2}{10}\right] \approx 35.567 \Rightarrow s \approx 5.96$

d. The property crime incidents averages 17.3 per 1,000 with 70% of all property crime incidents falling within one standard deviation of the mean property crime incidents.

33.   a. American: $\bar{x} = \dfrac{261}{8} = 32.625$; $s^2 = \dfrac{1}{7}\left[8571 - \dfrac{(261)^2}{8}\right] \approx 7.982 \Rightarrow s \approx 2.83$

Asian: $\overline{x} = \dfrac{463}{14} \approx 33.07$; $s^2 = \dfrac{1}{13}\left[15,629 - \dfrac{(463)^2}{14}\right] \approx 24.379 \Rightarrow s \approx 4.94$

Sample standard deviation was used as this does not include every car that could fall into this category.

b. Asian cars have the highest mpg for highway driving on the average, but it varies over 50% more than the average mpg varies for American cars.

4.4 The Normal Distribution

1. Yes, it looks like a bell curve.

3. No, it is right-tailed.

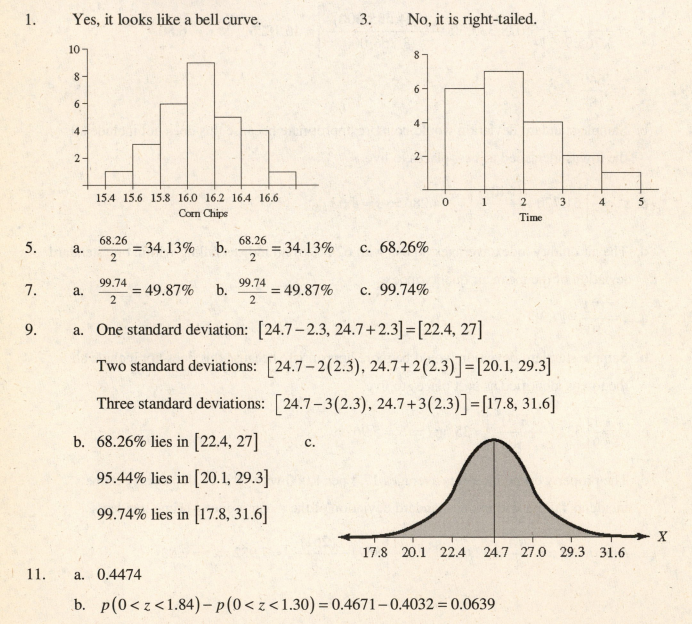

5. a. $\dfrac{68.26}{2} = 34.13\%$ b. $\dfrac{68.26}{2} = 34.13\%$ c. 68.26%

7. a. $\dfrac{99.74}{2} = 49.87\%$ b. $\dfrac{99.74}{2} = 49.87\%$ c. 99.74%

9. a. One standard deviation: $[24.7 - 2.3,\ 24.7 + 2.3] = [22.4,\ 27]$

    Two standard deviations: $[24.7 - 2(2.3),\ 24.7 + 2(2.3)] = [20.1,\ 29.3]$

    Three standard deviations: $[24.7 - 3(2.3),\ 24.7 + 3(2.3)] = [17.8,\ 31.6]$

    b. 68.26% lies in $[22.4,\ 27]$ c.

    95.44% lies in $[20.1,\ 29.3]$

    99.74% lies in $[17.8,\ 31.6]$

11. a. 0.4474

    b. $p(0 < z < 1.84) - p(0 < z < 1.30) = 0.4671 - 0.4032 = 0.0639$

c. $p(-0.37 < z < 0) + p(0 < z < 1.59) = p(0 < z < 0.37) + p(0 < z < 1.59)$

$$= 0.1443 + 0.4441 = 0.5884$$

d. $p(z > 1.91) = 0.5 - 0.4719 = 0.0281$

e. $p(0.88 < z < 1.32) = p(0 < z < 1.32) - p(0 < z < 0.88) = 0.4066 - 0.3106 = 0.0960$

f. $p(z < 0) + p(0 < z < 1.25) = p(z > 0) + p(0 < z < 1.25) = 0.5 + 0.3944 = 0.8944$

13. a. $c = 0.34$ b. $p(0 < z < -c) = 0.4812$ c. Solving for c:

$$-c = 2.08$$
$$c = -2.08$$

$$2p(0 < z < c) = 0.4648$$
$$p(0 < z < c) = 0.2324$$

$$c = 0.62$$

d. c must be negative since $p > 0.5$ e. $c = 1.64$ f. Solving for c:

$$p(c < z < 0) + p(z > 0) = 0.6064$$
$$p(c < z < 0) + .5 = 0.6064$$
$$p(c < z < 0) = 0.1064$$
$$p(0 < z < -c) = 0.1064$$
$$-c = 0.27$$
$$c = -0.27$$

$$p(z > -c) = 0.1003$$
$$-c = 1.28$$
$$c = -1.28$$

15. $u = 250$ $\sigma = 24$

a. $z = \dfrac{260 - 250}{24} \approx 0.42$ b. $z = \dfrac{240 - 250}{24} \approx -0.42$ c. $z = \dfrac{300 - 250}{24} \approx 2.08$

d. $z = \dfrac{215 - 250}{24} \approx -1.46$ e. $z = \dfrac{321 - 250}{24} \approx 2.96$ f. $z = \dfrac{197 - 250}{24} \approx -2.21$

17. a. $z_1 = \dfrac{36.8 - 36.8}{2.5} = 0$ $z_2 = \dfrac{39.3 - 36.8}{2.5} = 1$

$p(36.8 < x < 39.3) = p(0 < z < 1) = 0.3413$

b. $z_1 = \dfrac{34.2 - 36.8}{2.5} = -1.04$ $z_2 = \dfrac{38.7 - 36.8}{2.5} = 0.76$

$$
\begin{aligned}
p(34.2 < x < 38.7) &= p(-1.04 < z < 0.76) \\
&= p(-1.04 < z < 0) + p(0 < z < 0.76) \\
&= p(0 < z < 1.04) + p(0 < z < 0.76) \\
&= 0.3508 + 0.2764 = 0.6272
\end{aligned}
$$

c. $z = \dfrac{40 - 36.8}{2.5} = 1.28$

$p(x < 40.0) = p(z < 1.28) = .5 + p(0 < z < 1.28) = 0.5 + 0.3997 = 0.8997$

d. $z_1 = \dfrac{32.3 - 36.8}{2.5} = -1.8$ $z_2 = \dfrac{41.3 - 36.8}{2.5} = 1.8$

$p(32.3 < x < 41.3) = p(-1.8 < z < 1.8) = 2p(0 < z < 1.8) = 2 \cdot 0.4641 = 0.9282$

e. 0

f. $z = \dfrac{37.9 - 36.8}{2.5} = 0.44$ $p(x > 37.9) = p(z > .44) = 0.3300$

19. a. $z = \dfrac{15.5 - 16.0}{0.3} \approx -1.67$

$p(x < 15.5) = p(z < -1.67) = p(z > 1.67) = 0.0475 = 4.75\%$

b. $z_1 = \dfrac{15.8 - 16.0}{0.3} \approx -0.67$ $z_2 = \dfrac{16.2 - 16.0}{0.3} \approx 0.67$

$p(15.8 < x < 16.2) = p(-0.67 < z < 0.67) = 2p(0 < z < 0.67)$

$$= 2 \cdot 0.2486 = 0.4972 = 49.72\%$$

21. a. mean = 2 hours 36 minutes = 156 minutes

2 hours 15 minutes = 135 minutes

$z = \dfrac{135 - 156}{24} = -0.875$

$$
\begin{aligned}
p(x < 135) &= p(z < -0.875) \\
&= p(z > 0.875) \\
&= \dfrac{0.1922 + 0.1894}{2} = 0.1908
\end{aligned}
$$

b. 2 hours = 120 minutes, 3 hours = 180 minutes

$$z_1 = \frac{120-156}{24} = -1.5 \qquad z_2 = \frac{180-156}{24} = 1$$

$$
\begin{aligned}
p(120 < x < 180) &= p(-1.5 < z < 1) \\
&= p(-1.5 < z < 0) + p(0 < z < 1) \\
&= p(0 < z < 1.5) + p(0 < z < 1) \\
&= 0.4332 + 0.3413 = 0.7745
\end{aligned}
$$

23. a. $z = \dfrac{0.9-1.0}{0.06} \approx -1.67$

$$
\begin{aligned}
p(x > 0.9) &= p(z > -1.67) \\
&= p(-1.67 < z < 0) + 0.5 \\
&= p(0 < z < 1.67) + 0.5 \\
&= 0.4525 + 0.5 = 0.9525 = 95.25\%
\end{aligned}
$$

b. $z = \dfrac{1.05-1.0}{0.06} \approx 0.83$

$$p(x < 1.05) = p(z < 0.83) = 0.5 + p(0 < z < 0.83) = 0.5 + 0.2967 = 0.7967 = 79.67\%$$

25. a. $p(z > c) = .10$. Using the table, we see that $c = 1.28$

$$
\begin{aligned}
\frac{x-72}{12} &= 1.28 \\
x-72 &= 15.36 \\
x &= 87.36
\end{aligned}
$$

About 87.

b. Using the table:

$$
\begin{aligned}
p(z < c) &= 0.20 \\
p(z > -c) &= 0.20 \\
-c &= 0.84 \\
c &= -0.84
\end{aligned}
\qquad\qquad
\begin{aligned}
\frac{x-72}{12} &= -0.84 \\
x-72 &= -10.08 \\
x &= 61.92
\end{aligned}
$$

62 or less

27. $p(z > c) = 0.34$. Using the table, we see that $c = 0.41$

$$\frac{x - 8.5}{1.5} = 0.41$$

$$x - 8.5 = 0.615$$

$$x = 9.115$$

9.1 minutes or more

4.5 Polls and Margin of Error

1.   a. $z = 0.68$ b. $z = 0.31$ c. $z = 1.39$ d. $z = 2.65$

3.   a. $z = 1.44$ b. $z = 1.28$ c. $z = 1.15$

d. $z = 2.575$ since it is halfway between 0.4949 and 0.4951

5. $\alpha = 0.92$ 7. $\alpha = 0.75$

$z_{\alpha/2} = z_{.46} = 1.75$ $z_{\alpha/2} = z_{.375} = 1.15$

9.   a. $\alpha = 0.90$ $z_{\alpha/2} = z_{.45} = 1.645$ Since .45 is halfway between 0.4495 and 0.4505

$$MOE = \frac{1.645}{2\sqrt{2,034}} \approx 0.018 \text{ or } 1.8\%$$

b. $\alpha = 0.98$ $z_{\alpha/2} = z_{.49} = 2.33$ $MOE = \frac{2.33}{2\sqrt{2,034}} \approx 0.026 \text{ or } 2.6\%$

11.   a. The sample proportion is $\frac{x}{n} = \frac{450}{600} = 0.75$ or 75%

$\alpha = 0.90$ $z_{\alpha/2} = z_{.45} = 1.645$ $MOE = \frac{1.645}{2\sqrt{600}} \approx 0.034 \text{ or } 3.4\%;\ 75.0\% \pm 3.4\%$

b. The sample proportion is 0.75 or 75%

$\alpha = 0.95$ $z_{\alpha/2} = z_{.475} = 1.96$ $MOE = \frac{1.96}{2\sqrt{600}} \approx 0.040 \text{ or } 4.0\%;\ 75.0\% \pm 4.0\%$

13.   a. $\frac{x}{n} = \frac{2,141}{2,710} \approx 0.790$ or 79.0%   b. $\frac{x}{n} = \frac{569}{2,710} \approx 0.210$ or 21.0%

c. $\alpha = 0.90$ $z_{\alpha/2} = z_{.45} = 1.645$

$$MOE = \frac{1.645}{2\sqrt{2,710}} \approx 0.016 \text{ or } \pm 1.6\%$$

15. a. $\dfrac{x}{n} = \dfrac{1,049}{1,220} \approx 0.860$ or 86.0% b. $\dfrac{x}{n} = \dfrac{1,043}{1,490} \approx 0.700$ or 70.0%

    c. $\alpha = 0.95 \qquad z_{\alpha/2} = z_{.475} = 1.96$

    For men: For women:

    $\text{MOE} = \dfrac{1.96}{2\sqrt{1,220}} \approx 0.028$ or $\pm 2.8\%$ $\text{MOE} = \dfrac{1.96}{2\sqrt{1,490}} \approx 0.025$ or $\pm 2.5\%$

17. a. $\dfrac{x}{n} = \dfrac{92,011}{129,593} \approx 0.710$ or 71% b. $\dfrac{x}{n} = \dfrac{37,582}{129,593} \approx 0.290$ or 29%

    c. $\alpha = 0.95 \qquad z_{\alpha/2} = z_{.475} = 1.96$

    $\text{MOE} = \dfrac{1.96}{2\sqrt{129,593}} \approx 0.003$ or $\pm 0.3\%$

19. a. $\dfrac{x}{n} = \dfrac{58,317}{129,593} \approx 0.450$ or 45.0% b. $\dfrac{x}{n} = \dfrac{60,908}{129,593} \approx 0.470$ or 47.0%

    c. $\dfrac{x}{n} = \dfrac{10,368}{129,593} \approx 0.080$ or 8.0% d. $\alpha = 0.90 \qquad z_{\alpha/2} = z_{.45} = 1.645$

    $\text{MOE} = \dfrac{1.645}{2\sqrt{129,593}} \approx 0.002$ or $\pm 0.2\%$

21. a. $\dfrac{x}{n} = \dfrac{902}{2,503} \approx 0.360$ or 36.0%

    b. $\alpha = 0.90$ c. $\alpha = 0.98$

    $z_{\alpha/2} = z_{.45} = 1.645$ $z_{\alpha/2} = z_{.49} = 2.33$

    $\text{MOE} = \dfrac{1.645}{2\sqrt{2,503}} \approx 0.016$ or $\pm 1.6\%$ $\text{MOE} = \dfrac{2.33}{2\sqrt{2,503}} \approx 0.023$ or $\pm 2.3\%$

    d. The answer in (c) is larger than the answer in (b). In order to guarantee 98% accuracy, we must increase the error.

23. a. $\alpha = 0.95$

    $z_{\alpha/2} = z_{.475} = 1.96$

    $\text{MOE} \dfrac{1.96}{2\sqrt{430}} \approx 0.047$ or $\pm 4.7\%$

b. $\text{MOE} = \dfrac{1.96}{2\sqrt{765}} \approx 0.035$ or $\pm 3.5\%$ \qquad c. combined: $n = 430 + 765 = 1{,}195$

$$\text{MOE} = \dfrac{1.96}{2\sqrt{1{,}195}} \approx 0.028 \text{ or } \pm 2.8\%$$

25. $\dfrac{z_{\alpha/2}}{2\sqrt{1{,}763}} = 0.025$ \qquad\qquad 27. $n = 640 + 820 = 1{,}460$

$z_{\alpha/2} \approx 2.10$ \qquad\qquad\qquad\qquad $\dfrac{z_{\alpha/2}}{2\sqrt{1{,}460}} = 0.026$

$\alpha/2 = 0.4821$ \qquad\qquad\qquad\qquad $z_{\alpha/2} \approx 1.99$

$\alpha = 0.9642$ about 96.4% \qquad\qquad $\alpha/2 = 0.4767$

$\qquad\qquad\qquad\qquad\qquad\qquad\qquad\qquad \alpha = 0.9534$ about 95.3%

29. 95% level of confidence: $\alpha/2 = 0.475 \Rightarrow z_{\alpha/2} = 1.96$

a. $n = \left(\dfrac{1.96}{2(0.03)}\right)^2 \approx 1067.11 \Rightarrow 1{,}068$ \quad b. $n = \left(\dfrac{1.96}{2(0.02)}\right)^2 = 2{,}401$ \quad c. $n = \left(\dfrac{1.96}{2(0.01)}\right)^2 = 9{,}604$

d. Smaller margin of errors require a larger sample size, and larger margin of errors may have a smaller sample size.

31. 98% level of confidence: $\alpha/2 = 0.49 \Rightarrow z_{\alpha/2} = 2.33$

a. $n = \left(\dfrac{2.33}{2(0.03)}\right)^2 \approx 1508.027 \Rightarrow 1{,}509$ \qquad b. $n = \left(\dfrac{2.33}{2(0.02)}\right)^2 \approx 3393.06 = 3{,}394$

c. $n = \left(\dfrac{2.33}{2(0.01)}\right)^2 \approx 13{,}572.25 = 13{,}573$

d. Smaller margin of errors require a larger sample size, and larger margin of errors may have a smaller sample size.

4.6 Linear Regression

1. a. $m = \dfrac{6(1,039) - (64)(85)}{6(814) - (64)^2} = \dfrac{794}{788} \approx 1.00761421$

$b = \bar{y} - m\bar{x}$

$\bar{y} = \dfrac{85}{6} \approx 14.17$

$\bar{x} = \dfrac{64}{6} \approx 10.67$

$b = \dfrac{85}{6} - \dfrac{64}{6} = 3.5$

$y = 1.0x + 3.5$

  b. $y = 11 + 3.5 = 14.5$ c. $19 = x + 3.5$

                                        $x = 15.5$

  d. $r = \dfrac{6(1,039) - (64)(85)}{\sqrt{6(814) - (64)^2}\sqrt{6(1,351) - (85)^2}} = \dfrac{794}{\sqrt{788} \cdot \sqrt{881}} \approx 0.9529485$

  e. Yes, they are reliable since r is close to 1.

3. a. $m = \dfrac{5(279) - (37)(38)}{5(299) - (37)^2} = -\dfrac{11}{126} \approx -0.0873015873$

$\bar{y} = \dfrac{38}{5} = 7.6$ $\bar{x} = \dfrac{37}{5} = 7.4$ $b = \bar{y} - m\bar{x} = 7.6 - (-0.087)(7.4) = 8.246$

$y = -0.087x + 8.246$

  b. $y = -0.087(5) + 8.246 = 7.811$

  c. $\quad 7 = -0.087x + 8.246$

$-1.246 = -0.087x$

$\quad\quad x = 14.322$

  d. $r = \dfrac{5(279) - (37)(38)}{\sqrt{5(299) - (37)^2} \cdot \sqrt{5(310) - (38)^2}} = \dfrac{-11}{\sqrt{126} \cdot \sqrt{106}} \approx -0.095$

  e. No, they are not reliable since r is close to 0.

5.   a. Yes, the ordered pairs seem to exhibit a linear trend.

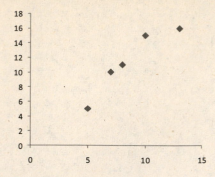

b.

x	y	x^2	y^2	xy
5	5	25	25	25
7	10	49	100	70
8	11	64	121	88
10	15	100	225	150
13	16	169	256	208
$\Sigma x = 43$	$\Sigma y = 57$	$\Sigma x^2 = 407$	$\Sigma y^2 = 727$	$\Sigma xy = 541$

$n = 5$

$$m = \frac{5(541)-(43)(57)}{5(407)-(43)^2} = \frac{254}{186} \approx 1.365591398$$

$$\bar{y} = \frac{57}{5} = 11.4 \qquad\qquad \bar{x} = \frac{43}{5} = 8.6$$

$$b = \bar{y} - m\bar{x} = 11.4 - (1.365591398)(8.6) \approx -0.3$$

$$y = 1.4x - 0.3$$

c. $y = 1.4(9) - 0.3 = 12.3$

d.

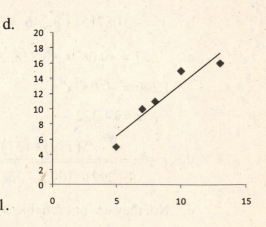

e. $r = \dfrac{5(541)-(43)(57)}{\sqrt{5(407)-(43)^2} \cdot \sqrt{5(727)-(57)^2}}$

$$= \frac{254}{\sqrt{186} \cdot \sqrt{386}} \approx 0.9479459$$

f. Yes, the prediction is reliable since r is close to 1.

116

7. a. No, the ordered pairs do not seem to exhibit a linear trend.

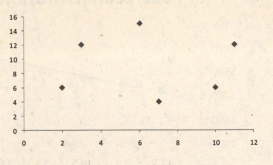

b.

x	y	x^2	y^2	xy
2	6	4	36	12
3	12	9	144	36
6	15	36	225	90
7	4	49	16	28
10	6	100	36	60
11	12	121	144	132
$\Sigma x = 39$	$\Sigma y = 55$	$\Sigma x^2 = 319$	$\Sigma y^2 = 601$	$\Sigma xy = 358$

$n = 6$

$$m = \frac{6(358) - (39)(55)}{6(319) - (39)^2} = \frac{3}{393} \approx 0.0076335878$$

$$\bar{y} = \frac{55}{6} \approx 9.17 \qquad \bar{x} = \frac{39}{6} = 6.5$$

$$b = \bar{y} - m\bar{x} = \frac{55}{6} - (0.0076335878)(6.5) \approx 9.1$$

$$y = 0.008x + 9.1$$

c. $y = 0.008(8) + 9.1 = 9.164$ d.

e. $$r = \frac{6(358) - (39)(55)}{\sqrt{6(319) - (39)^2} \cdot \sqrt{6(601) - (55)^2}}$$

$$= \frac{3}{\sqrt{393} \cdot \sqrt{581}} \approx 0.0062782255$$

f. The prediction is not reliable since r is close to 0.

9.   a. Yes, the data exhibit a linear trend.

Four-year Institutions

b.

x = Average Hourly Wage	y = Average Tuition at 4-year Institutions	x^2	y^2	xy
12.49	3,110	156.000	9,672,100	38,843.9
13.00	3,229	169.000	10,426,441	41,977.0
13.47	3,349	181.441	11,215,801	45,111.0
14.00	3,501	196.000	12,257,001	49,014.0
14.53	3,735	211.121	13,950,225	54,269.6
14.95	4,059	223.503	16,475,481	60,682.1
$\Sigma x = 82.44$	$\Sigma y = 20,983$	$\Sigma x^2 = 1,137.0644$	$\Sigma y^2 = 73,997,049$	$\Sigma xy = 289,897.53$

$n = 6$

$$m = \frac{6(289,897.53) - (82.44)(20,983)}{6(1,137.0644) - (82.44)^2} \approx 366.7166037$$

$$\bar{y} = \frac{20,983}{6} = 3,497.1\bar{6} \qquad \bar{x} = \frac{82.44}{6} = 13.74$$

$$b = \bar{y} - m\bar{x} = \frac{20,983}{6} - (366.7166037)(13.74) \approx -1,548.52$$

$$y = 366.72x - 1,541.52$$

c. $y = (366.72)(14.75) - 1,541.52 = \$3,867.55$

d. $\quad 4000 = 366.72x - 1,541.52$

$\quad 5,541.52 = 366.72x$

$\quad\quad x = \$15.11$

e. $r = \dfrac{6(289,898) - (82.44)(20,983)}{\sqrt{6(1,137.06) - (82.44)^2} \cdot \sqrt{6(73,997,049) - (20,983)^2}} \approx 0.9732507334$

f. Yes, r is close to 1.

11. a. Yes, the data exhibit a linear trend.

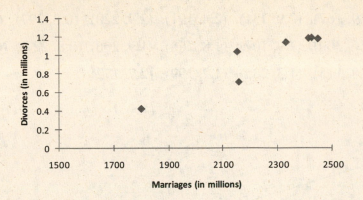

b.

x = Marriages	y = Divorces	x^2	y^2	xy
1.800	0.479	3.24	0.2294	0.8622
2.158	0.708	4.6570	0.5013	1.5279
2.152	1.036	4.6311	1.0733	2.2295
2.413	1.182	5.8226	1.3971	2.8522
2.425	1.187	5.8806	1.4090	2.8785
2.448	1.175	5.9927	1.3806	2.8764
2.329	1.135	5.4242	1.2882	2.6434
$\Sigma x = 15.725$	$\Sigma y = 6.902$	$\Sigma x^2 = 35.648207$	$\Sigma y^2 = 7.278944$	$\Sigma xy = 15.869992$

$n = 7$

$$m = \frac{7(15.869992) - (15.725)(6.902)}{7(35.648207) - (15.725)^2} = \frac{2.555994}{2.261824} \approx 1.130058749$$

$$\bar{y} = \frac{6.902}{7} \approx 0.99 \qquad \bar{x} = \frac{15.725}{7} \approx 2.25$$

$$b = \bar{y} - m\bar{x} = \frac{6.902}{7} - (1.130058749)\left(\frac{15.725}{7}\right) \approx -1.553$$

$$y = 1.130x - 1.553$$

c. $y = 1.130(2.75) - 1.553 = 1.5545$ million

d. $1.5 = 1.130x - 1.553$

$3.053 = 1.130x$

$x = 2.702$ million

e. $$r = \frac{7(15.8701) - (15.725)(6.902)}{\sqrt{7(35.6482) - (15.725)^2} \cdot \sqrt{7(7.2789) - (6.902)^2}} \approx 0.93344427$$

f. Yes, they are reliable since r is close to 1.

19.   a. $y = -24.0x + 269$; $r = -0.86$

   b. Yes   (2.5, 211), (3.9, 167), (2.9, 131), (2.4, 191), (2.9, 220), (0.8, 297), (9.1, 71),

   (0.8, 211), (0.6, 300), (7.9, 107), (1.8, 266), (1.9, 266), (0.8, 227), (6.5, 86),

   (1.6, 207), (5.8, 115), (1.3, 285), (1.2, 199), (2.7, 172)

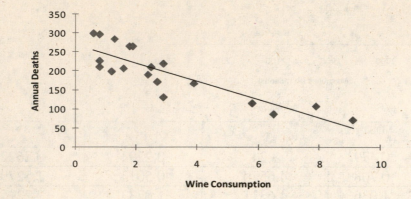

Chapter 4 Review

1.   Sorting the data:

x	x^2
4	16
5	25
5	25
6	36
7	49
8	64
8	64
8	64
10	100
10	100
$\Sigma x = 71$	$\Sigma x^2 = 543$

a. $\bar{x} = \dfrac{71}{10} = 7.1$

b. median $= \dfrac{7+8}{2} = \dfrac{15}{2} = 7.5$

c. mode $= 8$

d. $s^2 = \dfrac{1}{9}\left[543 - \dfrac{(71)^2}{10} \right] \approx 4.3222$

   $s \approx 2.1$

3.   a. $n = 23 + 45 + 53 + 31 + 17 = 169$

   $\dfrac{23+45}{169} = \dfrac{68}{169} \approx 0.40$ or 40%

   b. $\dfrac{31+17}{169} = \dfrac{48}{169} \approx 0.28$ or 28%   c. $\dfrac{45+53+31+17}{169} = \dfrac{146}{169} \approx 0.86$ or 86%

d. $\dfrac{23}{169} \approx 0.14$ or 14%

e. $\dfrac{53+31}{169} = \dfrac{84}{169} \approx 0.50$ or 50%

f. Cannot determine.

5. Let x = next score

$$\frac{74+65+85+76+x}{5} = 80$$

$$300 + x = 400$$

$$x = 100$$

7. a. Timo: mean = $\dfrac{103+99+107+93+92}{5} = \dfrac{494}{5} = 98.8$

Henke: mean = $\dfrac{101+92+83+96+111}{5} = \dfrac{483}{5} = 96.6$ Henke has a lower mean.

b. Timo:

$$s^2 = \frac{(103-98.8)^2 + (99-98.8)^2 + (107-98.8)^2 + (93-98.8)^2 + (92-98.8)^2}{4} = \frac{164.8}{4} = 41.2$$

$s \approx 6.4$

Henke:

$$s^2 = \frac{(101-96.6)^2 + (92-96.6)^2 + (83-96.6)^2 + (96-96.6)^2 + (111-96.6)^2}{4}$$

$$= \frac{433.2}{4} = 108.3$$

$s \approx 10.4$

c. Timo is more consistent because his standard deviation is lower.

9. a. Continuous   b. Neither   c. Discrete

d. Neither   e. Discrete   f. Continuous

11. a. One standard deviation $[78-7, 78+7] = [71, 85]$

Two standard deviations $[78-2(7), 78+2(7)] = [64, 92]$

Three standard deviations $[78-3(7), 78+3(7)] = [57, 99]$

b. 68.26% of the data lies in $[71, 85]$ c.

95.44% of the data lies in [64, 92]

99.74% of the data lies in [57, 99]

121

13. $p(z < c) = 0.34$. Using the tables, we find $c = -0.41$

$$\frac{x - 420}{45} = -0.41$$

$$x - 420 = -18.45$$

$$x = 401.55$$

Cutoff at 401.

15.   a. $\frac{x}{n} = \frac{800}{1200} \approx 0.67$ $\alpha = 0.90$

$$\text{MOE} = \frac{z_{\alpha/2}}{2\sqrt{n}} = \frac{z_{.45}}{2\sqrt{1200}} = \frac{1.645}{2\sqrt{1200}} \approx 0.0237435298 \quad \text{about } 2.4\%; \ 66.7\% \pm 2.4\%$$

   b. $\frac{x}{n} = 0.67$ $\alpha = 0.95$

$$\text{MOE} = \frac{z_{.475}}{2\sqrt{1200}} = \frac{1.96}{2\sqrt{1200}} \approx 0.0282901632 \quad \text{about } 2.8\%; \ 66.7\% \pm 2.8\%$$

17.   a. $\alpha = 0.95$ $\frac{z_{.475}}{2\sqrt{580}} = \frac{1.96}{2\sqrt{580}} \approx 0.0406922851 \quad \text{about } \pm 4.1\%$

   b. $\frac{z_{.475}}{2\sqrt{970}} = \frac{1.96}{2\sqrt{970}} \approx 0.0314659037 \quad \text{about } \pm 3.1\%$

   c. $\frac{z_{.475}}{2\sqrt{580+970}} = \frac{1.96}{2\sqrt{1,550}} \approx 0.0248920249 \quad \text{about } \pm 2.5\%$

19. 94% level of confidence: $\alpha/2 = 0.47 \Rightarrow z_{\alpha/2} = 1.88$

$$n = \left(\frac{1.88}{2(0.025)}\right)^2 = 1413.76 \Rightarrow 1,414$$

21.

x	y	x^2	y^2	xy
5	38	25	1,444	190
10	30	100	900	300
20	33	400	1,089	660
21	25	441	625	525
24	18	576	324	432
30	20	900	400	600
$\Sigma x = 110$	$\Sigma y = 164$	$\Sigma x^2 = 2,442$	$\Sigma y^2 = 4,782$	$\Sigma xy = 2,707$

a. Yes, the ordered pairs seem to exhibit a linear trend.

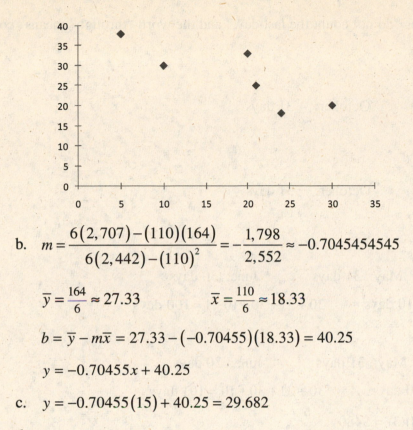

b. $m = \dfrac{6(2,707)-(110)(164)}{6(2,442)-(110)^2} = -\dfrac{1,798}{2,552} \approx -0.7045454545$

$\bar{y} = \dfrac{164}{6} \approx 27.33$ $\qquad\qquad$ $\bar{x} = \dfrac{110}{6} \approx 18.33$

$b = \bar{y} - m\bar{x} = 27.33 - (-0.70455)(18.33) = 40.25$

$y = -0.70455x + 40.25$

c. $y = -0.70455(15) + 40.25 = 29.682$

d.

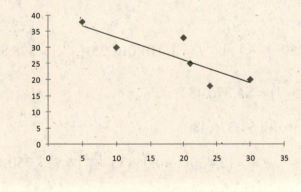

e. $r = \dfrac{6(2,707)-(110)(164)}{\sqrt{6(2,442)-(110)^2}\,\sqrt{6(4,782)-(164)^2}}$

$= \dfrac{-1.798}{\sqrt{2,552}\,\sqrt{1,796}} \approx -0.839839286$

f. Yes, the prediction is reliable since r is close to -1.

5.1 Simple Interest

Remember, the word "to" means "do not count the last day," and the word "through" means "count the last day."

1. a. 9/1 to 10/31

    September: 30 days   October: 31 days

    $30 + 31 - 1 = 60$ days

    b. 9/1 through 10/31

    September: 30 days   October: 31 days

    $30 + 31 = 61$ days

3. a. 4/1 to 7/10

    April: 30 days   May: 31 days   June: 30 days

    July 1st to July 10th: 10 days $30 + 31 + 30 + 10 - 1 = 100$ days

    b. 4/1 through 7/10

    April: 30 days   May: 31 days   June: 30 days

    July 1st to July 10th: 10 days $30 + 31 + 30 + 10 = 101$ days

5. $I = Prt = 2,000(0.08)(3) = \$480$

7. $I = 420\,(0.0675)\left(\frac{325}{365}\right) \approx \$25.24$

9. a. $I = 1,410(0.1225)\left(\frac{60}{365}\right) \approx \$28.39$ b. $I = 1,410(0.1225)\left(\frac{61}{365}\right) \approx \$28.87$

11. $FV = P(1 + rt) = 3,670(1 + 0.0275 \cdot 7) = \$4,376.48$

    In 7 years, the investment will be worth \$4,376.48.

13. $t = 660 \text{ days} \cdot \dfrac{1 \text{ year}}{365 \text{ days}} = \dfrac{132}{73}$ years $FV = 12,430\left(1 + 0.05875 \cdot \frac{132}{73}\right) \approx \$13,750.47$

    In 660 days, the investment will be worth \$13,750.47.

15. $FV = P(1 + rt) = 1,400(1 + 0.07125 \cdot 3) = \$1,699.25$

    The borrower will pay back a total of \$1,699.25.

17. $FV = 5,900\left(1 + 0.145 \cdot \frac{112}{365}\right) \approx \$6,162.51$ The borrower will pay back a total of \$6,162.51.

19.   April 1 through July 10 = 30+31+30+10 = 101 days

$$FV = 16,500\left(1 + 0.11875 \cdot \frac{101}{365}\right) \approx \$17,042.18 \quad \text{The borrower will pay back a total of } \$17,042.18.$$

21.   Solving for P:

$$FV = P(1+rt)$$
$$8,600 = P(1+0.095 \cdot 3)$$
$$P = \$6,692.61$$

One would have to invest $6,692.61 now to have the future value in the given time.

23. $t = 512 \text{ days} \cdot \dfrac{1 \text{ year}}{365 \text{ days}} = \dfrac{512}{365} \text{ years}$

Solving for P:

$$1,112 = P\left(1 + 0.03625 \cdot \frac{512}{365}\right)$$

$$1,112 \approx P(1.05085)$$

$$P \approx \$1,058.19$$

One would have to invest $1,058.19 now to have the future value in the given time.

25.   February 10 to October 15 = 19 + 31 + 30 + 31 + 30 + 31 + 31 + 30 + 15 − 1 = 247

$$t = 247 \text{ days} \cdot \dfrac{1 \text{ year}}{365 \text{ days}} = \dfrac{247}{365} \text{ year}$$

Solving for P:

$$1,311 = P\left(1 + 0.065 \cdot \frac{247}{365}\right)$$

$$1,311 \approx P(1.04399)$$

$$P \approx \$1,255.76$$

One would have to invest $1,255.76 now to have the future value in the given time.

27.   Solve for t:

$$FV = P(1+rt)$$
$$1,235.84 = 1,000(1+0.085t)$$
$$1.23584 = 1 + 0.085t$$
$$0.23584 = 0.085t$$
$$2.7746 \approx t$$
$$t = 2.7746(365) \approx 1,013 \text{ days}$$

29.   days = March + April + May + June + July

$$= 31 + 30 + 31 + 30 + 31 = 153$$ $$FV = P(1 + rt)$$

$$\text{years} = \frac{153}{365}$$ $$= 226,500\left(1 + 0.06875 \cdot \frac{153}{365}\right)$$

$$\approx 233,027.39$$

The lump sum payment is $233,027.39.

31. $$FV = 18,500(1 + 0.13 \cdot 3) = \$25,715$$ 33. $$FV = 6,200(1 + 0.098 \cdot 3) = \$8,022.80$$

Monthly payment $$= \frac{25,715}{36} \approx \$714.31$$ Monthly payment $$= \frac{8,022.80}{36} \approx \$222.86$$

35.   a. $$FV = P(1 + rt)$$

$$= 110,000\left(1 + 0.0775 \cdot \frac{90}{365}\right)$$

$$\approx 112,102.05$$

They would have to write a check for $112,102.05.

b. $$112,102.05 - 110,000 = \$2,102.05$$

37.

Time Interval	Days	Daily Balance
April 11 – April 14	4	$126.38
April 15 – April 21	7	$126.38 – $15.00 = $111.38
April 22 – April 30	9	$111.38 + $25.52 = $136.90
May 1 – May 10	10	$136.90 + $32.18 = $169.08

Average daily balance $$= \frac{4 \cdot (126.38) + 7(111.38) + 9(136.90) + 10(169.08)}{4 + 7 + 9 + 10}$$

$$= \frac{4208.08}{30} \approx 140.2693333 \approx \$140.27$$

$$I = Prt = 140.2693333 \cdot 0.18 \cdot \frac{30}{365} \approx \$2.08$$

39.

Time Interval	Days	Daily Balance
March 1 – March 4	4	$157.14
March 5 – March 16	12	$157.14 – $25.00 = $132.14
March 17 – March 31	15	$132.14 + $36.12 = $168.26

126

$$\text{Average daily balance} = \frac{4(157.14) + 12(132.14) + 15(168.26)}{4 + 12 + 15}$$

$$= \frac{4,738.14}{31} \approx 152.8432258 \approx \$152.84$$

$$I = Prt = 152.8432258 \cdot 0.21 \cdot \frac{31}{365} \approx \$2.73$$

41.   a. $\$162,500 \cdot 0.05 = \$8,125.00$

      b. $\$162,500 \cdot 0.90 = \$146,250$

      c. $\$162,500 \cdot 0.05 = \$8,125.00$

      d.   Total Interest $I = Prt = 8,125 \cdot 0.10 \cdot 4 = \$3,250$

        Monthly Interest $= \frac{\$3,250}{48} \approx \$67.71$

      e.   Down Payment $= \$8,125.00$

        Promissory Note $= \$8,125.00 + (48)(\$67.71) = \$11,375.08$

        Total: $\$8,125 + \$11,375.08 = \$19,500.08$

      f. $\$146,250 - 0.06(162,500) = \$136,500$

      g. $\$19,500.08 + \$136,500 = \$156,000.08$

43.   a. $\$389,400 \cdot 0.10 = \$38,940$

      b. $\$389,400 \cdot 0.80 = \$311,520$

      c. $\$389,400 \cdot 0.10 = \$38,940$

      d.   Total interest $I = Prt = \$38,940 \cdot 0.11 \cdot 4 = \$17,133.60$

        Monthly Interest $= \frac{\$17,133.60}{48} \approx \$356.95$

      e.   Down Payment $= \$38,940$

        Promissory Note $= \$38,940 + (48)(\$356.95) = \$56,073.60$

        Total $= \$38,940 + \$56,073.60 = \$95,013.60$

      f. $\$311,520 - 0.06(\$389,400) = \$288,156$

      g. $\$288,156 + \$95,013.60 = \$383,169.60$

45.

Time Interval	Days	Daily Balance ($)
Jan 10 – Jan 24	$24 - 9 = 15$	1,000
Jan 25 – Feb 9	$31 - 24 = 7$ $7 + 9 = 16$	$1,000 - 20 = 980$
Feb 10 – Feb 24	$24 - 9 = 15$	997.65
Feb 25 – Mar 9	$28 - 24 = 4$ $4 + 9 = 13$	$997.65 - 20 = 977.65$
Mar 10 – Mar 24	$24 - 9 = 15$	993.57
Mar 25 – Apr 9	$31 - 24 = 7$ $7 + 9 = 16$	$993.57 - 20 = 973.57$

a. Average daily balance $= \dfrac{15 \cdot 1,000 + 16 \cdot 980}{15 + 16} \approx 989.6774 = \$989.68$

$I = Prt$

$= 989.6774 \cdot 0.21 \cdot \dfrac{31}{365}$

$\approx \$17.65$

New Balance $= 980 + 17.65 = \$997.65$

b. Average daily balance $= \dfrac{15 \cdot 997.65 + 13 \cdot 977.65}{15 + 13} \approx 988.364285 = \$988.36$

$I = Prt = 988.364285 \cdot 0.21 \cdot \dfrac{28}{365} \approx \$15.92$

New Balance $= 977.65 + 15.92 = \$993.57$

c. Average daily balance $= \dfrac{15 \cdot 993.57 + 16 \cdot 973.57}{31} \approx 983.2474194 \approx \$983.25$

$I = Prt = 983.2474194 \cdot 0.21 \cdot \dfrac{31}{365} \approx \$17.54$

New Balance $= 973.57 + 17.54 = \$991.11$

d. The minimum $20 payment is reducing your debt by about $3 each month. It will take you roughly 28 years to pay off your debt. This means you will have paid the credit card company roughly $6,720 on a $1,000 debt. (572% of your debt was total interest.)

47.

Time Interval	Days	Daily Balance ($)
Jan 10 – Jan 24	$24 - 9 = 15$	1,000
Jan 25 – Feb 9	$31 - 24 = 7$ $7 + 9 = 16$	$1,000 - 40 = 960$
Feb 10 – Feb 24	$24 - 9 = 15$	977.47
Feb 25 – Mar 9	$28 - 24 = 4$ $4 + 9 = 13$	$977.47 - 40 = 937.47$
Mar 10 – Mar 24	$24 - 9 = 15$	952.92
Mar 25 – Apr 9	$31 - 24 = 7$ $7 + 9 = 16$	$952.92 - 40 = 912.92$

a. Average daily balance $= \dfrac{15 \cdot 1,000 + 16 \cdot 960}{15 + 16} \approx 979.3548387 \approx \$979.35$

$I = Prt = 979.35 \cdot 0.21 \cdot \dfrac{31}{365} \approx \$17.47$

New Balance $= 960 + 17.47 = \$977.47$

b. Average daily balance $= \dfrac{15 \cdot 977.47 + 13 \cdot 937.47}{15 + 13} \approx 958.8985714 \approx \$958.90$

$I = Prt = 958.8985714 \cdot 0.21 \cdot \dfrac{28}{365} \approx \$15.45$

New Balance $= 937.47 + 15.45 = \$952.92$

c. Average daily balance $= \dfrac{15 \cdot 952.92 + 16 \cdot 912.92}{15 + 16} \approx 932.2748387 \approx \$932.27$

$I = Prt = 932.2748387 \cdot 0.21 \cdot \dfrac{31}{365} \approx \$16.63$

New Balance $= 912.92 + 16.63 = \$929.55$

d. The change in the minimum required payment policy has had some impact,

because you're reducing your debt by approximately \$23 per month.

5.2 Compound Interest

1. a. $\dfrac{0.12}{4} = 0.03$
    b. $\dfrac{0.12}{12} = 0.01$
    c. $\dfrac{0.12}{365} \approx 0.000328767$

    d. $\dfrac{0.12}{26} \approx 0.0046153846$
    e. $\dfrac{0.12}{24} = 0.005$

3. a. $\dfrac{0.031}{4} = 0.00775$
    b. $\dfrac{0.031}{12} \approx 0.002583333$
    c. $\dfrac{0.031}{365} \approx 0.000084932$

d. $\frac{0.031}{26} \approx 0.001192307$ e. $\frac{0.031}{24} \approx 0.001291667$

5.   a. $\frac{0.097}{4} = 0.02425$ b. $\frac{0.097}{12} \approx 0.008083333$ c. $\frac{0.097}{365} \approx 0.000265753$

     d. $\frac{0.097}{26} \approx 0.003730769$ e. $\frac{0.097}{24} \approx 0.004041667$

7.   a. (8.5 years)(4 quarters/year) = 34 quarters

     b. (8.5 years)(12 months/year) = 102 months

     c. (8.5 years)(365 days/year) = 3,102.5 days

9.   a. (30 years)(4 quarters/year) = 120 quarters

     b. (30 years)(12 months/year) = 360 months

     c. (30 years)(365 day/year) = 10,950 days

11.   a. $FV = P(1+i)^n = 3,000(1+0.06)^{15} \approx \$7,189.67$

     b. After 15 years, the investment is worth $7,189.67.

13.   a. $n = (8.5 \text{ years})(4 \text{ quarters/year}) = 34 \text{ quarters}$ $FV = 5,200\left(1+\frac{0.0675}{4}\right)^{34} \approx \$9,185.46$

     b. After 8.5 years, the investment is worth $9,185.46.

15.   a. $n = (17 \text{ years})(365 \text{ days/year}) = 6,205 \text{ days}$ $FV = 1,960\left(1+\frac{0.04125}{365}\right)^{6,205} \approx \$3,951.74$

     b. After 17 years, the investment is worth $3,951.74.

17.   a. $(1+i)^n = 1 + rt$

       $\left(1+\frac{0.08}{12}\right)^{12} = 1+r$

           $r \approx 0.0829995068 \approx 8.30\%$

     b. The given compound rate is equivalent to 8.30% simple interest.

19.   a. $\left(1+\frac{0.0425}{365}\right)^{365} = 1+r$

       $r \approx 0.0434134749 \approx 4.34\%$

     b. The given compound rate is equivalent to 4.34% simple interest.

21. a. $\left(1+\frac{0.1}{4}\right)^4 = 1+r$ b. $\left(1+\frac{0.1}{12}\right)^{12} = 1+r$

$r \approx 0.1038128906 \approx 10.38\%$ $r \approx 0.1047130674 \approx 10.47\%$

The given compound rate is   The given compound rate is

equivalent to 10.38% simple interest.   equivalent to 10.47% simple interest.

c. $\left(1+\frac{0.1}{365}\right)^{365} = 1+r$

$r \approx 0.1051557816 \approx 10.52\%$

The given compound rate is equivalent to 10.52% simple interest.

23. a. $1{,}000 = P(1+0.08)^7$

$1{,}000 \approx P(1.713824269)$

$P \approx \$583.49$

b. One would have to invest \$583.49 now to have the future value in the given time.

25. a. $3{,}758 = P\left(1+\frac{0.11875}{12}\right)^{211}$

$3{,}758 \approx P(7.986536102)$

$P \approx \$470.54$

b. One would have to invest \$470.54 now to have the future value in the given time.

27. a.

Month	$FV = P(1+rt)$
1	$10{,}000\left(1+0.10\cdot\frac{1}{12}\right) \approx \$10{,}083.33$
2	$10{,}083.33\left(1+0.10\cdot\frac{1}{12}\right) \approx \$10{,}167.36$
3	$10{,}167.36\left(1+0.10\cdot\frac{1}{12}\right) \approx \$10{,}252.09$
4	$10{,}252.09\left(1+0.10\cdot\frac{1}{12}\right) \approx \$10{,}337.52$
5	$10{,}337.52\left(1+0.10\cdot\frac{1}{12}\right) \approx \$10{,}423.67$
6	$10{,}423.67\left(1+0.10\cdot\frac{1}{12}\right) \approx \$10{,}510.53$

b. $FV = P(1+i)^n = 10{,}000\left(1+0.10\cdot\frac{1}{12}\right)^6 \approx \$10{,}510.53$

29.   a.

Year	$FV = P(1 + rt)$
1	$15{,}000(1+0.06) = \$15{,}900$
2	$15{,}900(1+0.06) = \$16{,}854$
3	$16{,}854(1+0.06) = \$17{,}865.24$

   b. $FV = P(1+i)^n = 15{,}000(1+0.06)^3 = \$17{,}865.24$

31. $FV = P(1+i)^n = 10{,}000\left(1+0.06\cdot\dfrac{1}{365}\right)^{365\cdot100} \approx \$4{,}032{,}299.13$

33.   2 children; $2\cdot2=4$ grandchildren; $4\cdot2=8$ great-grandchildren; $8\cdot2=16$ great-great

   grandchildren. Total $\$ = 16\cdot\$1{,}000{,}000 = \$16{,}000{,}000$

$$FV = P(1+i)^n$$

   $16{,}000{,}000 = P\left(1+0.075\cdot\dfrac{1}{12}\right)^{12\cdot100}$

   $9{,}058.33 \approx P$

He would have to invest at least \$9,058.33.

35.   a. $FV = 3{,}000\left(1+\dfrac{0.065}{365}\right)^{365\cdot18} \approx \$9{,}664.97$

   b. $FV = 3{,}000\left(1+\dfrac{0.065}{365}\right)^{365\cdot18+30} \approx \$9{,}716.74$

      The difference is the interest: $\$9{,}716.74 - \$9{,}664.97 = \$51.77$

   c. $FV = 3{,}000\left(1+\dfrac{0.065}{365}\right)^{365\cdot8} \approx \$5{,}045.85$

37.   a. number of years $= 65 - 27 = 38$ b. number of years $= 65 - 35 = 30$

   $FV = 1{,}000\left(1+\dfrac{0.07875}{12}\right)^{12\cdot38} \approx \$19{,}741.51$ $FV = 1{,}000\left(1+\dfrac{0.07875}{12}\right)^{12\cdot30} \approx \$10{,}535.83$

39.   a. number of years $= 65 - 35 = 30$

   $100{,}000 = P\left(1+\dfrac{0.08375}{365}\right)^{365\cdot30}$

   $100{,}000 \approx P(12.33217616)$

   $P \approx \$8{,}108.87$

b. $FV = 8,108.87\left(1+\dfrac{0.08375}{365}\right)^{365\cdot30+30} \approx \$100,690.65$

The difference is the interest: $\$100,690.65 - \$100,000.00 = \$690.65$

When Marlene retires, her account will not have exactly $100,000 in it, so we will not compute the monthly interest on this amount.

41. FNB: $(1+i)^n = 1+rt$

$$\left(1+\dfrac{0.0912}{365}\right)^{365} = 1+r$$

$$r \approx 0.0954756014 \approx 9.55\%$$

CS: $\left(1+\dfrac{0.0913}{4}\right)^4 = 1+r$

$$r \approx 0.0944737207 \approx 9.45\%$$

First National Bank has the better offer.

43. a. $\left(1+\dfrac{0.0339}{365}\right)^{365} = 1+r$

$$r \approx 0.034479525 \approx 3.45\%$$

Does not verify.

b. $\left(1+\dfrac{0.0339}{360}\right)^{360} = 1+r$

$$r \approx 0.0344795024 \approx 3.45\%$$

Does not verify.

c. $\left(1+\dfrac{0.0339}{360}\right)^{365} = 1+r$

$$r \approx 0.034966662 \approx 3.50\%$$

The annual yield for the 5-year certificate verifies if you pro-rate the interest over a 360-day year, and then pay interest for 365 days.

45. $\left(1+\dfrac{0.093}{365}\right)^{365} = 1+r$

$$r \approx 0.097448735 \approx 9.74\%$$

Does not verify.

47. a. Solving for r:

$$\left(1+\dfrac{0.05}{12}\right)^{12} = 1+r$$

$$r \approx 0.0511618979 \approx 5.12\%$$

b. $FV = P(1+i)^n$

$$= 1,000\left(1+\dfrac{0.05}{12}\right)^6$$

$$\approx \$1,025.26$$

c. $\$1,025.26 - \$1,000 = \$25.26$

d. $\dfrac{25.56}{1,000} = 0.02526 \approx 2.53\%$

e. Part (d) is for 6 months; part (a) is for 1 year.

f. You earn interest on interest; that is what compounding means.

49. a. $FV = P(1+i)^n$ b. $FV = P(1+i)^n$

$= 1,000 \left(1 + \frac{0.167}{4}\right)^{2 \cdot 4}$ $= 1,000 \left(1 + \frac{0.081}{4}\right)^{2 \cdot 4}$

$\approx \$1,387.10$ $\approx \$1,173.96$

$\$1,387.10 - \$1,000 = \$387.10$ $\$1,173.96 - \$1,000 = \$173.96$

c. $FV = P(1+i)^n$

$= 1,000 \left(1 + \frac{0.02}{4}\right)^{2 \cdot 4}$

$\approx \$1,040.71$

$\$1,040.71 - \$1,000 = \$40.71$

51. $(1+i)^n = 1 + rt$ Let $t = 1$ year and solve for r in the formula:

$$(1+i)^n = 1 + r$$
$$r = (1+i)^n - 1$$

53. $r = \left(1 + \frac{0.0725}{4}\right)^4 - 1 \approx 0.0744950191$ or 7.45%

55. a. $r = \left(1 + \frac{0.05625}{2}\right)^2 - 1 \approx 0.0570410156$ or 5.70%

b. $r = \left(1 + \frac{0.05625}{4}\right)^4 - 1 \approx 0.0574476862$ or 5.74%

c. $r = \left(1 + \frac{0.05625}{12}\right)^{12} - 1 \approx 0.0577230954$ or 5.77%

d. $r = \left(1 + \frac{0.05625}{365}\right)^{365} - 1 \approx 0.0578575315$ or 5.79%

e. $r = \left(1 + \frac{0.05625}{26}\right)^{26} - 1 \approx 0.0577978427$ or 5.78%

f. $r = \left(1 + \frac{0.05625}{24}\right)^{24} - 1 \approx 0.0577924951$ or 5.78%

57. $i = 5.00\%$ because $£1,000 \cdot (1 + 0.05)^{100} \approx £131,000$

59. $i \approx 4.57\%$ because $4,500(1 + 0.0457)^{100} \approx \$391,000$

134

61. $\dfrac{\$391,000}{\$4,500} \approx 8689\%$

75.   a. $FV = 1,000\left(1+\dfrac{0.05}{365}\right)^{5061} \approx \$2,000.19$ b. $FV = 1,000\left(1+\dfrac{0.05}{365}\right)^{2\cdot 5061} \approx \$4,000.74$

   c. $FV = 1,000\left(1+\dfrac{0.05}{365}\right)^{3\cdot 5061} \approx \$8,002.23$ d. $FV = 1,000\left(1+\dfrac{0.05}{365}\right)^{4\cdot 5061} \approx \$16,005.95$

   e. Every extra period of 5,061 days doubles the money, approximately.

77.   a. $(1+0.06)^t = 2$

   $n \approx 11.90$ years, $t \approx 11.90$ years

   b. $(1+0.07)^t = 2$

   $n \approx 10.24$ years, $t \approx 10.24$ years

   c. $(1+0.1)^t = 2$

   $n \approx 7.27$ years, $t \approx 7.27$ years

   d. The larger i is (the larger the interest rate), the shorter the doubling time.

79.   To accumulate $25,000:

   $15,000\left(1+\dfrac{0.09375}{365}\right)^{365t} = 25,000$

   $n \approx 1,989.07$ days, $t \approx 5.45$ years

   To accumulate $100,000:

   $15,000\left(1+\dfrac{0.09375}{365}\right)^{365t} = 100,000$

   $n \approx 7,387.07 \approx 7,387$ days,

   $t \approx 20.24$ years

5.3 Annuities

1. $FV = 120 \cdot \dfrac{\left(1+\dfrac{0.0575}{12}\right)^{12} - 1}{\dfrac{0.0575}{12}} \approx \$1,478.56$

3. $FV = 100 \cdot \dfrac{\left(1+\dfrac{0.05875}{12}\right)^{12\cdot 4} - 1}{\dfrac{0.05875}{12}} \cdot \left(1+\dfrac{0.05875}{12}\right) \approx \$5,422.51$

5.   a. $FV(\text{ord}) = 75 \cdot \dfrac{\left(1+\dfrac{0.07}{12}\right)^{3} - 1}{\dfrac{0.07}{12}} \approx \$226.32$

135

b. $FV_{total} = FV_{Oct} + FV_{Nov} + FV_{Dec}$

$$= 75\left(1 + \frac{0.07}{12}\right)^2 + 75\left(1 + \frac{0.07}{12}\right) + 75$$

$$\approx \$226.32$$

c. $3 \cdot 75 = \$225$

d. $226.32 - 225 = \$1.32$

7. a. $FV(due) = 150 \cdot \dfrac{\left(1 + \frac{0.0725}{12}\right)^3 - 1}{\frac{0.0725}{12}}\left(1 + \frac{0.0725}{12}\right) \approx \$455.46$

b. $FV_{total} = FV_{Sept} + FV_{Oct} + FV_{Nov}$

$$= 150\left(1 + \frac{0.0725}{12}\right)^3 + 150\left(1 + \frac{0.0725}{12}\right)^2 + 150\left(1 + \frac{0.0725}{12}\right)$$

$$\approx \$455.46$$

c. $3 \cdot 150 = \$450$

d. $455.46 - 450 = \$5.46$

9. a. $FV = 175 \cdot \dfrac{\left(1 + \frac{0.105}{12}\right)^{12 \cdot 26} - 1}{\frac{0.105}{12}} \approx \$283,037.86$

b. $12 \cdot 26 \cdot 175 = \$54,600.00$ c. $\$283,037.86 - \$54,600.00 = \$228,437.86$

11. a. $FV = 290 \cdot \dfrac{\left(1 + \frac{0.11}{12}\right)^{12 \cdot 20} - 1}{\frac{0.11}{12}} \approx \$251,035.03$

b. $12 \cdot 20 \cdot 290 = \$69,600.00$ c. $\$251,035.03 - \$69,600.00 = \$181,435.03$

13. a. $P\left(1 + \frac{0.0575}{12}\right)^{12} = \$1,478.56$ (use non-rounded answer from problem 1)

$$P \approx \$1,396.14$$

15. a. $P\left(1 + \frac{0.07}{12}\right)^3 = \$226.32$

$$P \approx \$222.40$$

17. a. $P\left(1 + \frac{0.105}{12}\right)^{12 \cdot 26} = \$283,037.86$

$$P \approx \$18,680.03$$

19. $\displaystyle pymt \cdot \frac{\left(1+\frac{0.0925}{12}\right)^{12\cdot30}-1}{\frac{0.0925}{12}} = \$100,000$ 21. $\displaystyle pymt \cdot \frac{\left(1+\frac{0.105}{12}\right)^{12\cdot40}-1}{\frac{0.105}{12}} = \$250,000$

$$pymt \approx \$51.84 \qquad\qquad pymt \approx \$33.93$$

23. $\displaystyle pymt \cdot \frac{\left(1+\frac{0.105}{12}\right)^{12\cdot40}-1}{\frac{0.105}{12}} \cdot \left(1+\frac{.105}{12}\right) = \$250,000$

$$pymt \approx \$33.63$$

25. a. $\displaystyle FV = 100 \cdot \frac{\left(1+\frac{0.08125}{26}\right)^{26\cdot35.5}-1}{\frac{0.08125}{26}} \approx \$537,986.93$

  b.

	Beginning Balance	Interest	Withdrawal	Ending Balance
1	$537,986.93	$2,734.77	$650	$540,071.70
2	$540,071.70	$2,745.36	$650	$542,167.06
3	$542,167.06	$2,756.02	$650	$544,273.08
4	$544,273.08	$2,766.72	$650	$546,389.80
5	$546,389.80	$2,777.48	$650	**$548,517.28**

$$I_1 = 537,986.93 \cdot \frac{0.061}{12} \approx \$2,734.77 \qquad I_2 = 540,071.70 \cdot \frac{0.061}{12} \approx \$2,745.36$$

$$I_3 = 542,167.06 \cdot \frac{0.061}{12} \approx \$2,756.02 \qquad I_4 = 544,273.03 \cdot \frac{0.061}{12} \approx \$2,766.72$$

$$I_5 = 546,389.70 \cdot \frac{0.061}{12} \approx \$2,777.48$$

27. a. $\displaystyle P \cdot \frac{0.065}{12} = 950$

$$P \approx 175,384.62$$

  b. $\displaystyle pymt \cdot \frac{\left(1+\frac{0.0825}{12}\right)^{12\cdot30}-1}{\frac{0.0825}{12}} = \$175,384.62$

$$pymt \approx \$111.84$$

29. a. $\displaystyle FV = 200 \cdot \frac{\left(1+\frac{0.01}{12}\right)^{12}-1}{\frac{0.01}{12}} \approx \$2,411.03$

b. $FV = 200 \cdot \dfrac{\left(1 + \frac{0.0225}{12}\right)^{12} - 1}{\frac{0.0225}{12}} \approx \$2,424.91$

$FV = 2,411.03\left(1 + \frac{0.0225}{12}\right)^{12} \approx \$2,465.84$

$\$2,424.91 + \$2,465.84 = \$4,890.75$

c. $FV = 200 \cdot \dfrac{\left(1 + \frac{0.045}{12}\right)^{12} - 1}{\frac{0.045}{12}} \approx \$2,450.12$

$FV = 4,890.75\left(1 + \frac{0.045}{12}\right)^{12} \approx \$5,115.43$

$\$2,450.12 + \$5,115.43 = \$7,565.55$

31. a. $FV = 2000 \cdot \dfrac{(1 + 0.085)^{51} - 1}{0.085} \approx \$1,484,909.47$

b. Shannon: $1000(11) + 2000(40) = \$91,000$

Parents: $1000(11) = \$11,000$

Total contribution is $\$102,000$.

c. $\$1,484,909.47 - \$91,000 - \$11,000 = \$1,382,909.47$

d. $FV = 2000 \cdot \dfrac{(1 + 0.085)^{46} - 1}{0.085}$

$\approx \$979,650.96$

e. $FV = 2,000 \cdot \dfrac{(1 + 0.085)^{41} - 1}{0.085}$

$\approx \$643,631.10$

33. $pymt \cdot \dfrac{(1 + 0.085)^{30} - 1}{0.085} = \$1,484,909.47$

$pymt \approx \$11,954.38$ per year

35. $pymt \cdot \dfrac{\left(1 + \frac{0.09}{12}\right)^{12 \cdot 2} - 1}{\frac{0.09}{12}} = \$1,200$

$pymt \approx \$45.82$

37. a. $I = Prt = 65,000 \cdot \frac{0.11}{12} \approx \$595.83$

b. $pymt \cdot \dfrac{\left(1 + \frac{0.08375}{2}\right)^{2 \cdot 5} - 1}{\frac{0.08375}{2}} = \$65,000$

$pymt \approx \$5,367.01$

138

c.

Period	Starting Balance	Interest	Deposit	Ending Balance
1	0	0	$5,367.01	$5,367.01
2	$5,367.01	$224.74	$5,367.01	$10,958.76
3	$10,958.76	$458.90	$5,367.01	$16,784.67
4	$16,784.67	$702.86	$5,367.01	$22,854.54
5	$22,854.54	$957.03	$5,367.01	$29,178.58
6	$29,178.58	$1,221.85	$5,367.01	$35,767.44
7	$35,767.44	$1,497.76	$5,367.01	$42,632.21
8	$42,632.21	$1,785.22	$5,367.01	$49,784.44
9	$49,784.44	$2,084.72	$5,367.01	$57,236.17
10	$57,236.17	$2,396.76	$5,367.01	$64,999.94

39. $$P = pymt \cdot \frac{(1+i)^n - 1}{i(1+i)^n} = pymt \cdot \frac{1-(1+i)^{-n}}{i}$$

41. $$P = 120 \cdot \frac{1-\left(1+\frac{0.0575}{12}\right)^{-12}}{\left(\frac{0.0575}{12}\right)} \approx \$1,396.14$$

43. $$P = 75 \cdot \frac{1-\left(1+\frac{0.07}{12}\right)^{-3}}{\left(\frac{0.07}{12}\right)} \approx \$222.40$$

53. Initial TI-83/84 screen should hightlight BEGIN for an annuity due.

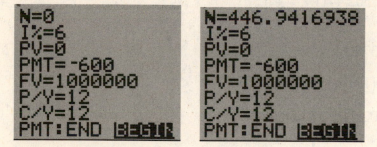

The amount of time is 446.9416938 months = 37.24514115 years or about 37 years 3 months.

5.4 Amortized Loans

1. a. $$pymt \cdot \frac{\left(1+\frac{0.095}{12}\right)^{4\cdot12} - 1}{\frac{0.095}{12}} = 5,000\left(1+\frac{0.095}{12}\right)^{4\cdot12}$$

$$pymt \approx \$125.62$$

139

b. Total payments $= (125.62)(12)(4) = \$6029.76$

Interest $= 6,029.76 - 5,000 = \$1,029.76$

3.   a. $pymt \cdot \dfrac{\left(1 + \frac{0.06125}{12}\right)^{5 \cdot 12} - 1}{\frac{0.06125}{12}} = 10,000\left(1 + \frac{0.06125}{12}\right)^{5 \cdot 12}$

$$pymt \approx \$193.91$$

b. Total payments $= (193.91)(12)(5) = \$11,634.60$

Interest $= 11,634.60 - 10,000 = \$1,634.60$

5.   a. $pymt \cdot \dfrac{\left(1 + \frac{0.095}{12}\right)^{30 \cdot 12} - 1}{\frac{0.095}{12}} = 155,000\left(1 + \frac{0.095}{12}\right)^{30 \cdot 12}$

$$pymt \approx \$1,303.32$$

b. Total payments $= (1,303.32)(12)(30) = \$469,195.20$

Interest $= 469,195.20 - 155,000 = \$314,195.20$

7.   10% down $= \$1,611.38$

Loan amount $= 16,113.82 - 1,611.38 = \$14,502.44$

a. $pymt \cdot \dfrac{\left(1 + \frac{0.115}{12}\right)^{4 \cdot 12} - 1}{\frac{0.11}{12}} = 14,502.44\left(1 + \frac{0.115}{12}\right)^{4 \cdot 12}$

$$pymt \approx \$378.35$$

b. Total payments $= (378.35)(12)(4) = \$18,160.80$

Interest $= 18,160.80 - 14,502.44 = \$3,658.36$

c.

Payment Number	Principal Portion	Interest Portion	Total Payment	Balance
0				$14,502.44
1	$239.37	$138.98	$378.35	$14,263.07
2	$241.66	$136.69	$378.35	$14,021.41

$I_1 = Prt = 14,502.44(0.115)\left(\frac{1}{12}\right) \approx \$138.98$

$I_2 = Prt = 14,263.07(0.115)\left(\frac{1}{12}\right) \approx \$136.69$

9. 20% down = $42,500

    Loan amount = $170,000

    a. $pymt \cdot \dfrac{\left(1+\frac{0.10875}{12}\right)^{30\cdot12}-1}{\frac{0.10875}{12}} = 170,000\left(1+\frac{0.10875}{12}\right)^{30\cdot12}$

    $$pymt \approx \$1,602.91$$

    b. Total payments = $(1,602.91)(12)(30) = \$577,047.60$

    Interest = $577,047.60 - 170,000 = \$407,047.60$

    c.

Payment Number	Principal Portion	Interest Portion	Total Payment	Balance
0				$170,000.00
1	$62.28	$1,540.63	$1,602.91	$169,937.72
2	$62.85	$1,540.06	$1,602.91	$169,874.87

    d. If he has no other monthly payments

    $0.38x = 1,602.91$

    $x \approx \$4,218.18$ per month (assuming only home loan payment)

11. a. Dealer: $FV = 14,829.32(1+0.0775\cdot4) = \$19,426.41$

    Since there are 48 payments: $\dfrac{\$19,426.41}{48} = \$404.72$

    Bank: $pymt \cdot \dfrac{\left(1+\frac{0.0775}{12}\right)^{4\cdot12}-1}{\frac{0.0775}{12}} = 14,829.32\left(1+\frac{0.0775}{12}\right)^{4\cdot12} = \$360.29$

    b. Dealer: Interest = $404.72(48) - 14,829.32 = \$4,597.24$

    Bank: Interest = $360.29(48) - 14,829.32 = \$2,464.60$

    c. Choose the bank loan since the interest is less.

13. a. $pymt \cdot \dfrac{\left(1+\frac{0.09875}{12}\right)^{3\cdot12}-1}{\frac{0.09875}{12}} = 11,000\left(1+\frac{0.09875}{12}\right)^{3\cdot12}$

    $$pymt \approx \$354.29$$

    Total payments = $(354.29)(12)(3) = \$12,754.44$

    Interest = $12,754.44 - 11,000 = \$1,754.44$

141

b. $pymt \cdot \dfrac{\left(1+\frac{0.09875}{12}\right)^{4\cdot12}-1}{\frac{0.09875}{12}} = 11,000\left(1+\frac{0.09875}{12}\right)^{4\cdot12}$

$pymt \approx \$278.33$

Total payments $= (278.33)(12)(4) = \$13,359.84$

Interest $= 13,359.84 - 11,000 = \$2,359.84$

c. $pymt \cdot \dfrac{\left(1+\frac{0.09875}{12}\right)^{5\cdot12}-1}{\frac{0.09875}{12}} = 11,000\left(1+\frac{0.09875}{12}\right)^{5\cdot12}$

$pymt \approx \$233.04$

Total payments $= (233.04)(12)(5) = \$13,982.40$

Interest $= 13,982.40 - 11,000 = \$2,982.40$

15. a. $pymt \cdot \dfrac{\left(1+\frac{0.06}{12}\right)^{30\cdot12}-1}{\frac{0.06}{12}} = 100,000\left(1+\frac{0.06}{12}\right)^{30\cdot12}$

$pymt \approx \$599.55$

Total payments $= (599.55)(12)(30) = \$215,838$

Interest $= 215,838 - 100,000 = \$115,838$

b. $pymt \cdot \dfrac{\left(1+\frac{0.07}{12}\right)^{30\cdot12}-1}{\frac{0.07}{12}} = 100,000\left(1+\frac{0.07}{12}\right)^{30\cdot12}$

$pymt \approx \$665.30$

Total payments $= (665.30)(12)(30) = \$239,508$

Interest $= \$139,508$

c. $pymt \cdot \dfrac{\left(1+\frac{0.08}{12}\right)^{30\cdot12}-1}{\frac{0.08}{12}} = 100,000\left(1+\frac{0.08}{12}\right)^{30\cdot12}$

$pymt \approx \$733.76$

Total payments $= (773.76)(12)(30) = \$264,153.60$

Interest $= \$164,153.60$

d. $pymt \cdot \dfrac{\left(1 + \frac{0.09}{12}\right)^{30 \cdot 12} - 1}{\frac{0.09}{12}} = 100{,}000\left(1 + \dfrac{0.09}{12}\right)^{30 \cdot 12}$

$$pymt \approx \$804.62$$

Total payments $= (804.62)(12)(30) = \$289{,}663.20$ $\qquad$ Interest $= \$189{,}663.20$

e. $pymt \cdot \dfrac{\left(1 + \frac{0.1}{12}\right)^{30 \cdot 12} - 1}{\frac{0.1}{12}} = 100{,}000\left(1 + \dfrac{0.1}{12}\right)^{30 \cdot 12}$

$$pymt \approx \$877.57$$

Total payments $= (877.57)(12)(30) = \$315{,}925.20$

Interest $= \$215{,}925.20$

f. $pymt \cdot \dfrac{\left(1 + \frac{0.11}{12}\right)^{30 \cdot 12} - 1}{\frac{0.11}{12}} = 100{,}000\left(1 + \dfrac{0.11}{12}\right)^{30 \cdot 12}$

$$pymt \approx \$952.32$$

Total payments $= (952.32)(12)(30) = \$342{,}835.20$

Interest $= \$242{,}835.20$

17. a. $pymt \cdot \dfrac{\left(1 + \frac{0.1}{12}\right)^{30 \cdot 12} - 1}{\frac{0.1}{12}} = 100{,}000\left(1 + \dfrac{0.1}{12}\right)^{30 \cdot 12}$

$$pymt \approx \$877.57$$

Total payments $= (877.57)(12)(30) = \$315{,}925.20$ $\qquad$ Interest $= \$215{,}925.20$

b. $pymt \cdot \dfrac{\left(1 + \frac{0.1}{26}\right)^{26 \cdot 30} - 1}{\frac{0.1}{26}} = 100{,}000\left(1 + \dfrac{0.1}{26}\right)^{26 \cdot 30}$

$$pymt \approx \$404.89$$

Total payments $= (404.89)(26)(30) = \$315{,}814.20$

Interest $= \$215{,}814.20$

19. a. 30 year: $pymt \cdot \dfrac{\left(1+\frac{0.105}{12}\right)^{30\cdot12}-1}{\frac{0.105}{12}} = 100,000\left(1+\frac{0.105}{12}\right)^{30\cdot12}$

$$pymt \approx \$914.74$$

15 year: $pymt \cdot \dfrac{\left(1+\frac{0.09875}{12}\right)^{15\cdot12}-1}{\frac{0.09875}{12}} = 100,000\left(1+\frac{0.09875}{12}\right)^{15\cdot12}$

$$pymt \approx \$1,066.97$$

Both verify, but not exactly. Their computations are actually more accurate than ours, since they are using more accurate round-off rules than we do.

b. 30 year: Total payments $= (914.74)(12)(30) = \$329,306.40$

15 year: Total payments $= (1066.97)(12)(15) = \$192,054.60$

Savings $= 329,306.40 - 192,054.60 = \$137,251.80$

21. a. $pymt \cdot \dfrac{\left(1+\frac{0.0775}{12}\right)^{4}-1}{\frac{0.0775}{12}} = 75,000\left(1+\frac{0.0775}{12}\right)^{4}$

$$pymt \approx \$19,053.71$$

b.

Payment Number	Principal Portion	Interest Portion	Total Payment	Balance Due
0				$75,000.00
1	$18,569.33	$484.38	$19,053.71	$56,430.67
2	$18,689.26	$364.45	$19,053.71	$37,741.41
3	$18,809.96	$243.75	$19,053.71	$18,931.45
4	$18,931.45	$122.27	$19,053.72	$0

23. a. $pymt \cdot \dfrac{\left(1+\frac{0.09125}{12}\right)^{4}-1}{\frac{0.09125}{12}} = 93,000\left(1+\frac{0.09125}{12}\right)^{4}$

$$pymt \approx \$23,693.67$$

b.

Payment Number	Principal Portion	Interest Portion	Total Payment	Balance Due
0				$93,000.00
1	$22,986.48	$707.19	$23,693.67	$70,013.52
2	$23,161.28	$532.39	$23,693.67	$46,852.24
3	$23,337.40	$356.27	$23,693.67	$23,514.84
4	$23,514.85	$178.81	$23,693.66	$0

25.   a. $I = Prt = 6,243(0.0575)\left(\frac{1}{12}\right) \approx \$29.91$

b. $pymt \cdot \dfrac{\left(1+\frac{0.0575}{12}\right)^8 - 1}{\frac{0.0575}{12}} = 6,243\left(1+\frac{0.0575}{12}\right)^8$

$pymt \approx \$797.30$

c.

Payment Number	Principal Portion	Interest Portion	Total Payment	Balance
0				$6,243.00
1	$767.39	$29.91	$797.30	$5,475.61
2	$771.06	$26.24	$797.30	$4,704.55
3	$774.76	$22.54	$797.30	$3,929.79
4	$778.47	$18.83	$797.30	$3,151.32
5	$782.20	$15.10	$797.30	$2,369.12
6	$785.95	$11.35	$797.30	$1,583.17
7	$789.71	$7.59	$797.30	$793.46
8	$793.46	$3.80	$797.26	$0

d.   Adding the interest from the schedule: $135.36

27.   a. $I = Prt$ b. $pymt \cdot \dfrac{\left(1+\frac{0.0825}{12}\right)^7 - 1}{\frac{0.0825}{12}} = 12,982\left(1+\frac{0.0825}{12}\right)^7$

$= 12,982(0.0825)\left(\frac{1}{12}\right)$ $pymt \approx \$1,905.92$

$\approx \$89.25$

c.

Payment Number	Principal Portion	Interest Portion	Total Payment	Balance
0				$12,982.00
1	$1,816.67	$89.25	$1,905.92	$11,165.33
2	$1,829.16	$76.76	$1,905.92	$9,336.17
3	$1,841.73	$64.19	$1,905.92	$7,494.44
4	$1,854.40	$51.52	$1,905.92	$5,640.04
5	$1,867.14	$38.78	$1,905.92	$3,772.90
6	$1,879.98	$25.94	$1,905.92	$1,892.92
7	$1,892.92	$13.01	$1,905.93	$0

d. Adding the interest from the schedule: $359.45

29. Loan amount $= 16,113.82 - 1,611.38 = \$14,502.44$

$$pymt \cdot \frac{\left(1+\frac{0.115}{12}\right)^{4\cdot12}-1}{\frac{0.115}{12}} = 14,502.44\left(1+\frac{0.115}{12}\right)^{4\cdot12}$$

$$pymt \approx \$378.35$$

3 years, 2 months = 38 months

$$\text{unpaid balance} = 14,502.44\left(1+\frac{0.115}{12}\right)^{38} - 378.35\cdot\frac{\left(1+\frac{0.115}{12}\right)^{38}-1}{\frac{0.115}{12}} \approx \$3,591.73$$

31. Loan amount $= 212,500 - 42,500 = \$170,000$

$$pymt \cdot \frac{\left(1+\frac{0.10875}{12}\right)^{30\cdot12}-1}{\frac{0.10875}{12}} = 170,000\left(1+\frac{0.10875}{12}\right)^{30\cdot12}$$

$$pymt \approx \$1,602.91$$

8 years, 2 months = 98 months

$$\text{unpaid balance} = 170,000\left(1+\frac{0.10875}{12}\right)^{98} - 1,602.91\cdot\frac{\left(1+\frac{0.10875}{12}\right)^{98}-1}{\frac{0.10875}{12}} \approx \$160,234.64$$

33. a. $$pymt \cdot \frac{\left(1+\frac{0.13375}{12}\right)^{30\cdot12}-1}{\frac{0.13375}{12}} = 152,850\left(1+\frac{0.13375}{12}\right)^{30\cdot12}$$

$$pymt \approx \$1,735.74$$

146

b. 3 years = 36 months

$$\text{unpaid balance} = 152{,}850\left(1+\frac{0.13375}{12}\right)^{36} - 1{,}735.74 \cdot \frac{\left(1+\frac{0.13375}{12}\right)^{36}-1}{\frac{0.13375}{12}} \approx \$151{,}437.74$$

c. $$pymt \cdot \frac{\left(1+\frac{0.08875}{12}\right)^{30\cdot12}-1}{\frac{0.08875}{12}} = 151{,}437.74\left(1+\frac{0.08875}{12}\right)^{30\cdot12}$$

$$pymt \approx \$1{,}204.91$$

d. Total payments $= (1735.74)(360) = \$624{,}866.40$

Interest $= 624{,}866.40 - 152{,}850 = \$472{,}016.40$

e. Total interest = interest on old loan + interest on new loan.

Old loan: Total payments $= (1735.74)(36) = \$62{,}486.64$

Principal paid $= 152{,}850 - 151{,}437.74 = \$1{,}412.26$

Interest paid $62{,}486.64 - 1{,}412.26 = \$61{,}074.38$

New loan: Total payments $= (1{,}204.91)(360) = \$433{,}767.60$

Interest $= 433{,}767.60 - 151{,}437.74 = \$282{,}329.86$

Total interest $= 61{,}074.38 + 282{,}329.86 = \$343{,}404.24$

35. a. $$pymt \cdot \frac{\left(1+\frac{0.105}{12}\right)^{30\cdot12}-1}{\frac{0.105}{12}} = 187{,}900\left(1+\frac{0.105}{12}\right)^{30\cdot12}$$

$$pymt \approx \$1{,}718.80$$

b. 10 years = 120 months

$$\text{Unpaid balance} = 187{,}900\left(1+\frac{0.105}{12}\right)^{120} - 1{,}718.80 \cdot \frac{\left(1+\frac{0.105}{12}\right)^{120}-1}{\frac{0.105}{12}} \approx \$172{,}157.40$$

c. 20 years = 240 months

Total payments remaining $= (1{,}718.80)(240) = \$412{,}512.00$

Interest remaining $= 412{,}512 - 172{,}157.40 = \$240{,}354.60$

d. $FV = 859.4 \dfrac{\left(1 + \frac{0.09}{12}\right)^{20 \cdot 12} - 1}{\frac{0.09}{12}} \approx \$573{,}981.98$

e. $FV = 172{,}157.40 \left(1 + \frac{0.0975}{12}\right)^{240} \approx \$1{,}200{,}543.86$

37. a. 10% down = \$18,950

    b. Loan amount (bank) = \$151,600

    c. \$18,950

    d. $pymt \cdot \dfrac{\left(1 + \frac{0.115}{12}\right)^{30 \cdot 12} - 1}{\frac{0.115}{12}} = 151{,}600 \left(1 + \frac{0.115}{12}\right)^{30 \cdot 12}$

    $$pymt \approx \$1{,}501.28$$

    e. $I = Prt = 18{,}950 \left(\frac{0.12}{12}\right) = \$189.50$

39. a. $18{,}950 = pymt \cdot \dfrac{\left(1 + \frac{0.06}{12}\right)^{5 \cdot 12} - 1}{\frac{0.06}{12}}$

    $$pymt \approx \$271.61$$

    b. $1{,}501.28 + 189.50 + 271.61 = \$1{,}962.39$

    c. \$1,501.28

41. Down payment: $\$275{,}400(0.2) = \$55{,}080$

    Amount to be financed: $\$275{,}400 - \$55{,}080 = \$220{,}320$

    $pymt \cdot \dfrac{\left(1 + \frac{0.06}{12}\right)^{30 \cdot 12} - 1}{\frac{0.06}{12}} = 220{,}320 \left(1 + \frac{0.06}{12}\right)^{30 \cdot 12}$

    $$pymt \approx \$1{,}320.93$$

Payment Number	Principal Portion	Interest Portion	Total Payment	Balance Due
0				$220,320
1	$219.33	$1,101.60	$1,320.93	$220,100.67
2	$220.43	$1,100.50	$1,320.93	$219,880.24
.				
.				
47	$275.89	$1,045.04	$1,320.93	$208,731.99
48	$277.27	$1,043.66	$1,320.93	$208,454.72

Using TI 83/84: Use 1: Finance under APPLICATIONS to store values of I%, PMT, PV, and

P/Y, and to find the balance.

Receiving $142,000 was not enough to pay off the loan. They did not make a profit but will

have to pay $208,454.72 - 142,000 = $66,454.72$.

43. a. Down payment: $189,000(0.2) = $37,800$

   Amount to be financed: $189,000 - $37,800 = $151,200$

$$pymt \cdot \frac{\left(1+\frac{0.0575}{12}\right)^{30\cdot12} - 1}{\frac{0.0575}{12}} = 151,200\left(1+\frac{0.0575}{12}\right)^{30\cdot12}$$

$$pymt \approx \$882.36$$

   b. $$\$982.36 \cdot \frac{\left(1+\frac{0.0575}{12}\right)^{12n} - 1}{\frac{0.0575}{12}} = 151,200\left(1+\frac{0.0575}{12}\right)^{12n}$$

To solve directly, use logarithms: $n \approx 23.32$ years. The loan will be paid off early in

approximately 6.69 years.

To solve by trial and error using the calculator:

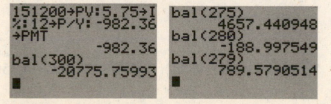

The loan will be paid off sometime between the 279th and 280th payment, or

149

$360 - 280 = 80$ payments early, which is approximately 6.67 years.

5.5 Annual Percentage Rate on a Graphing Calculator

1. Loan amount $= 16,113.82 - 1,611.38 = \$14,502.44$

    $$pymt \cdot \frac{\left(1 + \frac{0.115}{12}\right)^{4 \cdot 12} - 1}{\frac{0.115}{12}} = 14,502.44 \left(1 + \frac{0.115}{12}\right)^{4 \cdot 12}$$

    $$pymt \approx \$378.35$$

    Since finance charges are $\$814.14$, $P = 14,502.44 - 814.14 = \$13,688.30$

    $$378.35 \cdot \frac{(1 + i)^{4 \cdot 12} - 1}{i} = 13,688.30(1 + i)^{4 \cdot 12}$$

    Solving, $i \approx 0.0121865$ $\qquad APR = (0.0121865)(12) = 0.146238 \approx 14.6\%$

3. Loan amount $= 212,500 - 42,500 = \$170,000$

    $$pymt \cdot \frac{\left(1 + \frac{0.10875}{12}\right)^{30 \cdot 12} - 1}{\frac{0.10875}{12}} = 170,000 \left(1 + \frac{0.10875}{12}\right)^{30 \cdot 12}$$

    $$pymt \approx \$1,602.91$$

    2 points $= (0.02)(170,000) = \$3,400$

    $P = 170,000 - 3,400 - 1,318.10 = \$165,281.90$

    $$1,602.91 \cdot \frac{(1 + i)^{30 \cdot 12} - 1}{i} = 165,281.90(1 + i)^{30 \cdot 12}$$

    Solving, $i \approx 0.009359$ $\qquad APR = (0.009359)(12) = 0.112308 \approx 11.2\%$

5. a. $I = Prt = 4,600(0.08)(4) = \$1,472$ $\qquad$ Total $= 4,600 + 1,472 = \$6,072$

    Since there are 48 payments, $pymt = \$126.50$

    b. $126.5 \cdot \frac{(1 + i)^{4 \cdot 12} - 1}{i} = 4,600(1 + i)^{4 \cdot 12}$

    Solving, $i \approx 0.0119544$ $\qquad APR = (0.0119544)(12) = 0.1434528 \approx 14.35\%$

    Verifies; this is within the tolerance.

7. a. Loan amount $= 10,340 - 1,034 = \$9,306$

    $I = Prt = 9,306(0.095)(5) = \$4,420.35$ $\qquad$ Total $= 9,306 + 4,420.35 = \$13,726.35$

    Since there are 60 payments, $pymt = \$228.77$

    b. $228.77 \cdot \dfrac{(1+i)^{5 \cdot 12} - 1}{i} = 9,306(1+i)^{5 \cdot 12}$

    Solving, $i \approx 0.0137477$ $\qquad$ $APR = (0.0137477)(12) = 0.1649724 \approx 16.50\%$

    Doesn't verify; the advertised APR is incorrect.

9. Either one could be less expensive, depending on the A.P.R.

11. Really Friendly S and L will have lower payments but higher fees and/or more points.

13. Loan amount $= 119,000 - 23,800 = \$95,200$

    $pymt \cdot \dfrac{\left(1 + \frac{0.0825}{12}\right)^{30 \cdot 12} - 1}{\frac{0.0825}{12}} = 95,200\left(1 + \frac{0.0825}{12}\right)^{30 \cdot 12}$

    $pymt \approx \$715.21$

    Find the legal principal:

    $715.21 \cdot \dfrac{\left(1 + \frac{0.0923}{12}\right)^{360} - 1}{\frac{0.0923}{12}} = P\left(1 + \frac{0.0923}{12}\right)^{360}$

    $P \approx \$87,090.47$

    Fees $= 95,200 - 87,090.47 = \$8,109.53$

15. a. Loan amount $= 124,500 - 24,900 = \$99,600$

    $pymt \cdot \dfrac{\left(1 + \frac{0.0925}{12}\right)^{360} - 1}{\frac{0.0925}{12}} = 99,600\left(1 + \frac{0.0925}{12}\right)^{360}$

    $pymt \approx \$819.38$

    b. $pymt \cdot \dfrac{\left(1 + \frac{0.09}{12}\right)^{360} - 1}{\frac{0.09}{12}} = 99,600\left(1 + \frac{0.09}{12}\right)^{360}$

    $pymt \approx \$801.40$

c. $819.38 \cdot \dfrac{\left(1+\frac{0.1023}{12}\right)^{360}-1}{\frac{0.1023}{12}} = P\left(1+\frac{0.1023}{12}\right)^{360}$

$$P \approx \$91,590.25$$

Fees $= 99,600 - 91,590.25 = \$8,009.75$

d. $801.40 \cdot \dfrac{\left(1+\frac{0.1016}{12}\right)^{360}-1}{\frac{0.1016}{12}} = P\left(1+\frac{0.1016}{12}\right)^{360}$

$$P \approx \$90,103.88$$

Fees $= 99,600 - 90,103.88 = \$9,496.12$

e. The RTC loan has a lower monthly payment but has higher fees.

17. The fees and points must be \$0.

18. They were both designed to make comparisons easier. APR is to stop the confusion that results when fees and points are charged. Annual yield is to stop the confusion that results from varying compounding schemes.

5.6 Payout Annuities

1. a. $P\left(1+\frac{0.08}{12}\right)^{20\cdot12} = 1,200 \cdot \dfrac{\left(1+\frac{0.08}{12}\right)^{20\cdot12}-1}{\frac{0.08}{12}}$

    b. $(1,200)(12)(20) = \$288,000$

$$P = \$143,465.15$$

3. a. $P\left(1+\frac{0.073}{12}\right)^{25\cdot12} = 1,300 \cdot \dfrac{\left(1+\frac{0.073}{12}\right)^{25\cdot12}-1}{\frac{0.073}{12}}$

    b. $(1,300)(12)(25) = \$390,000$

$$P \approx \$179,055.64$$

5. a. $143,465.15 = pymt\,\dfrac{\left(1+\frac{0.08}{12}\right)^{30\cdot12}-1}{\frac{0.08}{12}}$

    b. Pay in $= (96.26)(12)(30) = \$34,653.60$

$$pymt \approx \$96.26$$

Receive $= \$288,000$

Therefore, Suzanne receives \$253,346.40 more than she paid.

152

7.   a. $179,055.64 = pymt \dfrac{\left(1+\frac{0.073}{12}\right)^{30\cdot12}-1}{\frac{0.073}{12}}$ b. $179,055.64 = pymt \dfrac{\left(1+\frac{0.073}{12}\right)^{20\cdot12}-1}{\frac{0.073}{12}}$

         $pymt \approx \$138.30$ $pymt \approx \$331.39$

9.   a. $P = 14,000 \cdot \dfrac{1-\left(\frac{1+0.04}{1+0.08}\right)^{20}}{0.08-0.04} \approx \$185,464.46$ b. $\$14,000$

     c. $14,000+(0.04)(14,000) = \$14,560$ d. $14,000(1+0.04)^{19} = \$29,495.89$

11.   a. $13,000(1+0.035)^{30} = \$36,488.32$

     b. $P = 36,488.32 \cdot \dfrac{1-\left(\frac{1+0.035}{1+0.072}\right)^{25}}{0.072-0.035} \approx \$576,352.60$

     c. $576,352.60 = pymt \dfrac{\left(1+\frac{0.072}{12}\right)^{30\cdot12}-1}{\frac{0.072}{12}}$ d. $576,352.60 = pymt \dfrac{\left(1+\frac{0.072}{12}\right)^{20\cdot12}-1}{\frac{0.072}{12}}$

         $pymt \approx \$454.10$ $pymt \approx \$1,079.79$

13.   a. $FV = 1,000 \cdot \dfrac{\left(1+\frac{0.10}{12}\right)^{12}-1}{\frac{0.10}{12}} \approx \$12,565.57$ b. $(1+i)^{n} = 1+rt$

                                                  $\left(1+\frac{0.10}{12}\right)^{12} = 1+r$

                                                 $r \approx 10.47130674\%$

     c. $P = 12,565.57 \cdot \dfrac{1-\left(\frac{1+0.03}{1+0.1047130674}\right)^{25}}{0.1047130674-0.03} \approx \$138,977.90$

     d. $138,977.90 = pymt \cdot \dfrac{\left(1+\frac{0.10}{12}\right)^{360}-1}{\frac{0.10}{12}}$

         $pymt \approx \$61.48$

15. a. $FV = 1,400 \cdot \dfrac{\left(1+\frac{0.089}{12}\right)^{12} - 1}{\frac{0.089}{12}} \approx \$17,502.53$ b. $\left(1+\frac{0.089}{12}\right)^{12} = 1 + r$

$$r \approx 0.092721727 = 9.2721727\%$$

c. $P = 17,502.53 \cdot \dfrac{1 - \left(\frac{1+0.05}{1+0.092721727}\right)^{23}}{0.092721727 - 0.05} \approx \$245,972.94$

d. $245,972.94 = pymt \cdot \dfrac{\left(1+\frac{0.089}{12}\right)^{20 \cdot 12} - 1}{\frac{0.089}{12}}$

$$pymt \approx \$372.99$$

17. $P(1+0.08)^{20} = 50,000 \dfrac{(1+0.08)^{20} - 1}{0.08}$

$$P \approx \$490,907.37$$

Chapter 5 Review

1. $I = Prt = 8,140(0.0975)(11) = \$8,730.15$

3. $FV = P(1+rt) = 12,288(1+0.0425 \cdot 15) = \$20,121.60$

5. $FV = P(1+rt) = 3,550\left(1+0.125 \cdot \frac{14}{12}\right) \approx \$4,067.71$

7. $FV = P(1+rt)$

$84,120 = P(1+0.0725 \cdot 25)$

$P \approx \$29,909.33$

9. $FV = P(1+i)^n = 8,140\left(1+\frac{0.0975}{12}\right)^{12 \cdot 11} \approx \$23,687.72$

11. $\text{Interest} = P(1+i)^n - P$

$= 7,990\left(1+\frac{0.0475}{12}\right)^{12 \cdot 11} - 7,990 \approx 13,459.15 - 7,990 = \$5,469.15$

13. $FV = P(1+i)^n$

$$33,120 = P\left(1+\frac{0.0625}{365}\right)^{365 \cdot 25}$$

$$P \approx \$6,943.26$$

15. $FV\,(\text{simple interest}) = FV\,(\text{compounded daily})$

$$P(1+rt) = P(1+i)^n$$

$$1+rt = (1+i)^n$$

$$1+r \cdot 1 = \left(1+\frac{0.07}{365}\right)^{365}$$

$$r \approx 1.0725 - 1$$

$$r = 7.25\%$$

17. $FV\,(\text{ord}) = pymt \cdot \dfrac{(1+i)^n - 1}{i}$

$$= 230 \cdot \frac{\left(1+\frac{0.0625}{12}\right)^{12 \cdot 20} - 1}{\frac{0.0625}{12}}$$

$$\approx \$109,474.35$$

19. $FV\,(\text{due}) = FV\,(\text{ord}) \cdot (1+i)$

$$= pymt \cdot \frac{(1+i)^n - 1}{i}(1+i)$$

$$= 450 \cdot \frac{\left(1+\frac{0.0825}{12}\right)^{12 \cdot 20} - 1}{\frac{0.0825}{12}}\left(1+\frac{0.0825}{12}\right)$$

$$\approx \$275,327.05$$

21.   a. $pymt \cdot \dfrac{(1+i)^n - 1}{i} = P(1+i)^n$ b. Total payments: $525.05(12 \cdot 5) = \$31,503$

$$pymt \cdot \frac{\left(1+\frac{0.095}{12}\right)^{12 \cdot 5} - 1}{\frac{0.095}{12}} = 25,000\left(1+\frac{0.095}{12}\right)^{12 \cdot 5}$$ Interest: $31,503 - 25,000 = \$6,503$

$$pymt \approx \$525.05$$

23.   days = Feb + March + April + May + June + July + August

$$= 14 + 31 + 30 + 31 + 30 + 31 + 31 = 198$$

years $= \frac{198}{365}$

$$FV = P(1 + rt)$$

$$= 351,500\left(1 + 0.05875 \cdot \frac{198}{365}\right) \approx \$362,702.26$$

25.

Time Interval	Days	Daily Balance
Aug. 26 – Aug. 29	4	$3,472.38
Aug. 30 – Sept. 1	3	$3,472.38 – $100.00 = $3,372.38
Sept. 2 – Sept 9	8	$3,372.38 + $34.12 = $3,406.50
Sept. 10 – Sept 25	16	$3,406.50 + $62.00 = $3,468.50

Average daily balance

$$= \frac{4(3472.38) + 3(3372.38) + 8(3406.50) + 16(3468.50)}{4 + 3 + 8 + 16} \approx \$3,443.70$$

$$I = Prt = (3443.70)(0.195)\left(\frac{31}{365}\right) \approx \$57.03$$

27.   a. $FV = P(1 + i)^n$

$$250,000 = P\left(1 + \frac{0.08125}{4}\right)^{4\cdot40}$$

$$P \approx \$10,014.44$$

b. $P(1 + i)^n = 250,000\left(1 + \frac{0.08125}{12}\right)^1 \approx 251,692.71$

Interest: $251,692.71 - 250,000 = \$1,692.71$

29. $FV(\text{ord}) = pymt \cdot \dfrac{(1+i)^n - 1}{i}$

$$= 200 \cdot \frac{\left(1 + \frac{0.06125}{12}\right)^{12\cdot33} - 1}{\frac{0.06125}{12}} \approx \$255,048.53$$

31. $FV = 200 \cdot \dfrac{\left(1 + \frac{0.06125}{12}\right)^{11\cdot12} - 1}{\frac{0.06125}{12}} \cdot \left(1 + \frac{0.06125}{12}\right) \approx \$37,738.26$

156

33.   a. $FV(\text{ord}) = pymt \cdot \dfrac{(1+i)^n - 1}{i}$

$$= 100 \cdot \dfrac{\left(1 + \dfrac{0.06125}{12}\right)^{30 \cdot 12} - 1}{\dfrac{0.06125}{12}} \approx \$102{,}887.14$$

b.

Month	Beginning	Interest Earned	Withdrawal	End
1	$102,887.14	$493.00	$1,000	$102,380.14
2	$102,380.14	$490.57	$1,000	$101,870.71
3	$101,870.71	$488.13	$1,000	$101,358.84
4	$101,358.84	$485.68	$1,000	$100,844.52
5	$100,844.52	$483.21	$1,000	$100,327.73

35.   a. $pymt \cdot \dfrac{(1+i)^n - 1}{i} = P(1+i)^n$

$$pymt \cdot \dfrac{\left(1 + \dfrac{0.07375}{12}\right)^{12 \cdot 30} - 1}{\dfrac{0.07375}{12}} = 180{,}480\left(1 + \dfrac{0.07375}{12}\right)^{12 \cdot 30}$$

$$pymt \approx \$1{,}246.53$$

b. Total payments: $1{,}246.53(12 \cdot 30) = \$448{,}750.80$

Interest: $448{,}750.80 - 180{,}480 = \$268{,}270.80$

c.

Payment	Principal Portion	Interest Portion	Total Payment	Balance
0				$180,480
1	$137.33	$1,109.20	$1,246.53	$180,342.67
2	$138.17	$1,108.36	$1,246.53	$180,204.50

37.   a. $I = Prt = 41{,}519(0.0775)\left(\dfrac{1}{12}\right) \approx \$268.14$

b. $pymt \cdot \dfrac{(1+i)^n - 1}{i} = P(1+i)^n$

$$pymt \cdot \dfrac{\left(1 + \frac{0.0775}{12}\right)^7 - 1}{\frac{0.0775}{12}} = 41{,}519\left(1 + \frac{0.0775}{12}\right)^7$$

$$pymt \approx \$6{,}085.50$$

c.

Payment	Principal Portion	Interest Portion	Total Payment	Balance
0				$41,519.00
1	$5,817.36	$268.14	$6,085.50	$35,701.64
2	$5,854.93	$230.57	$6,085.50	$29,846.71
3	$5,892.74	$192.76	$6,085.50	$23,953.97
4	$5,930.80	$154.70	$6,085.50	$18,023.17
5	$5,969.10	$116.40	$6,085.50	$12,054.07
6	$6,007.65	$77.85	$6,085.50	$6,046.42
7	$6,046.42	$39.05	$6,085.47	$0.00

d. Total interest: $6{,}085.50(7) - 41{,}519 = \$1{,}079.50$

39. a. Loan amount $= 6{,}200 - 200 = \$6{,}000$

   $I = Prt = 6{,}000(0.099)(4) = \$2{,}376$

   Total payments $= 6{,}000 + 2{,}376 = \$8{,}376$

   Monthly payment $= \frac{8{,}376}{48} = \$174.50$

   b. $174.50 \cdot \dfrac{(1+i)^{48} - 1}{i} = 6{,}000(1+i)^{48}$

   Solving, $i \approx 0.01453357$

   $APR = (0.01453357)(12) = 0.17440284 \approx 17.44\%$

   Does not verify; the advertised APR is incorrect.

41. Loan amount $= (0.8)(198{,}000) = \$158{,}400$

   a. $pymt \cdot \dfrac{\left(1 + \frac{0.085}{12}\right)^{30 \cdot 12} - 1}{\frac{0.085}{12}} = 158{,}400\left(1 + \frac{0.085}{12}\right)^{30 \cdot 12}$

   $$pymt \approx \$1{,}217.96$$

b. $1,217.96 \cdot \dfrac{\left(1+\frac{0.0902}{12}\right)^{360}-1}{\frac{0.0902}{12}} = P\left(1+\frac{0.0902}{12}\right)^{360}$

$$P \approx \$151,100.02$$

Fees $= 158,400 - 151,100.02 = \$7,299.98$

43. a. $17,000\left(1+0.042\right)^{30} = \$58,409.10$

   b. $P = 58,409.10 \cdot \dfrac{1-\left(\frac{1+0.042}{1+0.083}\right)^{25}}{0.083-0.042} \approx \$881,764.23$

   c. $881,764.23 = pymt \cdot \dfrac{\left(1+\frac{0.083}{12}\right)^{360}-1}{\frac{0.083}{12}}$

$$pymt \approx \$556.55$$

6.1 Voting Systems

1. a. $314 + 155 + 1,052 + 479 = 2,000$ b. Cruz   c. Yes

3. a. $3,021 + 4,198 + 3,132 = 10,351$ b. Edelstein   c. No

5. a. $6 + 8 + 11 + 5 = 30$ b. Park   c. $\frac{6+8}{30} \approx 47\%$

   d. Deleting the lowest vote getter (W) and giving those votes to the second choice (B):

   P gets $6 + 8 = 14$ votes. B gets $11 + 5 = 16$ votes, which is a majority.

   Beach wins.

   e. $\frac{16}{30} \approx 53\%$

   f. $k = 3$. 3 pts for 1st place, 2 pts for 2nd, 1 pt for 3rd.   g. 63 points

   Score for:

   P: $6 \cdot 3 + 8 \cdot 3 + 11 \cdot 1 + 5 \cdot 1 = 18 + 24 + 11 + 5 = 58$

   B: $6 \cdot 2 + 8 \cdot 1 + 11 \cdot 3 + 5 \cdot 2 = 12 + 8 + 33 + 10 = 63$

   W: $6 \cdot 1 + 8 \cdot 2 + 11 \cdot 2 + 5 \cdot 3 = 6 + 16 + 22 + 15 = 59$

   Beach wins.

   h. W vs. P: P gets $6 + 18 = 14$, W gets $11 + 5 = 16$. W: 1 pt.   i. 2 points

   W vs. B: B gets $6 + 11 = 17$, W gets $8 + 5 = 13$. B: 1 pt.

   P vs. B: P gets $6 + 8 = 14$, B gets $11 + 5 = 16$. B: 1 pt.

   Beach wins.

7. a. $19 + 12 + 10 + 11 + 13 = 65$ b. C: $19 + 12 = 31$ c. $\frac{31}{65} \approx 48\%$

   P: 10

   R: $11 + 13 = 24$

   Coastline wins.

   d. Delete P, give votes to C.   e. $\frac{41}{65} \approx 63\%$

   C: $19 + 12 + 10 = 41$, a majority.

   R: $11 + 13 = 24$

   Coastline wins.

f. C: $19\cdot3+12\cdot3+10\cdot2+11\cdot1+13\cdot2=150$ g. 150 points

P: $19\cdot2+12\cdot1+10\cdot3+11\cdot2+13\cdot1=115$

R: $19\cdot1+12\cdot2+10\cdot1+11\cdot3+13\cdot3=125$

Coastline wins.

h. C vs. P: C gets $19+12+13=44$, P gets $10+11=21$. C: 1 pt   i. 2 points

C vs. R: C gets $19+12+10=41$, R gets $11+13=24$. C: 1 pt

P vs. R: P gets $19+10=29$, R gets $12+11+13=36$. R: 1 pt

Coastline wins.

9. a. $25+13+19+27+30+26=140$ b. B: $25+13=38$ c. $\frac{56}{140}=40\%$

N: $19+27=46$

S: $30+26=56$

Shattuck wins.

d. Delete B, give votes to 2nd place.   e. $\frac{71}{140}\approx51\%$

N: $25+19+27=71$, a majority.

S: $13+30+26=69$

Nirgiotis wins.

f. B: $25\cdot3+13\cdot3+19\cdot1+27\cdot2+30\cdot2+26\cdot1=273$

N: $25\cdot2+13\cdot1+19\cdot3+27\cdot3+30\cdot1+26\cdot2=283$

S: $25\cdot1+13\cdot2+19\cdot2+27\cdot1+30\cdot3+26\cdot3=284$ Shattuck wins.

g. 284 points

h. B vs. N: B gets $25+13+30=68$, N gets $19+27+26=72$. N: 1 pt.

B vs. S: B gets $25+13+27=65$, S gets $19+30+26=75$. S: 1 pt.

N vs. S: N gets $25+19+27=71$, S gets $13+30+26=69$. N: 1 pt.

Nirgiotis wins.

i. 2 points

11.   a. $225+134+382+214+81+197+109=1{,}342$ b. D: $225+134=359$

J: $382+214=596$

N: 81

T: $197+109=306$

Jones wins.

c. $\dfrac{596}{1{,}342}\approx44\%$ d. Delete N, give votes to 2nd place (J).   e. $\dfrac{677}{1{,}342}\approx50.4\%$

D: $225+134=359$

J: $382+214+81=677$, a majority.

T: $197+109=306$ Jones wins.

f. D: $225\cdot4+134\cdot4+382\cdot3+214\cdot2+81\cdot1+197\cdot3+109\cdot2=3{,}900$ g. 3,960 points

J: $225\cdot3+134\cdot1+382\cdot4+214\cdot4+81\cdot3+197\cdot1+109\cdot3=3{,}960$

T: $225\cdot2+134\cdot2+382\cdot2+214\cdot3+81\cdot2+197\cdot4+109\cdot4=3{,}510$

N: $225\cdot1+134\cdot3+382\cdot1+214\cdot1+81\cdot4+197\cdot2+109\cdot1=2{,}050$

Jones wins.

h. D vs. J: D gets $225+134+197=556$, J gets $382+214+81+109=786$.

J: 1 pt.

D vs N: D gets $225+134+382+214+197+109=1{,}261$, N gets 81.

D: 1 pt.

D vs. T: D gets $225+134+382=741$, T gets $214+81+197+109=601$.

D: 1 pt.

J vs N: J gets $225+382+214+109=930$, N gets $134+81+197=412$.

J: 1 pt.

J vs T: J gets $225+382+214+81=902$, T gets $134+197+109=440$.

J: 1 pt.

N vs. T: N gets $134+81=215$, T gets $225+382+214+197+109=1{,}127$.

T: 1 pt.

Jones wins.

i. 3 points

13.   a. $1,897 + 1,025 + 4,368 + 2,790 + 6,897 + 9,571 + 5,206 = 31,754$

      b. A: $1,897 + 1,025 = 2,922$

          B: 4,368   D: $6,897 + 9,571 = 16,468$

          C: 2,790   E: 5,206

          Darter wins.

      c. $\dfrac{16,468}{31,754} \approx 52\%$

      d. Darter wins (He has a majority, so no runoff is needed.)   e. 52%

      f. A: $1,897 \cdot 5 + 1,025 \cdot 5 + 4,368 \cdot 1 + 2,790 \cdot 1 + 6,897 \cdot 1 + 9,571 \cdot 2 + 5,206 \cdot 1 = 53,013$

          B: $1,897 \cdot 2 + 1,025 \cdot 1 + 4,368 \cdot 5 + 2,790 \cdot 2 + 6,897 \cdot 4 + 9,571 \cdot 3 + 5,206 \cdot 2 = 98,952$

          C: $1,897 \cdot 4 + 1,025 \cdot 3 + 4,368 \cdot 3 + 2,790 \cdot 5 + 6,897 \cdot 3 + 9,571 \cdot 1 + 5,206 \cdot 3 = 83,597$

          D: $1,897 \cdot 3 + 1,025 \cdot 4 + 4,368 \cdot 4 + 2,790 \cdot 3 + 6,897 \cdot 5 + 9,571 \cdot 5 + 5,206 \cdot 4 = 138,797$

          E: $1,897 \cdot 1 + 1,025 \cdot 2 + 4,368 \cdot 2 + 2,790 \cdot 4 + 6,897 \cdot 2 + 9,571 \cdot 4 + 5,206 \cdot 5 = 101,951$

          Darter wins.

      g. 138,797 points

      h. A vs. B: A gets $1,897 + 1,025 = 2,922$,

          B gets $4,368 + 2,790 + 6,897 + 9,571 + 5,206 = 28,832$. B: 1 pt.

          A vs. C: A gets $1,897 + 1,025 + 9,571 = 12,493$,

          C gets $4,368 + 2,790 + 6,897 + 5,206 = 19,261$. C: 1 pt.

          A vs. D: A gets $1,897 + 1,025 = 2,922$,

          D gets $4,368 + 2,790 + 6,897 + 9,571 + 5,206 = 28,832$. D: 1 pt.

          A vs. E: A gets $1,897 + 1,025 = 2,922$,

          E gets $4,368 + 2,790 + 6,897 + 9,571 + 5,206 = 28,832$. E: 1 pt.

          B vs. C: B gets $4,368 + 6,897 + 9,571 = 20,836$.

          C gets $1,897 + 1,025 + 2,790 + 5,206 = 10,918$. B: 1 pt.

          B vs. D: B gets 4,368,

          D gets $1,897 + 1,025 + 2,790 + 6,897 + 9,571 + 5,206 = 27,386$. D: 1 pt.

          B vs. E: B gets $1,897 + 4,368 + 6,897 = 13,162$,

E gets $1,025 + 2,790 + 9,571 + 5,206 = 18,592$. E: 1 pt.

C vs. D: C gets $1,897 + 2,790 = 4,687$,

D gets $1,025 + 4,368 + 6,897 + 9,571 + 5,206 = 27,067$. D: 1 pt.

C vs. E: C gets $1,897 + 1,025 + 4,368 + 2,790 + 6,897 = 16,977$

E gets $9,571 + 5,206 = 14,777$. C: 1 pt.

D vs. E: D gets $1,897 + 1,025 + 4,368 + 6,897 + 9,571 = 23,758$

E gets $2,790 + 5,206 = 7,996$. D: 1 pt.

Darter wins.

   i.   4 points

15. $6! = 6 \cdot 5 \cdot 4 \cdot 3 \cdot 2 \cdot 1 = 720$

17. $_6C_2 = \dfrac{6!}{2!4!} = \dfrac{6 \cdot 5}{2} = 15$

19.   a. Since $k = 3$, 3 is the maximum number of points per vote.

          The total score $= 3 \cdot 25 = 75$

      b. The smallest k value is 1, $1 \cdot 25 = 25$

21.   a. There will be $_7C_2 = \dfrac{7!}{2!5!} = \dfrac{7 \cdot 6}{2} = 21$ pair-wise comparisons. Each candidate is paired with

          each of the remaining 6 candidates, so most points for a candidate would be 6. For example,

          AB, AC, AD, AE, AF, AG.

      b. They could lose every one, score $= 0$.

23.   a. A: 7, B: 4, C: 2

          A has the majority of the first-place votes

          and therefore should win.

	A	B	C
First	7	4	2
Second	0	9	4
Third	6	0	7
Total	13	13	13

      b. A: $7 \cdot 3 + 0 \cdot 2 + 6 \cdot 1 = 21 + 0 + 6 = 27$

          B: $4 \cdot 3 + 9 \cdot 2 + 0 \cdot 1 = 12 + 18 + 0 = 30$

          C: $2 \cdot 3 + 4 \cdot 2 + 7 \cdot 1 = 6 + 8 + 7 = 21$

          Since 30 is the greatest points, B wins.

	A	B	C
First	21	12	6
Second	0	18	8
Third	6	0	7
Total	27	30	21

      c. Yes, A received 7 votes for first choice, which is a majority of the 13 votes, but A did not

          win.

25. a. Number of first-choice votes: A: $6 + 4 = 10$, B: 8, C: 9

    Majority of votes needed: $\frac{27}{2} = 13.5$

	6	8	4	9
First	A	C	A	C
Second	C	A	C	A

    Since none of the candidates received a
    majority, eliminate the candidate with the
    fewest first-choice votes, B.

    Now, find number of first-choice votes: A: $6 + 4 = 10$, C: $8 + 9 = 17$

    C wins by getting a majority of first-choice votes.

    b. Eliminate A. B wins by getting 14 out of 27 first-choice votes.

	6	8	13
First	B	B	C
Second	C	C	B

    c. Yes. C wins the original election. All changes are in favor of C, but C does not win again.

Section 6.2 Methods of Apportionment

1. a. $900,000 = 900$ thousand; $700,000 = 700$ thousand; $400,000 = 400$ thousand;

    total $= 900 + 700 + 400 = 2,000$

    b. Standard divisor: $d = \dfrac{\text{total population}}{\text{total number of seats}} = \dfrac{2,000}{40} = 50.00$

    c. Standard quota: $q = \dfrac{\text{each state's population}}{d}$, (round to two decimal places)

    Lower quota = standard quota truncated to whole number.

    Upper quota = standard quota rounded up to next whole number.

State	A	B	C	Total
Population (thousands)	900	700	400	2,000
Std q $(d = 50.00)$	18.00	14.00	8.00	—
Lower q	18	14	8	40
Upper q	19.00	15.00	9.00	43

3. a. $18,976,821 = 18.977$ million; $12,281,054 = 12.281$ million; $8,414,347 = 8.414$ million;

    total $= 18.977 + 12.281 + 8.414 = 39.672$

    b. Standard divisor: $d = \dfrac{\text{total population}}{\text{total number of seats}} = \dfrac{39.672}{15} = 2.64$

c. Standard quota: $q = \dfrac{\text{each state's population}}{d}$, (round to two decimal places)

Lower quota = standard quota truncated to whole number.

Upper quota = standard quota rounded up to next whole number.

State	NY	PA	NJ	Total
Population (millions)	18.977	12.281	8.414	39.672
Std q $(d=2.64)$	7.19	4.65	3.19	—
Lower q $(d=2.64)$	7	4	3	14
Upper q $(d=2.64)$	8	5	4	17

5. Since the sum of the lower quotas equals the total number of seats to be apportioned, the apportionment process is complete.

State	A	B	C	Total
Population (thousands)	900	700	400	2,000
Std q $(d=50.00)$	18.00	14.00	8.00	—
Lower q $(d=50.00)$	18	14	8	40
Additional seats	0	0	0	0
Seats Hamilton	**18**	**14**	**8**	**40**

7. Since the sum of the lower quotas is 1 less than the total number of seats, assign an additional seat to the state with the highest decimal part in its standard quota.

State	NY	PA	NJ	Total
Population (millions)	18.977	12.281	8.414	39.672
Std q $(d=2.64)$	7.19	4.65	3.19	—
Lower q $(d=2.64)$	7	4	3	14
Additional seats	0	1	0	1
Seats Hamilton	**7**	**5**	**3**	**15**

9. total = 1,768 + 1,357 + 1,091 + 893 = 5,109

Standard divisor: $d = \dfrac{5,109}{200} = 25.55$

Standard quota: $q = \dfrac{\text{each school's population}}{d}$, (round to two decimal places)

Lower quota = standard quota truncated to whole number.

Since the sum of the lower quotas is 2 less than the total number of calculators, assign an additional calculator to the two schools with the highest decimal part in its standard quota. See table.

166

School	A	B	C	D	Total
Population	1,768	1,357	1,091	893	5,109
Std q $(d = 25.55)$	69.20	53.11	42.70	34.95	—
Lower q $(d = 25.55)$	69	53	42	34	198
Additional calculators	0	0	1	1	2
Calculators Hamilton	**69**	**53**	**43**	**35**	**200**

11. total $= 2,680 + 6,550 + 2,995 + 8,475 = 20,700$ Standard divisor: $d = \dfrac{20,700}{24} = 862.50$

Standard quota: $q = \dfrac{\text{each region's population}}{d}$, (round to two decimal places)

Lower quota = standard quota truncated to whole number.

Since the sum of the lower quotas is 2 less than the total number of seats, assign an additional seat to the two regions with the highest decimal part in its standard quota.

Region	N	S	E	W	Total
Population	2,680	6,550	2,995	8,475	20,700
Std q $(d = 862.50)$	3.11	7.59	3.47	9.83	—
Lower q $(d = 862.50)$	3	7	3	9	22
Additional seats	0	1	0	1	2
Seats Hamilton	**3**	**8**	**3**	**10**	**24**

13. total $= 5,413 + 5,215 + 294 + 4,575 + 8,986 = 24,483$

Standard divisor: $d = \dfrac{24,483}{20} = 1,224.15$

Standard quota: $q = \dfrac{\text{each country's population}}{d}$, (round to two decimal places)

Lower quota = standard quota truncated to whole number.

Since the sum of the lower quotas is 2 less than the total number of seats, assign an additional seat to the two countries with the highest decimal part in its standard quota. See table next page.

Region	D	F	I	N	S	Total
Population (thousands)	5,413	5,215	294	4,575	8,986	24,483
Std q $(d = 1,224.15)$	4.42	4.26	0.24	3.74	7.34	—
Lower q $(d = 1,224.15)$	4	4	0	3	7	18
Additional seats	1	0	0	1	0	2
Seats Hamilton	**5**	**4**	**0**	**4**	**7**	**20**

15. $total = 3{,}957 + 6{,}588 + 14{,}281 + 6{,}824 + 5{,}360 + 3{,}000 = 40{,}010$

Standard divisor: $d = \dfrac{40{,}010}{25} = 1{,}600.4$

Standard quota: $q = \dfrac{\text{each country's population}}{d}$, (rounded to two decimal places)

Lower quota = standard quota truncated to whole number.

Since the sum of the lower quotas is 3 less than the total number of seats, assign an additional

seat to the countries with the highest decimal part in its standard quota.

Country	C	E	G	H	N	P	Total
Population (thousands)	3,957	6,588	14,281	6,824	5,360	3,000	40,010
Std q $(d = 1,600.4)$	2.47	4.12	8.92	4.26	3.35	1.87	—
Lower q $(d = 1,600.4)$	2	4	8	4	3	1	22
Additional seats	1	0	1	0	0	1	3
Seats Hamilton	**3**	**4**	**9**	**4**	**3**	**2**	**25**

17. a. For Jefferson's method, use the standard divisor d. Since the sum of the lower quotas

equals the total number of seats, the apportionment process is complete. See table.

b. For Adam's method, use the modified divisor $(d = 52)$, greater than d, to obtain the

modified quotas, upper modified quotas, and final apportionment. See table.

c. For Webster's method, use the standard divisor d. Since the sum of the rounded standard

quotas equals the total number of seats, the apportionment process is complete.

See table.

State	A	B	C	Total	Comments
Population (thousands)	900	700	400	2,000	
Std q $(d = 50.00)$	18.00	14.00	8.00	—	
Lower q $(d = 50.00)$	18	14	8	40	Jefferson
Upper q $(d = 50.00)$	19	15	9	43	Use $d_m > d$
Rounded q $(d = 50.00)$	18	14	8	40	Webster
Additional seats (Hamilton)	0	0	0	0	
Seats Hamilton	18	14	8	040	Hamilton
Modified std $(d = 52)$	17.31	13.46	7.69	—	Adams
Upper q	18	14	8	40	

19. a. For Jefferson's method, use the modified divisor $(d = 2.4)$, less than d, to obtain the modified quotas, lower modified quotas, and final apportionment. See table on next page.

b. For Adam's method, use the modified divisor $(d = 3)$, greater than d, to obtain the modified quotas, upper modified quotas, and final apportionment. See table on next page.

c. For Webster's method, use the standard divisor d. Since the sum of the rounded standard quotas equals the total number of seats, the apportionment process is complete. See table.

State	NY	PA	NJ	Total	Comments
Population (millions)	18.977	12.281	8.414	39.672	
Std q $(d = 2.64)$	7.19	4.65	3.19	—	
Lower q $(d = 2.64)$	7	4	3	14	Use $d_m < d$
Upper q $(d = 2.64)$	8	5	4	17	Use $d_m > d$
Rounded q $(d = 2.64)$	7	5	3	15	Webster
Additional seats (Hamilton)	0	1	0	1	
Seats Hamilton	7	5	3	15	Hamilton
Modified std $(d = 2.4)$	7.91	5.12	3.51	—	
Lower q	7	5	3	15	Jefferson
Modified std $(d = 3)$	6.33	4.09	2.80		
Upper q	7	5	3	15	Adams

21. a. For Jefferson's method, use the modified divisor $(d = 25.3)$, less than d, to obtain the modified quotas, lower modified quotas, and final apportionment. See table.

b. For Adam's method, use the modified divisor $(d = 25.7)$, greater than d, to obtain the modified quotas, upper modified quotas, and final apportionment. See table.

c. For Webster's method, use the standard divisor d. Since the sum of the rounded standard quotas equals the total number of seats, the apportionment process is complete. See table.

School	A	B	C	D	Total	Comments
Population	1,768	1,357	1,091	893	5,109	
Std q $(d = 25.55)$	69.20	53.11	42.70	34.95	—	
Lower q $(d = 25.55)$	69	53	42	34	198	Use $d_m < d$
Upper q $(d = 25.55)$	70	54	43	35	202	Use $d_m > d$
Rounded q $(d = 25.55)$	**69**	**53**	**43**	**35**	**200**	**Webster**
Additional seats (Hamilton)	0	0	1	1	2	
Seats Hamilton	**69**	**53**	**43**	**35**	**200**	**Hamilton**
Modified std $(d = 25.3)$	69.88	53.64	43.12	35.30		
Lower q	**69**	**53**	**43**	**35**	**200**	**Jefferson**
Modified std $(d = 25.7)$	68.79	52.8	42.45	34.75		
Upper q	**69**	**53**	**43**	**35**	**200**	**Adams**

23.   a. For Jefferson's method, use the modified divisor $(d = 800)$, less than d, to obtain the modified quotas, lower modified quotas, and final apportionment. See table on next page.

b. For Adam's method, use the modified divisor $(d = 940)$, greater than d, to obtain the modified quotas, upper modified quotas, and final apportionment. See table on next page.

c. For Webster's method, use the standard divisor d. Since the sum of the rounded standard quotas equals the total number of seats, the apportionment process is complete. See table.

Region	N	S	E	W	Total	Comments
Population	2,680	6,550	2,995	8,475	20,700	
Std q $(d = 862.50)$	3.11	7.59	3.47	9.83	—	
Lower q $(d = 862.50)$	3	7	3	9	22	Use $d_m < d$
Upper q $(d = 862.50)$	4	8	4	10	26	Use $d_m > d$
Rounded q $(d = 862.50)$	**3**	**8**	**3**	**10**	**24**	**Webster**
Additional seats (Hamilton)	0	1	0	1	2	
Seats Hamilton	**3**	**8**	**3**	**10**	**24**	**Hamilton**
Modified std $(d = 800)$	3.35	8.19	3.74	10.59		
Lower q	**3**	**8**	**3**	**10**	**24**	**Jefferson**
Modified std $(d = 940)$	2.85	6.97	3.19	9.02		
Upper q	**3**	**7**	**4**	**10**	**24**	**Adams**

25. a. For Jefferson's method, use the modified divisor ($d = 1,100$), less than d, to obtain the modified quotas, lower modified quotas, and final apportionment. See table on next page.

   b. For Adam's method, use the modified divisor ($d = 1,425$), greater than d, to obtain the modified quotas, upper modified quotas, and final apportionment. See table on next page.

   c. For Webster's method, use the modified divisor ($d = 1,200$), different from d, to obtain the modified quotas, rounded modified quotas, and final apportionment.

Country	D	F	I	N	S	Total	Comments
Population (thousands)	5,413	5,215	294	4,575	8,986	24,483	
Std q $(d = 1,224.15)$	4.42	4.26	0.24	3.74	7.34	—	
Lower q $(d = 1,224.15)$	4	4	0	3	7	18	Use $d_m < d$
Upper q $(d = 1,224.15)$	5	5	1	4	8	23	Use $d_m > d$
Additional seats	1	0	0	1	0	2	
Seats Hamilton	5	4	0	4	7	20	**Hamilton**
Modified std $(d = 1,100)$	4.92	4.74	0.27	4.16	8.17		
Lower q	4	4	0	4	8	20	**Jefferson**
Modified std $(d = 1,425)$	3.80	3.66	0.21	3.21	6.31		
Upper q	4	4	1	4	7	20	**Adams**
Modified std $(d = 1,200)$	4.51	4.35	0.25	3.81	7.49		
Rounded q	5	4	0	4	7	20	**Webster**

27. a. For Jefferson's method, use the modified divisor ($d = 1,400$), less than d, to obtain the modified quotas, lower modified quotas, and final apportionment. See table on next page.

   b. For Adam's method, use the modified divisor ($d = 1,786$), greater than d, to obtain the modified quotas, upper modified quotas, and final apportionment. See table.

   c. For Webster's method, use the modified divisor ($d = 1,550$), different from d, to obtain the modified quotas, rounded modified quotas, and final apportionment. See table on next page.

Country	C	E	G	H	N	P	Total	Comments
Population (thousands)	3,957	6,588	14,281	6,824	5,360	3,000	40,010	
Std q $(d=1,600.4)$	2.47	4.12	8.92	4.26	3.35	1.87	——	
Lower q $(d=1,600.4)$	2	4	8	4	3	1	22	Use $d_m < d$
Upper q $(d=1,600.4)$	3	5	9	5	4	2	28	Use $d_m > d$
Rounded q $(d=1,600.4)$	2	4	9	4	3	2	24	Use $d_m < d$
Additional seats	1	0	1	0	0	1	3	
Seats Hamilton	3	4	9	4	3	2	25	**Hamilton**
Modified std $(d=1,400)$	2.83	4.71	10.20	4.87	3.83	2.14		
Lower q	2	4	10	4	3	2	25	**Jefferson**
Modified std $(d=1,786)$	2.22	3.69	7.996	3.82	3.001	1.68		
Upper q	3	4	8	4	4	2	25	**Adams**
Modified std $(d=1,550)$	2.55	4.25	9.21	4.40	3.46	1.94		
Rounded q	3	4	9	4	3	2	25	**Webster**

29.   Use the standard divisor d. Find the standard, lower, and upper quotas, and calculate the geometric means. Since the sum of the rounded standard quotas equals the total number of seats, the apportionment process is complete.

State	A	B	C	Total	Comments
Population (thousands)	900	700	400	2,000	
Std q $(d=50.00)$	18.00	14.00	8.00	——	
Lower q $(d=50.00)$	18	14	8	40	
Upper q $(d=50.00)$	19	15	9	41	
Geometric mean	18.49	14.49	8.49	——	
Rounded q $(d=50.00)$	18	14	8	40	**H–H**

31.   Use the standard divisor d. Find the standard, lower, and upper quotas, and calculate the geometric means. Since the sum of the rounded standard quotas equals the total number of seats, the apportionment process is complete. See table.

172

State	NY	PA	NJ	Total	Comments
Population (millions)	18.977	12.281	8.414	39.672	
Std q $(d = 2.64)$	7.19	4.65	3.19	—	
Lower q $(d = 2.64)$	7	4	3	14	
Upper q $(d = 2.64)$	8	5	4	17	
Geometric mean	7.48	4.47	3.46	—	
Rounded q $(d = 2.64)$	**7**	**5**	**3**	**15**	**H–H**

33. Use the standard divisor d. Find the standard, lower, and upper quotas, and calculate the geometric means. Since the sum of the rounded standard quotas equals the total number of seats, the apportionment process is complete.

School	A	B	C	D	Total	Comments
Population	1,768	1,357	1,091	893	5,109	
Std q $(d = 25.55)$	69.20	53.11	42.70	34.95	—	
Lower q $(d = 25.55)$	69	53	42	34	198	
Upper q $(d = 25.55)$	70	54	43	35	202	
Geometric mean	69.50	53.50	42.50	34.50	—	
Rounded q $(d = 25.55)$	**69**	**53**	**43**	**35**	**200**	**H–H**

35. Using the modified divisor $(d = 870)$, different from d, obtain the modified quotas, lower, and upper modified quotas, and geometric means. The final apportionment is shown in the table.

Region	N	S	E	W	Total	Comments
Population	2,680	6,550	2,995	8,475	20,700	
Std q $(d = 862.50)$	3.11	7.59	3.47	9.83	—	
Lower q $(d = 862.50)$	3	7	3	9	22	
Upper q $(d = 862.50)$	4	8	4	10	26	
Geometric mean	3.46	7.48	3.46	9.49	—	
Rounded q $(d = 862.50)$	3	8	4	10	25	Use $d_m > d$
Modified std $(d = 870)$	3.08	7.53	3.44	9.74		
Rounded q	**3**	**8**	**3**	**10**	**24**	**H–H**

37. Use the standard divisor d. Find the standard, lower, and upper quotas, and calculate the geometric means. Since the sum of the rounded standard quotas equals the total number of seats, the apportionment process is complete. See table on next page.

173

Region	D	F	I	N	S	Total	Comments
Population (thousands)	5,413	5,215	294	4,575	8,986	24,483	
Std q $(d = 1,224.15)$	4.42	4.26	0.24	3.74	7.34	—	
Lower q $(d = 1,224.15)$	4	4	0	3	7	18	
Upper q $(d = 1,224.15)$	5	5	1	4	8	23	
Geometric mean	4.47	4.47	0	3.46	7.48	—	
Rounded q $(d = 1,224.15)$	**4**	**4**	**1**	**4**	**7**	**20**	**H–H**

39. Use the standard divisor d. Find the standard, lower, and upper quotas, and calculate the geometric means. Since the sum of the rounded standard quotas equals the total number of seats, the apportionment process is complete. See table.

Country	C	E	G	H	N	P	Total	Comments
Population (thousands)	3,957	6,588	14,281	6,824	5,360	3,000	40,010	
Std q $(d = 1,600.4)$	2.47	4.12	8.92	4.26	3.35	1.87	—	
Lower q $(d = 1,600.4)$	2	4	8	4	3	1	22	
Upper q $(d = 1,600.4)$	3	5	9	5	4	2	28	
Geometric mean	2.45	4.47	8.49	4.47	3.46	1.41	—	
Rounded q $(d = 1,600.4)$	**3**	**4**	**9**	**4**	**3**	**2**	**25**	**H–H**

41. $$\text{HHN} = \frac{(\text{population})^2}{n(n+1)}$$

	Main	Satellite
Population	25,000	2,600
Instructors	465	47
HHN	2,884.30	2,996.45

Since 2,996.45 > 2,884.30, the satellite campus should receive the new instructor.

174

43. $$HHN = \frac{(\text{population})^2}{n(n+1)}$$

	Agnesi	Banach	Cantor
Population	567	871	666
Teachers	21	32	25
HHN	695.86	718.41	682.39

Since 718.41 is the largest, Banach school should receive the new instructor.

45. The hypothetical Delaware has one more and Virginia has one fewer than the actual.

Total population = 3,615.92 in thousands; $d = 3{,}615.92/105 = 34.44$. See table.

State	Population (thousands)	Std q $d = 34.44$	Lower q	Additional Seats	Hamilton	Actual	Upper q	Std q_m $d_m = 36.5$	Upper q_m $d_m = 36.5$ Adams
VA	630.56	18.31	18	0	18	19	19	17.28	18
MA	475.327	13.8	13	1	14	14	14	13.02	14
PA	432.879	12.57	12	1	13	13	13	11.86	12
NC	353.523	10.26	10	0	10	10	11	9.69	10
NY	331.589	9.63	9	1	10	10	10	9.08	10
MD	278.514	8.09	8	0	8	8	9	7.63	8
CT	236.841	6.88	6	1	7	7	7	6.49	7
SC	206.236	5.99	5	1	6	6	6	5.65	6
NJ	179.57	5.21	5	0	5	5	6	4.92	5
NH	141.822	4.12	4	0	4	4	5	3.89	4
VT	85.533	2.48	2	0	2	2	3	2.34	3
GA	70.835	2.06	2	0	2	2	3	1.94	2
KY	68.705	1.99	1	1	2	2	2	1.88	2
RI	68.446	1.99	1	1	2	2	2	1.88	2
DE	55.54	1.61	1	1	2	1	2	1.52	2
Total	3,615.92	—	97	8	105	105	112	—	105

47. Virginia has one fewer and Delaware has one more than the actual.

State	Population (thousands)	Std q $d = 34.44$	Rounded q	Actual
VA	630.56	18.31	18	**19**
MA	475.327	13.8	14	**14**
PA	432.879	12.57	13	**13**
NC	353.523	10.26	10	**10**
NY	331.589	9.63	10	**10**
MD	278.514	8.09	8	**8**
CT	236.841	6.88	7	**7**
SC	206.236	5.99	6	**6**
NJ	179.57	5.21	5	**5**
NH	141.822	4.12	4	**4**
VT	85.533	2.48	2	**2**
GA	70.835	2.06	2	**2**
KY	68.705	1.99	2	**2**
RI	68.446	1.99	2	**2**
DE	55.54	1.61	2	**1**
Total	3,615.92	——	105	**105**

Section 6.3 Flaws of Apportionment

1. a. 3,500,000 = 3.5 million; 4,200,000 = 4.2 million; 16,800,000 = 16.8 million;

   Total = 3.5 + 4.2 + 16.8 = 24.5

   b. Standard divisor: $d = \dfrac{\text{total population}}{\text{total number of seats}} = \dfrac{24.5}{32} = 0.77$

   c. Standard quotas: $q = \dfrac{\text{each state's population}}{d}$ (rounded to two decimal places)

   Lower quota = standard quota truncated to whole number

   Upper quota = standard quota rounded up to next whole number; See table.

   d. Using the modified divisor $d_m = 0.73$ (less than d), we obtain the modified quotas, lower

   modified quotas, and final apportionment shown in the table.

176

State	A	B	C	Total	Comments
Population (millions)	3.5	4.2	16.8	24.5	Part (a)
Std q $(d = 0.77)$	4.55	5.45	21.82	——	Part (c)
Lower q $(d = 0.77)$	4	5	21	30	Part (c)
Upper q $(d = 0.77)$	5	6	22	33	Part (c)
Modified lower q $(d_m = 0.73)$	4	5	23	32	d. Jefferson

    e. Yes. C has 23 seats, which is neither upper nor lower quota.

3.   a. $3,500,000 = 3.5$ million; $4,200,000 = 4.2$ million; $16,800,000 = 16.8$ million;

        Total $= 3.5 + 4.2 + 16.8 = 24.5$

    b. Standard divisor: $d = \dfrac{\text{total population}}{\text{total number of seats}} = \dfrac{24.5}{79} = 0.31$

    c. Standard quotas: $q = \dfrac{\text{each state's population}}{d}$ (rounded to two decimal places)

        Lower quota = standard quota truncated to whole number

        Upper quota = standard quota rounded up to next whole number

        See table on next page.

    d. Using the modified divisor $d_m = 0.317$ (greater than d), we obtain the modified quotas,

        upper modified quotas, and final apportionment shown in the table.

State	A	B	C	Total	Comments
Population (millions)	3.5	4.2	16.8	24.5	Part (a)
Std q $(d = 0.31)$	11.29	13.55	54.19	——	Part (c)
Lower q $(d = 0.31)$	11	13	54	78	Part (c)
Upper q $(d = 0.31)$	12	14	55	81	Part (c)
Modified upper q $(d_m = 0.317)$	12	14	53	79	d. Adams

    e. Yes. C has 53 seats, which is neither upper nor lower quota.

5.   a. $1,200,000 = 1.2$ million; $3,400,000 = 3.4$ million; $17,500,000 = 17.5$ million;

        $19,400,000 = 19.4$ million; Total $= 1.2 + 3.4 + 17.5 + 19.4 = 41.5$

    b. Standard divisor: $d = \dfrac{\text{total population}}{\text{total number of seats}} = \dfrac{41.5}{200} = 0.21$

   c. Standard quotas: $q = \dfrac{\text{each state's population}}{d}$ (rounded to two decimal places)

      Lower quota = standard quota truncated to whole number

      Upper quota = standard quota rounded up to next whole number

      See table.

   d. Using the modified divisor $d_m = 0.20715$ (different from d), we obtain the modified

      quotas, rounded modified quotas, and final apportionment shown in the table.

State	A	B	C	D	Total	Comments
Population (millions)	1.2	3.4	17.5	19.4	41.5	Part (a)
Std q $(d = 0.21)$	5.71	16.19	83.33	92.38	—	Part (c)
Lower q $(d = 0.21)$	5	16	83	92	196	Part (c)
Upper q $(d = 0.21)$	6	17	84	93	200	Part (c)
Rounded q $(d = 0.21)$	6	16	83	92	197	
Modified rounded q $(d_m = 0.20715)$	6	16	84	94	200	d. Webster

   e. Yes. D has 94 seats, which is neither upper nor lower quota.

7.   a. 1,200,000 = 1.2 million; 3,400,000 = 3.4 million; 17,500,000 = 17.5 million;

      19,400,000 = 19.4 million; Total = 1.2 + 3.4 + 17.5 + 19.4 = 41.5

   b. Standard divisor: $d = \dfrac{\text{total population}}{\text{total number of seats}} = \dfrac{41.5}{200} = 0.21$

   c. Standard quotas: $q = \dfrac{\text{each state's population}}{d}$ (rounded to two decimal places)

      Lower quota = standard quota truncated to whole number

      Upper quota = standard quota rounded up to next whole number; See table.

   d. Using the modified divisor $d_m = 0.2085$ (different from d), we obtain the modified

      quotas, lower and upper modified quotas, geometric means, rounded quotas, and final

      apportionment shown in the table.

State	A	B	C	D	Total	Comments
Population (millions)	1.2	3.4	17.5	19.4	41.5	Part (a)
Std q $(d = 0.21)$	5.71	16.19	83.33	92.38	—	Part (c)
Lower q $(d = 0.21)$	5	16	83	92	196	Part (c)
Upper q $(d = 0.21)$	6	17	84	93	200	Part (c)
Geometric mean	5.48	16.49	83.50	92.50	—	
Rounded q $(d = 0.21)$	6	16	83	92	197	
Modified rounded q $(d_m = 0.2085)$	6	16	84	94	200	d. H–H

    e. Yes. D has 94 seats, which is neither upper nor lower quota.

9.   a. $690,000 = 690$ thousand; $5,700,000 = 5,700$ thousand; $6,410,000 = 6,410$ thousand;

        Total $= 690 + 5,700 + 6,410 = 12,800$

    b. Standard divisor: $d = \dfrac{\text{total population}}{\text{total number of seats}} = \dfrac{12,800}{120} = 106.67$

    c. Find each state's standard and lower quotas. Since the sum of the lower quotas is 1 less than the total number of seats, assign an additional seat to the state with the highest decimal part in it standard quota. See table.

    d. New standard divisor: $d = \dfrac{\text{total population}}{\text{total number of seats}} = \dfrac{12,800}{121} = 105.79$

        Find the new standard and lower quotas. Now the two states with the highest decimal point of their standard quotas receive and additional seat.

State	A	B	C	Total	Comments
Population (thousands)	690	5,700	6,410	12,800	Part (a)
Std q $(d = 106.67)$	6.47	53.44	60.09	—	
Lower q $(d = 106.67)$	6	53	60	119	
Additional Seats	1	0	0	1	
Seats	7	53	60	120	c. Hamilton
New std q $(d = 12800 / 121 = 105.79)$	6.52	53.88	60.59	—	
New lower q $(d = 105.79)$	6	53	60	119	
Additional seats	0	1	1	2	
Seats	6	54	61	121	d. Hamilton

    e. Yes. State A has lost a seat at the expense of the two larger states even though its population didn't change.

11. a. $1{,}056{,}000 = 1{,}056$ thousand; $1{,}844{,}000 = 1{,}844$ thousand; $2{,}100{,}000 = 2{,}100$ thousand;

    Total $= 1{,}056 + 1{,}844 + 2{,}100 = 5{,}000$

    b. Standard divisor: $d = \dfrac{\text{total population}}{\text{total number of seats}} = \dfrac{5{,}000}{110} = 45.45$

    c. Find each state's standard and lower quotas. Since the sum of the lower quotas is 1 less than the total number of seats, assign an additional seat to the state with the highest decimal part in it standard quota. See table.

State	A	B	C	Total	Comments
Population (thousands)	1,056	1,844	2,100	5,000	Part (a)
Std q $(d = 45.45)$	23.23	40.57	46.20	—	
Lower q $(d = 45.45)$	23	40	46	109	
Additional Seats	0	1	0	1	
Seats	23	41	46	110	c. Hamilton

    d. $110 \cdot \dfrac{2{,}440}{5{,}000} = 53.68$; Add 53 new seats.

    e. Total new seats: $110 + 53 = 163$

    Total new population: $5{,}000 + 2{,}440 = 7{,}440$

    New standard divisor: $d = \dfrac{\text{total population}}{\text{total number of seats}} = \dfrac{7{,}440}{163} = 45.64$

    Find the new standard and lower quotas. Now, once again, the state with the highest decimal part of its standard quota receives an additional seat.

State	A	B	C	D	Total
Population (thousands)	1,056	1,844	2,100	2,440	7,400
Std q $(d = 45.64)$	23.14	40.40	46.01	53.46	—
Lower q $(d = 45.64)$	23	40	46	53	162
Additional Seats	0	0	0	1	1
Seats	23	40	46	54	163

    f. Yes, apportionment changes. State B is altered.

13. a. 2005: total $= 780 + 1{,}700 + 3{,}520 = 6{,}000$; $d = \dfrac{6{,}000}{11} = 545.45$

    Find each campus' standard, lower, and upper quotas. Since the sum of the lower quotas is 1 less than the total number of specialists, assign an additional specialist to the campus with the highest decimal part in its standard quota. See table

180

	A	B	C	Total
2005	780	1,700	3,520	6,000
Std q $(d = 545.45)$	1.43	3.12	6.45	—
Lower q $(d = 545.45)$	1	3	6	10
Additional Seats	0	0	1	1
2005 seats	1	3	7	11

b. 2006: total $= 820 + 1,880 + 3,740 = 6,440$; $d = \dfrac{6,440}{11} = 585.45$

Find each campus' standard, lower, and upper quotas. Since the sum of the lower quotas is 1 less than the total number of specialists, assign an additional specialist to the campus with the highest decimal part in its standard quota. See table.

	A	B	C	Total
2006	820	1,880	3,740	6,440
Std q $(d = 585.45)$	1.40	3.21	6.39	—
Lower q $(d = 585.45)$	1	3	6	10
Additional Seats	1	0	0	1
2006 seats	2	3	6	11

c. A: $\dfrac{820-780}{780} = 0.051$; B: $\dfrac{1,880-1,700}{1,700} = 0.106$; C: $\dfrac{3,740-3,520}{3,520} = 0.0625$

	A	B	C	Total
2005	780	1,700	3,520	6,000
2006	820	1,880	3,740	6,440
Increase	40/780	180/1700	220/3520	
% increase	5.1	10.6	6.25	
Change in seats	+1	0	−1	

d. Yes. C lost a seat to A, but C grew at a faster rate.

Chapter 6 Review

1. a. $5 + 7 + 6 + 12 + 13 + 17 + 14 = 74$

b. B: $12 + 13 = 25$

M: $5 + 7 + 6 = 18$

T: 14

V: 17   Beethoven wins.

c. $\frac{25}{74} = 34\%$

d. Delete the one with the fewest first-choice votes, T.

B: 25; M: 18; V: 17 + 14 = 31

Vivaldi only received $\frac{31}{74} = 42\%$ of the votes, which is not a majority. Repeat process.

Delete the next one with the fewest first-choice votes, M.

B: 5 + 12 + 13 = 30; V: 7 + 6 + 17 + 14 = 44

Vivaldi wins.

e. $\frac{44}{74} = 59\%$

f. $k = 4$; 4 pts for 1st place, 3 pts for 2nd place, 2 pts for 3rd place, 1 pt for 4th place

Score for:

B: $5 \cdot 3 + 7 \cdot 2 + 6 \cdot 1 + 12 \cdot 4 + 13 \cdot 4 + 17 \cdot 2 + 14 \cdot 2 = 197$

M: $5 \cdot 4 + 7 \cdot 4 + 6 \cdot 4 + 12 \cdot 1 + 13 \cdot 3 + 17 \cdot 3 + 14 \cdot 1 = 188$

T: $5 \cdot 1 + 7 \cdot 1 + 6 \cdot 3 + 12 \cdot 3 + 13 \cdot 2 + 17 \cdot 1 + 14 \cdot 4 = 165$

V: $5 \cdot 2 + 7 \cdot 3 + 6 \cdot 2 + 12 \cdot 2 + 13 \cdot 1 + 17 \cdot 4 + 14 \cdot 3 = 190$

Beethoven wins.

g. Beethoven received 197 points.

h. B vs M: B gets 12 + 13 + 14 = 39, M gets 5 + 7 + 6 + 17 = 35; B 1 pt

B vs T: B gets 5 + 7 + 12 + 13 + 17 = 54, T gets 6 + 14 = 20; B 1 pt

B vs V: B gets 5 + 12 + 13 = 30, V gets 7 + 6 + 17 + 14 = 44; V 1 pt

M vs T: M gets 5 + 7 + 6 + 13 + 17 = 48, T gets 12 + 14 = 26; M 1 pt

M vs V: M gets 5 + 7 + 6 + 13 = 31, V gets 12 + 17 + 14 = 43; V 1 pt

T vs V: T gets 6 + 12 + 13 + 14 = 45, V gets 5 + 7 + 17 = 29; T 1pt

B: 1 + 1 = 2, V: 1 + 1 = 2, M: 1, T: 1

There is a tie between Beethoven and Vivaldi.

i. They each received 2 points.

3. Since the sum of the lower quotas is 2 less than the total number of seats, assign an additional seat to the two states with the highest decimal part in its standard quota.

State	A	B	C	Total	Comments
Population (thousands)	1,200	2,300	2,500	6,000	
Std q $(d = 78.95)$	15.20	29.13	31.67	—	
Lower q $(d = 78.95)$	15	29	31	75	
Additional seats	0	0	1	1	
Seats Hamilton	**15**	**29**	**32**	**76**	**Hamilton**

5. Using the modified divisor $d_m = 80.1$ (greater than d), we obtain the modified quotas, upper modified quotas, and final apportionment shown here.

State	A	B	C	Total	Comments
Population (thousands)	1,200	2,300	2,500	6,000	
Std q $(d = 78.95)$	15.20	29.13	31.67	—	
Lower q $(d = 78.95)$	15	29	31	75	
Upper q $(d = 78.95)$	16	30	32	78	Use $d_m > d$
Modified std q $(d_m = 80.1)$	14.98	28.71	31.21	—	
Upper q	**15**	**29**	**32**	**76**	**Adams**

7. Using the geometric mean, $d = 78.95$, and rounded standard quotas, the final apportionment is shown here.

State	A	B	C	Total	Comments
Population (thousands)	1,200	2,300	2,500	6,000	
Std q $(d = 78.95)$	15.20	29.13	31.67	—	
Lower q $(d = 78.95)$	15	29	31	75	
Upper q $(d = 78.95)$	16	30	32	78	
Geometric mean $(d = 78.95)$	15.49	29.50	31.50	—	
Rounded q $(d = 78.95)$	**15**	**29**	**32**	**76**	**H–H**

9. Using the modified divisor $d_m = 2,700$ (less than d), we obtain the modified quotas, lower modified quotas, and final apportionment shown here.

Country	A	B	C	P	U	Total	Comments
Population (thousands)	39,145	8,724	15,824	6,191	3,399	73,283	
Std q $(d = 2,931.32)$	13.35	2.98	5.40	2.11	1.16	——	
Lower q $(d = 2,931.32)$	13	2	5	2	1	23	Use $d_m < d$
Modified std q $(d_m = 2,700)$	14.50	3.23	5.86	2.29	1.26	——	
Lower q	**14**	**3**	**5**	**2**	**1**	**25**	**Jefferson**

11. Using the modified divisor $d_m = 2,900$ (different than d), we obtain the modified quotas, rounded modified quotas, and final apportionment shown here.

Country	A	B	C	P	U	Total	Comments
Population (thousands)	39,145	8,724	15,824	6,191	3,399	73,283	
Std q $(d = 2,931.32)$	13.35	2.98	5.40	2.11	1.16	——	
Lower q $(d = 2,931.32)$	13	2	5	2	1	23	
Upper q $(d = 2,931.32)$	14	3	6	3	2	28	
Rounded q $(d = 2,931.32)$	13	3	5	2	1	24	Use $d_m < d$
Modified std q $(d_m = 2,900)$	13.50	3.01	5.46	2.13	1.17	——	
Rounded q	**14**	**3**	**5**	**2**	**1**	**25**	**Webster**

13. $$HHN = \frac{(\text{population})^2}{n(n+1)}$$

	Leibniz	Maclaurin	Napier	Total
Population	987	1,242	1,763	3,992
Seats	25	31	44	100
HHN	1,498.72	1,555.00	1,569.78	

Since 1,569.78 is the largest, Napier school should get the new instructor.

15. a. $1,600,000 = 1.6$ million; $3,500,000 = 3.5$ million; $15,300,000 = 15.3$ million;

Total $= 1.6 + 3.5 + 15.3 = 20.4$

b. Standard divisor: $d = \dfrac{\text{total population}}{\text{total number of seats}} = \dfrac{20.4}{72} = 0.28$

c. Standard quotas: $q = \dfrac{\text{each states population}}{d}$, (rounded to two decimal places)

Lower quota = standard quota truncated to whole number

Upper quota = standard quota rounded up to next whole number

See table.

State	A	B	C	Total	Comments
Population (millions)	1.6	3.5	15.3	20.4	Part (a)
Std q $(d = 0.28)$	5.71	12.5	54.64	—	Part (c)
Lower q $(d = 0.28)$	5	12	54	71	Part (c)
Upper q $(d = 0.28)$	6	13	55	74	Part (c)
Modified std q $(d_m = 0.29)$	5.52	12.07	52.76	—	Use $d_m > d$
Upper q	**6**	**13**	**53**	**72**	**Adams**

d. Using the modified divisor $d_m = 0.29$ (greater than d), we obtain the modified quotas, upper modified quotas, and find the apportionment shown in the table.

e. Yes. C has 53 seats, which is neither an upper nor a lower quota.

17. a. $1{,}100{,}000 = 1.100$ million; $3{,}500{,}000 = 3.500$ million; $17{,}600{,}000 = 17.600$ million; $19{,}400{,}000 = 19.400$ million;

   Total $= 1.100 + 3.500 + 17.600 + 19.400 = 41.6$

   b. Standard divisor: $d = \dfrac{\text{total population}}{\text{total number of seats}} = \dfrac{41.6}{201} = 0.21$

   c. Standard quotas: $q = \dfrac{\text{each states population}}{d}$, (rounded to two decimal places)

   Lower quota = standard quota truncated to whole number

   Upper quota = standard quota rounded up to next whole number

   See table.

   d. Using the modified divisor $d_m = 0.208$ (different from d), to obtain the modified quotas, lower and upper modified quotas, geometric means, rounded quotas, and find the apportionment shown in the table on next page.

State	A	B	C	D	Total	Comments
Population (millions)	1.100	3.500	17.600	19.400	41.6	Part (a)
Std q $(d = 0.21)$	5.24	16.67	83.81	92.38	—	Part (c)
Lower q $(d = 0.21)$	5	16	83	92	196	Part (c)
Upper q $(d = 0.21)$	6	17	84	93	200	Part (c)
Geometric mean $(d = 0.21)$	5.48	16.49	83.50	92.50	—	
Rounded q $(d = 0.21)$	5	17	84	92	198	Use $d_m < d$
Modified std q $(d_m = 0.208)$	5.29	16.83	84.62	93.27	—	Part (d)
Rounded q	**5**	**17**	**85**	**94**	**201**	**H–H**

e. Yes. Neither C nor D has a number of seats that is either an upper or lower quota.

19. a. $1,057,000 = 1,057$ thousand; $1,942,000 = 1,942$ thousand; $2,001,000 = 2,001$ thousand;

   Total $= 1,057 + 1,942 + 2,001 = 5,000$

   b. Standard divisor: $d = \dfrac{\text{total population}}{\text{total number of seats}} = \dfrac{5,000}{110} = 45.45$

   c. Find each state's standard, lower, and upper quotas. Since the sum of the lower quotas is 1 less than the total number of seats, assign an additional seat to the state with the highest decimal part in its standard quota.

State	A	B	C	Total	Comments
Population (thousands)	1,057	1,942	2,001	5,000	
Std q $(d = 45.45)$	23.26	42.73	44.03	—	
Lower q $(d = 45.45)$	23	42	44	109	
Additional seats	0	1	0	1	
Seats Hamilton	**23**	**43**	**44**	**110**	**Hamilton part (c)**

   d. $110 \cdot \dfrac{2,450}{5,000} = 53.9$; Add 53 seats.

   e. Total $= 5,000 + 2,450 = 7,450$; seats $= 110 + 53 = 163$

   $d = \dfrac{7,450}{163} = 45.71$; Find each state's standard, lower, and upper quotas. Since the sum of

   the lower quotas is 2 less than the total number of seats, assign an additional seat to the two

   state with the highest decimal part in its standard quota. See table.

186

State	A	B	C	D	Total	Comments
Population (thousands)	1,057	1,942	2,001	2,450	7,450	
Std q ($d = 45.71$)	23.12	42.49	43.78	53.60	—	
Lower q ($d = 45.71$)	23	42	43	53	161	
Additional seats	0	0	1	1	2	
Seats Hamilton	**23**	**42**	**44**	**54**	**163**	**Hamilton part (e)**

f. Yes. The apportionment changed.

Section 7.1 Place Systems

1. $891 = 8 \cdot 10^2 + 9 \cdot 10^1 + 1 \cdot 10^0 \quad$ or $8 \cdot 10^2 + 9 \cdot 10 + 1$

3. $3,258 = 3 \cdot 10^3 + 2 \cdot 10^2 + 5 \cdot 10^1 + 8 \cdot 10^0 \quad$ or $3 \cdot 10^3 + 2 \cdot 10^2 + 5 \cdot 10 + 8$

5. $372_8 = 3 \cdot 8^2 + 7 \cdot 8^1 + 2 \cdot 8^0 \ $ or $3 \cdot 8^2 + 7 \cdot 8 + 2$

7. $3592_{16} = 3 \cdot 16^3 + 5 \cdot 16^2 + 9 \cdot 16^1 + 2 \cdot 16^0 \ $ or $3 \cdot 16^3 + 5 \cdot 16^2 + 9 \cdot 16 + 2$

9. $ABCDE0_{16} = 10 \cdot 16^5 + 11 \cdot 16^4 + 12 \cdot 16^3 + 13 \cdot 16^2 + 14 \cdot 16^1 + 0 \cdot 16^0$

   or $10 \cdot 16^5 + 11 \cdot 16^4 + 12 \cdot 16^3 + 13 \cdot 16^2 + 14 \cdot 16$

11. $1011001_2 = 1 \cdot 2^6 + 0 \cdot 2^5 + 1 \cdot 2^4 + 1 \cdot 2^3 + 0 \cdot 2^2 + 0 \cdot 2^1 + 1 \cdot 2^0$

    or $2^6 + 2^4 + 2^3 + 1$

13. 1324_2 does not exist because base two only uses numerals 0 and 1.

15. $5,32,85_{60}$ does not exist because base 60 only uses numerals 0 through 59.

17. $4312_5 = 4 \cdot 5^3 + 3 \cdot 5^2 + 1 \cdot 5^1 + 2 \cdot 5^0 \ $ or $4 \cdot 5^3 + 3 \cdot 5^2 + 1 \cdot 5 + 2$

19. $123_4 = 1 \cdot 4^2 + 2 \cdot 4^1 + 3 \cdot 4^0 \ $ or $1 \cdot 4^2 + 2 \cdot 4 + 3$

21. $372_8 = 3 \cdot 8^2 + 7 \cdot 8^1 + 2 \cdot 8^0 = 3 \cdot 64 + 7 \cdot 8 + 2 = 192 + 56 + 2 = 250$

23. $3592_{16} = 3 \cdot 16^3 + 5 \cdot 16^2 + 9 \cdot 16^1 + 2 \cdot 16^0 = 3 \cdot 4096 + 5 \cdot 256 + 9 \cdot 16 + 2$

    $= 12,288 + 1,280 + 144 + 2 = 13,714$

25. $ABCDE0_{16} = 10 \cdot 16^5 + 11 \cdot 16^4 + 12 \cdot 16^3 + 13 \cdot 16^2 + 14 \cdot 16^1$

    $= 10 \cdot 1,048,576 + 11 \cdot 65,536 + 12 \cdot 4096 + 13 \cdot 256 + 14 \cdot 16$

    $= 10,485,760 + 720,896 + 49,152 + 3,328 + 224 = 11,259,360$

27. $1011001_2 = 2^6 + 2^4 + 2^3 + 1$

    $= 64 + 16 + 8 + 1 = 89$

29. $5,32,85_{60}$ does not exist.

31. $4312_5 = 4 \cdot 5^3 + 3 \cdot 5^2 + 1 \cdot 5 + 2$ 33. $123_4 = 1 \cdot 4^2 + 2 \cdot 4 + 3$

    $= 4 \cdot 125 + 3 \cdot 25 + 5 + 2$ $= 1 \cdot 16 + 2 \cdot 4 + 3$

    $= 500 + 75 + 5 + 2 = 582$ $= 16 + 8 + 3 = 27$

35. $$1705_8 = 1 \cdot 8^3 + 7 \cdot 8^2 + 0 \cdot 8^1 + 5 \cdot 8^0$$
$$= 1 \cdot 512 + 7 \cdot 64 + 0 + 5$$
$$= 512 + 448 + 5 = 965$$

37. $$11011011_2 = 1 \cdot 2^7 + 1 \cdot 2^6 + 0 \cdot 2^5 + 1 \cdot 2^4 + 1 \cdot 2^3 + 0 \cdot 2^2 + 1 \cdot 2^1 + 1 \cdot 2^0$$
$$= 2^7 + 2^6 + 2^4 + 2^3 + 2 + 1$$
$$= 128 + 64 + 16 + 8 + 2 + 1 = 219$$

39. $$55,28,33,59_{60} = 55 \cdot 60^3 + 28 \cdot 60^2 + 33 \cdot 60^1 + 59 \cdot 60^0$$
$$= 55 \cdot 216,000 + 28 \cdot 3600 + 33 \cdot 60^1 + 59$$
$$= 11,880,000 + 100,800 + 1980 + 59 = 11,982,839$$

41. $$5798_{16} = 5 \cdot 16^3 + 7 \cdot 16^2 + 9 \cdot 16^1 + 8 \cdot 16^0$$
$$= 5 \cdot 4096 + 7 \cdot 256 + 9 \cdot 16 + 8$$
$$= 20,480 + 1792 + 144 + 8 = 22,424$$

43. $$7224_8 = 7 \cdot 8^3 + 2 \cdot 8^2 + 2 \cdot 8^1 + 4 \cdot 8^0$$
$$= 7 \cdot 512 + 2 \cdot 64 + 2 \cdot 8 + 4$$
$$= 3584 + 128 + 16 + 4 = 3,732$$

45. $$23,44,14_{50} = 23 \cdot 50^2 + 44 \cdot 50^1 + 14 \cdot 50^0$$
$$= 23 \cdot 2500 + 44 \cdot 50 + 14$$
$$= 57,500 + 2200 + 14 = 59,714$$

47. $$253_7 = 2 \cdot 7^2 + 5 \cdot 7^1 + 3 \cdot 7^0$$
$$= 2 \cdot 49 + 5 \cdot 7^1 + 3$$
$$= 98 + 35 + 3 = 136$$

49. $$4_8 = 4 = 1 \cdot 2^2 + 0 \cdot 2^1 + 0 \cdot 2^0 = 100_2$$
$$5_8 = 5 = 1 \cdot 2^2 + 0 \cdot 2^1 + 1 \cdot 2^0 = 101_2$$
$$2_8 = 2 = 0 \cdot 2^2 + 1 \cdot 2^1 + 0 \cdot 2^0 = 010_2$$
$$452_8 = 100101010_2$$

51. $5_8 = 5 = 1 \cdot 2^2 + 0 \cdot 2^1 + 1 \cdot 2^0 = 101_2$

         $2_8 = 2 = 0 \cdot 2^2 + 1 \cdot 2^1 + 0 \cdot 2^0 = 010_2$

         $6_8 = 6 = 1 \cdot 2^2 + 1 \cdot 2^1 + 0 \cdot 2^0 = 110_2$

         $0_8 = 0 = 0 \cdot 2^2 + 0 \cdot 2^1 + 0 \cdot 2^0 = 000_2$

         $5260_8 = 101010110000_2$

53. $001_2 = 0 \cdot 2^2 + 0 \cdot 2^1 + 1 \cdot 2^0 = 1_8$

         $010_2 = 0 \cdot 2^2 + 1 \cdot 2^1 + 0 \cdot 2^0 = 2_8$

         $1010_2 = 12_8$

55. $010_2 = 0 \cdot 2^2 + 1 \cdot 2^1 + 0 \cdot 2^0 = 2_8$

         $110_2 = 1 \cdot 2^2 + 1 \cdot 2^1 + 0 \cdot 2^0 = 6_8$

         $110_2 = 1 \cdot 2^2 + 1 \cdot 2^1 + 0 \cdot 2^0 = 6_8$

         $10110110_2 = 266_8$

57. $5_{16} = 5 = 0 \cdot 2^3 + 1 \cdot 2^2 + 0 \cdot 2^1 + 1 \cdot 2^0 = 0101_2$

         $3_{16} = 3 = 0 \cdot 2^3 + 0 \cdot 2^2 + 1 \cdot 2^1 + 1 \cdot 2^0 = 0011_2$

         $A_{16} = 10 = 1 \cdot 2^3 + 0 \cdot 2^2 + 1 \cdot 2^1 + 0 \cdot 2^0 = 1010_2$

         $2_{16} = 2 = 0 \cdot 2^3 + 0 \cdot 2^2 + 1 \cdot 2^1 + 0 \cdot 2^0 = 0010_2$

         $53A2_{16} = 101001110100010_2$

59. $3_{16} = 3 = 0 \cdot 2^3 + 0 \cdot 2^2 + 1 \cdot 2^1 + 1 \cdot 2^0 = 0011_2$

         $A_{16} = 10 = 1 \cdot 2^3 + 0 \cdot 2^2 + 1 \cdot 2^1 + 0 \cdot 2^0 = 1010_2$

         $B_{16} = 11 = 1 \cdot 2^3 + 0 \cdot 2^2 + 1 \cdot 2^1 + 1 \cdot 2^0 = 1011_2$

         $0_{16} = 0 = 0 \cdot 2^3 + 0 \cdot 2^2 + 0 \cdot 2^1 + 0 \cdot 2^0 = 0000_2$

         $3AB0_{16} = 11101010110000_2$

61. $0001_2 = 0 \cdot 2^3 + 0 \cdot 2^2 + 0 \cdot 2^1 + 1 \cdot 2^0 = 1_{16}$

         $0100_2 = 0 \cdot 2^3 + 1 \cdot 2^2 + 0 \cdot 2^1 + 0 \cdot 2^0 = 4_{16}$

         $10100_2 = 14_{16}$

63. $1011_2 = 1 \cdot 2^3 + 0 \cdot 2^2 + 1 \cdot 2^1 + 1 \cdot 2^0 = B_{16}$

         $1010_2 = 1 \cdot 2^3 + 0 \cdot 2^2 + 1 \cdot 2^1 + 0 \cdot 2^0 = A_{16}$

         $10111010_2 = BA_{16}$

65.   a. $2_{16} = 2 = 0 \cdot 2^3 + 0 \cdot 2^2 + 1 \cdot 2^1 + 0 \cdot 2^0 = 0010_2$

           $B_{16} = 11 = 1 \cdot 2^3 + 0 \cdot 2^2 + 1 \cdot 2^1 + 1 \cdot 2^0 = 1011_2$

           $A_{16} = 10 = 1 \cdot 2^3 + 0 \cdot 2^2 + 1 \cdot 2^1 + 0 \cdot 2^0 = 1010_2$

           $2BA_{16} = 1010111010_2$

         b. $001_2 = 0 \cdot 2^2 + 0 \cdot 2^1 + 1 \cdot 2^0 = 1_8$

           $010_2 = 0 \cdot 2^2 + 1 \cdot 2^1 + 0 \cdot 2^0 = 2_8$

           $111_2 = 1 \cdot 2^2 + 1 \cdot 2^1 + 1 \cdot 2^0 = 7_8$

           $010_2 = 0 \cdot 2^2 + 1 \cdot 2^1 + 0 \cdot 2^0 = 2_8$

           $1010111010_2 = 1272_8$

67.

$4_8 = 4 = 1 \cdot 2^2 + 0 \cdot 2^1 + 0 \cdot 2^0 = 100_2$

$2_8 = 2 = 0 \cdot 2^2 + 1 \cdot 2^1 + 0 \cdot 2^0 = 010_2$

$42_8 = 100010_2 \rightarrow 00100010_2$

$0010_2 = 0 \cdot 2^3 + 0 \cdot 2^2 + 1 \cdot 2^1 + 0 \cdot 2^0 = 2_{16}$

$0010_2 = 0 \cdot 2^3 + 0 \cdot 2^2 + 1 \cdot 2^1 + 0 \cdot 2^0 = 2_{16}$

$42_8 = 22_{16}$

69.

$5_8 = 5 = 1 \cdot 2^2 + 0 \cdot 2^1 + 1 \cdot 2^0 = 101_2$

$4_8 = 4 = 1 \cdot 2^2 + 0 \cdot 2^1 + 0 \cdot 2^0 = 100_2$

$0_8 = 0 = 0 \cdot 2^2 + 0 \cdot 2^1 + 0 \cdot 2^0 = 000_2$

$540_8 = 101100000_2 \rightarrow 000101100000_2$

$0001_2 = 0 \cdot 2^3 + 0 \cdot 2^2 + 0 \cdot 2^1 + 1 \cdot 2^0 = 1_{16}$

$0110_2 = 0 \cdot 2^3 + 1 \cdot 2^2 + 1 \cdot 2^1 + 0 \cdot 2^0 = 6_{16}$

$0000_2 = 0 \cdot 2^3 + 0 \cdot 2^2 + 0 \cdot 2^1 + 0 \cdot 2^0 = 0_{16}$

$540_8 = 160_{16}$

71.

$A_{16} = 10 = 1 \cdot 2^3 + 0 \cdot 2^2 + 1 \cdot 2^1 + 0 \cdot 2^0 = 1010_2$

$B_{16} = 11 = 1 \cdot 2^3 + 0 \cdot 2^2 + 1 \cdot 2^1 + 1 \cdot 2^0 = 1011_2$

$8_{16} = 8 = 1 \cdot 2^3 + 0 \cdot 2^2 + 0 \cdot 2^1 + 0 \cdot 2^0 = 1000_2$

$AB8_{16} = 101010111000_2$

$101_2 = 1 \cdot 2^2 + 0 \cdot 2^1 + 1 \cdot 2^0 = 5_8$

$010_2 = 0 \cdot 2^2 + 1 \cdot 2^1 + 0 \cdot 2^0 = 2_8$

$111_2 = 1 \cdot 2^2 + 1 \cdot 2^1 + 1 \cdot 2^0 = 7_8$

$000_2 = 0 \cdot 2^2 + 0 \cdot 2^1 + 0 \cdot 2^0 = 0_8$

$AB8_{16} = 5270_8$

73.

$8_{16} = 8 = 1 \cdot 2^3 + 0 \cdot 2^2 + 0 \cdot 2^1 + 0 \cdot 2^0 = 1000_2$

$0_{16} = 0 = 0 \cdot 2^3 + 0 \cdot 2^2 + 0 \cdot 2^1 + 0 \cdot 2^0 = 0000_2$

$1_{16} = 0 = 0 \cdot 2^3 + 0 \cdot 2^2 + 0 \cdot 2^1 + 1 \cdot 2^0 = 0001_2$

$D_{16} = 13 = 1 \cdot 2^3 + 1 \cdot 2^2 + 0 \cdot 2^1 + 1 \cdot 2^0 = 1101_2$

$801D_{16} = 1000000000011101_2$

$\qquad = 001000000000011101_2$

$001_2 = 0 \cdot 2^2 + 0 \cdot 2^1 + 1 \cdot 2^0 = 1_8$

$000_2 = 0 \cdot 2^2 + 0 \cdot 2^1 + 0 \cdot 2^0 = 0_8$

$000_2 = 0 \cdot 2^2 + 0 \cdot 2^1 + 0 \cdot 2^0 = 0_8$

$000_2 = 0 \cdot 2^2 + 0 \cdot 2^1 + 0 \cdot 2^0 = 0_8$

$011_2 = 0 \cdot 2^2 + 1 \cdot 2^1 + 1 \cdot 2^0 = 3_8$

$101_2 = 1 \cdot 2^2 + 0 \cdot 2^1 + 1 \cdot 2^0 = 5_8$

$801D_{16} = 100035_8$

75.

$22,12_{60} = 22 \cdot 60^1 + 12 \cdot 60^0$

$\qquad = 1320 + 12$

$\qquad = 1,332$

77.

$33,14,25_{60} = 33 \cdot 60^2 + 14 \cdot 60^1 + 25 \cdot 60^0$

$\qquad = 118,800 + 840 + 25$

$\qquad = 119,665$

79.
$$60\overline{)364}^{\;6}$$
$$360$$
$$4$$

$$364 = 6 \cdot 60^1 + 4 \cdot 60^0 = 6,4_{60}$$

81.
$$3600\overline{)129845}^{\;36} \qquad 60\overline{)245}^{\;6}$$
$$129600 \qquad\qquad 240$$
$$245 \qquad\qquad\quad 5$$

$$129,845 = 36 \cdot 60^2 + 4 \cdot 60^1 + 5 \cdot 60^0$$
$$= 36,4,5_{60}$$

83. $\quad 534_9 = 5 \cdot 9^2 + 3 \cdot 9^1 + 4 \cdot 9^0$

$$= 405 + 27 + 4 = 436$$

$$343\overline{)436}^{\;1} \qquad 49\overline{)93}^{\;1} \qquad 7\overline{)44}^{\;6}$$
$$343 \qquad\quad 49 \qquad\quad 42$$
$$93 \qquad\quad 44 \qquad\quad 2$$

$$534_9 = 1 \cdot 7^3 + 1 \cdot 7^2 + 6 \cdot 7^1 + 2 \cdot 7^0$$
$$= 1162_7$$

85. $\quad 331_4 = 3 \cdot 4^2 + 3 \cdot 4^1 + 1 \cdot 4^0$

$$= 48 + 12 + 1 = 61$$

$$25\overline{)61}^{\;2} \qquad 5\overline{)11}^{\;2}$$
$$50 \qquad\quad 10$$
$$11 \qquad\quad 1$$

$$331_4 = 2 \cdot 5^2 + 2 \cdot 5^1 + 1 \cdot 5^0$$
$$= 221_5$$

87. – 89. *Student instructed to use trial-and-error process.*

87.   a. $121_x = 16$ b. $121_x = 36$

$$1 \cdot x^2 + 2 \cdot x^1 + 1 \cdot x^0 = 16 \qquad\qquad x^2 + 2x + 1 = 36$$
$$x^2 + 2x + 1 = 16 \qquad\qquad (x+1)^2 = 36$$
$$(x+1)^2 = 16 \qquad\qquad x + 1 = 6$$
$$x + 1 = 4 \qquad\qquad\qquad x = 5$$
$$x = 3$$

The base is 3.   The base is 5.

89.　a.
$$370_x = 248$$

$$3 \cdot x^2 + 7 \cdot x^1 + 0 \cdot x^0 = 248$$

$$3x^2 + 7x = 248$$

$$3x^2 + 7x - 248 = 0$$

$$(3x + 31)(x - 8) = 0$$

$$3x + 31 = 0 \quad \text{or} \quad x - 8 = 0$$

$$x = -\frac{31}{3} \quad \text{or} \quad x = 8$$

The base is 8.

　b.
$$370_x = 306$$

$$3x^2 + 7x = 306$$

$$3x^2 + 7x - 306 = 0$$

$$(3x + 34)(x - 9) = 0$$

$$3x + 34 = 0 \quad \text{or} \quad x - 9 = 0$$

$$x = -\frac{34}{3} \quad \text{or} \quad x = 9$$

The base is 9.

91.　a. 0, 1, 2, 3, 4

　b. $5^0 \ 5^1 \ 5^2 \ 5^3$

$$0_5 \quad 1_5 \quad 2_5 \quad 3_5 \quad 4_5$$
$$10_5 \quad 11_5 \quad 12_5 \quad 13_5 \quad 14_5$$

　c. $20_5 \quad 21_5 \quad 22_5 \quad 23_5 \quad 24_5$
$$30_5 \quad 31_5 \quad 32_5 \quad 33_5 \quad 34_5$$
$$40_5 \quad 41_5 \quad 42_5 \quad 43_5 \quad 44_5$$

93.　a. 0, 1, 2, 3, 4, 5, 6, 7, 8, 9, A

　b. $11^0 \ 11^1 \ 11^2 \ 11^3$

$$0_{11} \quad 1_{11} \quad 2_{11} \quad 3_{11} \quad 4_{11} \quad 5_{11} \quad 6_{11} \quad 7_{11} \quad 8_{11} \quad 9_{11} \quad A_{11}$$

　c. $10_{11} \quad 11_{11} \quad 12_{11} \quad 13_{11} \quad 14_{11} \quad 15_{11} \quad 16_{11} \quad 17_{11} \quad 18_{11} \quad 19_{11} \quad 1A_{11}$
$$20_{11} \quad 21_{11} \quad 22_{11}$$

95.　16-bit color has $2^{16} = 65{,}536$ shades, and 24-bit color has $2^{24} = 16{,}777{,}216$ shades.

Section 7.2 Addition and Subtraction in Different Bases

1.　$3_8 + 7_8 = 1 \cdot 8^1 + 2 \cdot 8^0 = 12_8$

3.　$4_8 + 5_8 = 1 \cdot 8^1 + 1 \cdot 8^0 = 11_8$

5.　$9_{16} + 7_{16} = 1 \cdot 16^1 + 0 \cdot 16^0 = 10_{16}$

7.　$B_{16} + 9_{16} = 1 \cdot 16^1 + 4 \cdot 16^0 = 14_{16}$

9.　$1_2 + 1_2 = 1 \cdot 2^1 + 0 \cdot 2^0 = 10_2$

11.　
$$11_2$$
$$+ \ \underline{11_2}$$
$$110_2$$

13.　a.　13_8
$$\underline{+\ 72_8}$$
$$105_8$$

　　　b.　$13_8 = 1 \cdot 8^1 + 3 \cdot 8^0 = 11$
$$72_8 = 7 \cdot 8^1 + 2 \cdot 8^0 = 58$$
$$105_8 = 1 \cdot 8^2 + 0 \cdot 8^1 + 5 \cdot 8^0 = 69$$
Check: $11 + 25 = 69$

15.　a.　475_8
$$\underline{+\ 254_8}$$
$$751_8$$

　　　b.　$475_8 = 4 \cdot 8^2 + 7 \cdot 8^1 + 5 \cdot 8^0 = 317$
$$254_8 = 2 \cdot 8^2 + 5 \cdot 8^1 + 4 \cdot 8^0 = 172$$
$$751_8 = 7 \cdot 8^2 + 5 \cdot 8^1 + 1 \cdot 8^0 = 489$$
Check: $317 + 172 = 489$

17.　a.　$A9_{16}$
$$\underline{+\ 7B_{16}}$$
$$124_{16}$$

　　　b.　$A9_{16} = 10 \cdot 16^1 + 9 \cdot 16^0 = 169$
$$7B_{16} = 7 \cdot 16^1 + 11 \cdot 16^0 = 123$$
$$124_{16} = 1 \cdot 16^2 + 2 \cdot 16^1 + 4 \cdot 16^0 = 292$$
Check: $169 + 123 = 292$

19.　a.　BBC_{16}
$$\underline{+\ CCD_{16}}$$
$$1889_{16}$$

　　　b.　$BBC_{16} = 11 \cdot 16^2 + 11 \cdot 16^1 + 12 \cdot 16^0 = 3004$
$$CCD_{16} = 12 \cdot 16^2 + 12 \cdot 16^1 + 13 \cdot 16^0 = 3277$$
$$1889_{16} = 1 \cdot 16^3 + 8 \cdot 16^2 + 8 \cdot 16^1 + 9 \cdot 16^0 = 6281$$
Check: $3004 + 3277 = 6281$

21.　a.　110_2
$$\underline{+\ 101_2}$$
$$1011_2$$

　　　b.　$110_2 = 1 \cdot 2^2 + 1 \cdot 2^1 + 0 \cdot 2^0 = 6$
$$101_2 = 1 \cdot 2^2 + 0 \cdot 2^1 + 1 \cdot 2^0 = 5$$
$$1011_2 = 1 \cdot 2^3 + 0 \cdot 2^2 + 1 \cdot 2^1 + 1 \cdot 2^0 = 11$$
Check: $6 + 5 = 11$

23.　a.　1100110_2
$$\underline{+\ 1101101_2}$$
$$11010011_2$$

b. $1100110_2 = 1 \cdot 2^6 + 1 \cdot 2^5 + 0 \cdot 2^4 + 0 \cdot 2^3 + 1 \cdot 2^2 + 1 \cdot 2^1 + 0 \cdot 2^0 = 102$

$1101101_2 = 1 \cdot 2^6 + 1 \cdot 2^5 + 0 \cdot 2^4 + 1 \cdot 2^3 + 1 \cdot 2^2 + 0 \cdot 2^1 + 1 \cdot 2^0 = 109$

$11010011_2 = 1 \cdot 2^7 + 1 \cdot 2^6 + 0 \cdot 2^5 + 1 \cdot 2^4 + 0 \cdot 2^3 + 0 \cdot 2^2 + 1 \cdot 2^1 + 1 \cdot 2^0 = 211$

Check: $102 + 109 = 211$

25.
$$
\begin{array}{r}
\overset{0\ 1}{\cancel{1}}3_8 \\
-\ 7_8 \\
\hline
4_8
\end{array}
$$

27.
$$
\begin{array}{r}
194_{16} \\
-\ 52_{16} \\
\hline
142_{16}
\end{array}
$$

29.
$$
\begin{array}{r}
\overset{0\ 1}{1\cancel{1}}01_2 \\
-\ 1010_2 \\
\hline
0011_2 \quad \text{or } 11_2
\end{array}
$$

31. a.
$$
\begin{array}{r}
\overset{0\ 1}{\cancel{1}}5_8 \\
-\ 6_8 \\
\hline
7_8
\end{array}
$$

b. $15_8 = 1 \cdot 8^1 + 5 \cdot 8^0 = 13$

$6_8 = 6 \cdot 8^0 = 6$

$7_8 = 7 \cdot 8^0 = 7$

Check: $13 - 6 = 7$

33. a.
$$
\begin{array}{r}
1\overset{A\ 1}{\cancel{B}}3_{16} \\
-\ 105_{16} \\
\hline
AE_{16}
\end{array}
$$

b. $1B3_{16} = 1 \cdot 16^2 + 11 \cdot 16^1 + 3 \cdot 16^0 = 435$

$105_{16} = 1 \cdot 16^2 + 0 \cdot 16^1 + 5 \cdot 16^0 = 261$

$AE_{16} = 10 \cdot 16^1 + 14 \cdot 16^0 = 174$

Check: $435 - 261 = 174$

35. a.
$$
\begin{array}{r}
1\overset{0\ 1}{\cancel{1}}011_2 \\
-\ 10110_2 \\
\hline
00101_2 \quad \text{or } 101_2
\end{array}
$$

b. $11011_2 = 1 \cdot 2^4 + 1 \cdot 2^3 + 0 \cdot 2^2 + 1 \cdot 2^1 + 1 \cdot 2^0 = 27$

$10110_2 = 1 \cdot 2^4 + 0 \cdot 2^3 + 1 \cdot 2^2 + 1 \cdot 2^1 + 0 \cdot 2^0 = 22$

$101_2 = 1 \cdot 2^2 + 0 \cdot 2^1 + 1 \cdot 2^0 = 5$ Check: $27 - 22 = 5$

37. a.
$$
\begin{array}{r}
\overset{2\ \overset{1}{3}\ 1}{\cancel{3}\,\cancel{4}}1_5 \\
-\ 243_5 \\
\hline
43_5
\end{array}
$$

b. $341_5 = 3 \cdot 5^2 + 4 \cdot 5^1 + 1 \cdot 5^0 = 96$

$243_5 = 2 \cdot 5^2 + 4 \cdot 5^1 + 3 \cdot 5^0 = 73$

$43_5 = 4 \cdot 5^1 + 3 \cdot 5^0 = 23$ Check: $96 - 73 = 23$

Section 7.3 Multiplication and Division in Different Bases

1. $\overset{1}{1}2_8$

   $\underline{\times\quad 6_8}$

   74_8

   $2_8 \cdot 6_8 = 12_{10} = 1 \cdot 8^1 + 4 \cdot 8^0 = 14_8$

   $\left(1_8 \cdot 6_8\right) + 1_8 = 7_{10} = 0 \cdot 8^1 + 7 \cdot 8^0 = 7_8$

3. $C1_{16}$

   $\underline{\times\quad 5_{16}}$

   $3C5_{16}$

   $1_{16} \cdot 5_{16} = 5_{10} = 0 \cdot 16^1 + 5 \cdot 16^0 = 5_{16}$

   $C_{16} \cdot 5_{16} = 60_{10} = 3 \cdot 16^1 + C \cdot 16^0 = 3C_{16}$

5. 101_2

   $\underline{\times\quad 11_2}$

   101

   $\underline{101\quad}$

   1111_2

7. $\overset{1}{3}2_4$

   $\underline{\times\quad 23_4}$

   222

   $\underline{130\quad}$

   2122_4

   $2_4 \cdot 3_4 = 6_{10} = 1 \cdot 4^1 + 2 \cdot 4^0 = 12_4$

   $\left(3_4 \cdot 3_4\right) + 1_4 = 10_{10} = 2 \cdot 4^1 + 2 \cdot 4^0 = 22_4$

   $2_4 \cdot 2_4 = 4_{10} = 1 \cdot 4^1 + 0 \cdot 4^0 = 10_4$

   $\left(3_4 \cdot 2_4\right) + 1_4 = 7_{10} = 1 \cdot 4^1 + 3 \cdot 4^0 = 13_4$

9. a. $\overset{3}{2}5_8$

   $\underline{\times\quad 5_8}$

   151_8

   $5_8 \cdot 5_8 = 25_{10} = 3 \cdot 8^1 + 1 \cdot 8^0 = 31_8$

   $\left(5_8 \cdot 2_8\right) + 3_{10} = 13_{10} = 1 \cdot 8^1 + 5 \cdot 8^0 = 15_8$

   b. $25_8 = 2 \cdot 8^1 + 5 \cdot 8^0 = 21$

   $5_8 = 5$

   $151_8 = 1 \cdot 8^2 + 5 \cdot 8^1 + 1 \cdot 8^0 = 105$

   Check: $21 \cdot 5 = 105$

11. a. $\overset{3\ 3}{1}BA_{16}$

    $\underline{\times\quad 15_{16}}$

    $8A2$

    $\underline{1BA\quad}$

    2442_{16}

$$5_{16} \cdot A_{16} = 50_{10} = 3 \cdot 16^1 + 2 \cdot 16^0 = 32_{16}$$

$$\left(5_{16} \cdot B_{16}\right) + 3_{10} = 58_{10} = 3 \cdot 16^1 + 10 \cdot 16^0 = 3A_{16}$$

b. $1BA_{16} = 1 \cdot 16^2 + 11 \cdot 16^1 + 10 \cdot 16^0 = 442$

$15_{16} = 1 \cdot 16^1 + 5 \cdot 16^0 = 21$

$2442_{16} = 2 \cdot 16^3 + 4 \cdot 16^2 + 4 \cdot 16^1 + 2 \cdot 16^0 = 9,282$

Check: $442 \cdot 21 = 9,282$

13. a.
$$
\begin{array}{r}
1101_2 \\
\times\ 10110_2 \\
\hline
11010 \\
\overset{1}{1101} \\
\overset{1\ 1}{\underline{1101}} \\
\hline
100011110_2
\end{array}
$$

b. $1101_2 = 1 \cdot 2^3 + 1 \cdot 2^2 + 0 \cdot 2^1 + 1 \cdot 2^0 = 13$

$10110_2 = 1 \cdot 2^4 + 0 \cdot 2^3 + 1 \cdot 2^2 + 1 \cdot 2^1 + 0 \cdot 2^0 = 22$

$100011110_2 = 1 \cdot 2^8 + 1 \cdot 2^4 + 1 \cdot 2^3 + 1 \cdot 2^2 + 1 \cdot 2^1 = 286$

Check: $13 \cdot 22 = 286$

15. a.
$$
\begin{array}{r}
31_5 \\
\times\ 23_5 \\
\hline
\overset{1}{143} \\
\underline{112} \\
1313_5
\end{array}
$$

b. $31_5 = 3 \cdot 5^1 + 1 \cdot 5^0 = 16$

$23_5 = 2 \cdot 5^1 + 3 \cdot 5^0 = 13$

$1313_5 = 1 \cdot 5^3 + 3 \cdot 5^2 + 1 \cdot 5^1 + 3 \cdot 5^0 = 208$

$3_5 \cdot 3_5 = 9_{10} = 1 \cdot 5^1 + 4 \cdot 5^0 = 14_5$

$2_5 \cdot 3_5 = 6_{10} = 1 \cdot 5^1 + 1 \cdot 5^0 = 11_5$

17.
$$
\begin{array}{r}
23_8 \\
2_8 \overline{)\ 46_8} \\
\underline{-4} \\
06 \\
\underline{-6} \\
0
\end{array}
$$

19. a.
$$
\begin{array}{r}
6F_{16} \\
3_{16} \overline{)\ 14D_{16}} \\
\underline{-12} \\
2D \\
\underline{-2D} \\
0
\end{array}
$$

$$14D_{16} = 1 \cdot 16^2 + 4 \cdot 16^1 + 13 \cdot 16^0 = 333$$

b. $3_{16} = 3 \cdot 16^0 = 3$

$6F_{16} = 6 \cdot 16^1 + 15 \cdot 16^0 = 111$

Check: $333 \div 3 = 111$

21. a. $31_8 \overline{) \, 144_8}$ with quotient 4_8

$$-144_8$$
$$0$$

$$31_8 \times 4_8 = 144_8$$

b. $144_8 = 1 \cdot 8^2 + 4 \cdot 8^1 + 4 \cdot 8^0 = 100$

$31_8 = 3 \cdot 8^1 + 1 \cdot 8^0 = 25$

$4_8 = 4 \cdot 8^0 = 4$

Check: $100 \div 25 = 4$

23. a. $15_8 \overline{) \, 153_8}$ with quotient 10_8 10_8 R 3_8

$$-15$$
$$3_8$$

b. 15_8

$\times \, 10_8$

150_8

Check: $150_8 + 3_8 = 153_8$

25. a. $152_8 \overline{) \, 277_8}$ with quotient 1_8 1_8 R 125_8

$$-152_8$$
$$125_8$$

b. 152_8

$\times \quad 1_8$

152_8

Check: $152_8 + 125_8 = 277_8$

27. a. $38_{16} \overline{) \, 79AB_{16}}$ with quotient $22C_{16}$ $22C_{16}$ R B_{16}

$$-70$$
$$9A$$
$$-70$$
$$2AB$$
$$-2A0$$
$$B_{16}$$

b. $\overset{1 \ \overset{2}{6}}{2 \, 2 C_{16}}$

$\times \, 38_{16}$

1160

664

$79A0_{16}$

Check: $79A0_{16} + B_{16} = 79AB_{16}$

29. a. $218_{16} \overline{) \, 790B_{16}}$ with quotient 39_{16} 39_{16} R $1B3_{16}$

$$-648$$
$$148B$$
$$-12D8$$
$$1B3_{16}$$

b. 218_{16}

$\times \, 39_{16}$

$12D8$

648

7758_{16}

Check: $7758_{16} + 1B3_{16} = 790B_{16}$

31. a. $110_2 \overline{)1100111_2}$ 10001_2

$$\underline{-110_2}$$

$$0111$$

$$\underline{-110}$$

$$1_2$$

10001_2 R 1_2

b. 10001_2

$$\times \underline{110_2}$$

$$100010$$

$$\underline{10001}$$

$$1100110_2$$

Check: $1100110_2 + 1_2 = 1100111_2$

33. a. $1100_2 \overline{)1101111_2}$ 1001_2 R 11_2

$$\underline{-1100}$$

$$1111$$

$$\underline{-1100}$$

$$11_2$$

b. 1100_2

$$\times \underline{1001_2}$$

$$1100$$

$$\underline{1100}$$

$$1101100_2$$

Check: $1101100_2 + 11_2 = 1101111_2$

35.

	1_5	2_5	3_5	4_5	10_5	11_5
1_5	1_5	2_5	3_5	4_5	10_5	11_5
2_5	2_5	4_5	11_5	13_5	20_5	22_5
3_5	3_5	11_5	14_5	22_5	30_5	33_5
4_5	4_5	13_5	22_5	31_5	40_5	44_5
10_5	10_5	20_5	30_5	40_5	100_5	110_5
11_5	11_5	22_5	33_5	44_5	110_5	121_5

37.

	1_{14}	2_{14}	3_{14}	4_{14}	5_{14}	6_{14}	7_{14}	8_{14}	9_{14}	A_{14}	B_{14}	C_{14}	D_{14}	10_{14}	11_{14}
1_{14}	1_{14}	2_{14}	3_{14}	4_{14}	5_{14}	6_{14}	7_{14}	8_{14}	9_{14}	A_{14}	B_{14}	C_{14}	D_{14}	10_{14}	11_{14}
2_{14}	2_{14}	4_{14}	6_{14}	8_{14}	A_{14}	C_{14}	10_{14}	12_{14}	14_{14}	16_{14}	18_{14}	$1A_{14}$	$1C_{14}$	20_{14}	22_{14}
3_{14}	3_{14}	6_{14}	9_{14}	C_{14}	11_{14}	14_{14}	17_{14}	$1A_{14}$	$1D_{14}$	22_{14}	25_{14}	28_{14}	$2B_{14}$	30_{14}	33_{14}
4_{14}	4_{14}	8_{14}	C_{14}	12_{14}	16_{14}	$1A_{14}$	20_{14}	24_{14}	28_{14}	$2C_{14}$	32_{14}	36_{14}	$3A_{14}$	40_{14}	44_{14}
5_{14}	5_{14}	A_{14}	11_{14}	16_{14}	$1B_{14}$	22_{14}	27_{14}	$2C_{14}$	33_{14}	38_{14}	$3D_{14}$	44_{14}	49_{14}	50_{14}	55_{14}
6_{14}	6_{14}	C_{14}	14_{14}	$1A_{14}$	22_{14}	28_{14}	30_{14}	36_{14}	$3C_{14}$	44_{14}	$4A_{14}$	52_{14}	58_{14}	60_{14}	66_{14}
7_{14}	7_{14}	10_{14}	17_{14}	20_{14}	27_{14}	30_{14}	37_{14}	40_{14}	47_{14}	50_{14}	57_{14}	60_{14}	67_{14}	70_{14}	77_{14}
8_{14}	8_{14}	12_{14}	$1A_{14}$	24_{14}	$2C_{14}$	36_{14}	40_{14}	48_{14}	52_{14}	$5A_{14}$	64_{14}	$6C_{14}$	76_{14}	80_{14}	88_{14}
9_{14}	9_{14}	14_{14}	$1D_{14}$	28_{14}	33_{14}	$3C_{14}$	47_{14}	52_{14}	$5B_{14}$	66_{14}	71_{14}	$7A_{14}$	85_{14}	90_{14}	99_{14}
A_{14}	A_{14}	16_{14}	22_{14}	$2C_{14}$	38_{14}	44_{14}	50_{14}	$5A_{14}$	66_{14}	72_{14}	$7C_{14}$	88_{14}	94_{14}	$A0_{14}$	AA_{14}
B_{14}	B_{14}	18_{14}	25_{14}	32_{14}	$3D_{14}$	$4A_{14}$	57_{14}	64_{14}	71_{14}	$7C_{14}$	89_{14}	96_{14}	$A3_{14}$	$B0_{14}$	BB_{14}
C_{14}	C_{14}	$1A_{14}$	28_{14}	36_{14}	44_{14}	52_{14}	60_{14}	$6C_{14}$	$7A_{14}$	88_{14}	96_{14}	$A4_{14}$	$B2_{14}$	$C0_{14}$	CC_{14}
D_{14}	D_{14}	$1C_{14}$	$2B_{14}$	$3A_{14}$	49_{14}	58_{14}	67_{14}	76_{14}	85_{14}	94_{14}	$A3_{14}$	$B2_{14}$	$C1_{14}$	$D0_{14}$	DD_{14}
10_{14}	10_{14}	20_{14}	30_{14}	40_{14}	50_{14}	60_{14}	70_{14}	80_{14}	90_{14}	$A0_{14}$	$B0_{14}$	$C0_{14}$	$D0_{14}$	100_{14}	110_{14}
11_{14}	11_{14}	22_{14}	33_{14}	44_{14}	55_{14}	66_{14}	77_{14}	88_{14}	99_{14}	AA_{14}	BB_{14}	CC_{14}	DD_{14}	110_{14}	121_{14}

39.
$$\begin{array}{r} 133_5 \\ 23_5 \overline{)\ 4230_5} \\ -23 \\ \hline 143 \\ -124 \\ \hline 140 \\ -124 \\ \hline 11_5 \end{array}$$

133_5 R 11_5

41.
$$\begin{array}{r} B9_{14} \\ 102_{14} \overline{)\ BADC_{14}} \\ -B18 \\ \hline 95C \\ -914 \\ \hline 48_{14} \end{array}$$

$B9_{14}$ R 48_{14}

Section 7.4 Prime Numbers and Perfect Numbers

1.   a. $42 = 2 \cdot 21 = 2 \cdot 3 \cdot 7$

     b.   composite

3.   a.   23

     b.   prime

5. a. $54 = 2 \cdot 27$ b.   composite

         $= 2 \cdot 3 \cdot 9$

         $= 2 \cdot 3 \cdot 3 \cdot 3$

         $= 2 \cdot 3^3$

7. Note: $\sqrt{299} = 17.29\ldots$

    $\dfrac{299}{2} = 149.5$ $\dfrac{299}{7} = 42.7\ldots$

    $\dfrac{299}{3} = 99.6\ldots$ $\dfrac{299}{11} = 27.18\ldots$

    $\dfrac{299}{5} = 59.8$ $\dfrac{299}{13} = 23$

    $299 = 13 \cdot 23$, so it is composite.

9. Note: $\sqrt{207} = 14.38\ldots$

    $\dfrac{207}{2} = 103.5$

    $\dfrac{207}{3} = 69$

    $207 = 3 \cdot 69$, so it is composite.

11. Note: $\sqrt{233} = 15.26\ldots$

    $\dfrac{233}{2} = 116.5$

    $\dfrac{233}{3} = 77.66\ldots$

    $\dfrac{233}{5} = 46.6$

    $\dfrac{233}{7} = 33.28\ldots$

    $\dfrac{233}{11} = 21.18\ldots$

    $\dfrac{233}{13} = 17.92\ldots$

    No need to go further; 233 is prime.

13. Since $\sqrt{49} = 7$, eliminate multiples of 2, then 3, then 4, etc., through 7.

    ~~26~~ ~~27~~ ~~28~~ 29 ~~30~~ 31
    ~~32~~ ~~33~~ ~~34~~ ~~35~~ ~~36~~ 37
    ~~38~~ ~~39~~ ~~40~~ 41 ~~42~~ 43
    ~~44~~ ~~45~~ ~~46~~ 47 ~~48~~ ~~49~~

    The prime numbers are 29, 31, 37, 41, 43, 47.

15. $2^n + 1 = 2^4 + 1 = 16 + 1 = 17$

    Since $17 = 17 \cdot 1$, it is prime.

17. $2^n + 1 = 2^6 + 1 = 64 + 1 = 65$

    Since $65 = 5(13)$, it is composite.

19. $2^n - 1 = 2^4 - 1 = 16 - 1 = 15$

    Since $\frac{15}{3} = 5$, it is composite.

21. $2^n - 1 = 2^9 - 1 = 512 - 1 = 511$

    Since $\frac{511}{7} = 73$, it is composite.

23. Note: $\sqrt{42} = 6.4\ldots$

    $\frac{42}{2} = 21 \Rightarrow 2$ and 21 are factors.

    $\frac{42}{3} = 14 \Rightarrow 3$ and 14 are factors.

    $\frac{42}{4} = 10.5$

    $\frac{42}{5} = 8.4$

    $\frac{42}{6} = 7 \Rightarrow 6$ and 7 are factors.

    The proper factors are $1, 2, 3, 6, 7, 14, 21$.

25. Note: $\sqrt{54} = 7.3\ldots$

    $\frac{54}{2} = 27 \Rightarrow 2$ and 27 are factors.

    $\frac{54}{3} = 18 \Rightarrow 3$ and 18 are factors.

    $\frac{54}{4} = 13.5$

    $\frac{54}{5} = 10.8$

    $\frac{54}{6} = 9 \Rightarrow 6$ and 9 are factors.

    $\frac{54}{7} = 7.7\ldots$

    The proper factors are $1, 2, 3, 6, 9, 18, 27$.

27. Since $1 + 2 + 3 + 6 + 7 + 14 + 21 = 54 > 42$, it is abundant.

29. Since $1 + 2 + 3 + 6 + 9 + 18 + 27 = 66 > 54$, it is abundant.

31. Note: $\sqrt{61} = 7.8\ldots$

    $\frac{61}{2} = 30.5$

    $\frac{61}{3} = 20.3\ldots$

    $\frac{61}{5} = 12.2$

    $\frac{61}{7} = 8.7\ldots$

    The proper factor is 1.

    Since $1 < 61$, it is deficient.

33. Note: $\sqrt{62} = 7.8\ldots$

    $\frac{62}{2} = 31 \Rightarrow 2$ and 31 are factors.

    $\frac{62}{3} = 20.6\ldots$

    $\frac{62}{4} = 15.5$

    $\frac{62}{5} = 12.4$

    $\frac{62}{7} = 8.8\ldots$

    The proper factors are $1, 2, 31$.

    Since $1 + 2 + 31 = 34 < 62$, it is deficient.

35. a. 13

    b. $2^{12} = 4,096$

    c. $2^{13} - 1 = 8,191$

    d. $4096 \cdot 8191 = 33,550,336$

37. a. 19

    b. $2^{18} = 262,144$

    c. $2^{19} - 1 = 524,287$

    d. $262,144 \cdot 524,287 = 1.374386913 \times 10^{11}$

39.   a. Since $2^{127} - 1 = 2^n - 1$, $n = 127$.   b. 1.45×10^{76}

$$\left(2^{n-1}\right)\left(2^n - 1\right) = \left(2^{127-1}\right)\left(2^{127} - 1\right)$$

41. $pq = 35$

     $5 \cdot 7 = 35$

     The private key is 5, 7.

43. $pq = 143$

     $11 \cdot 13 = 143$

     The private key is 11, 13.

45.   a. $6 = 1 \cdot 2^2 + 1 \cdot 2^1 = 110$

      $28 = 1 \cdot 2^4 + 1 \cdot 2^3 + 1 \cdot 2^2 = 11100$

      $496 = 1 \cdot 2^8 + 1 \cdot 2^7 + 1 \cdot 2^6 + 1 \cdot 2^5 + 1 \cdot 2^4 = 111110000$

      $8128 = 1 \cdot 2^{12} + 1 \cdot 2^{11} + 1 \cdot 2^{10} + 1 \cdot 2^9 + 1 \cdot 2^8 + 1 \cdot 2^7 + 1 \cdot 2^6 = 1111111000000$

  b. The nth perfect number is a sequence of m ones, where m is the nth prime, followed by $2(n-1)$ zeros, $n \neq 1$. (When $n = 1$, there is 1 zero.)

Section 7.5 Fibonacci Numbers and the Golden Ratio

1.   The pattern is 1, 1, 2, 3, 5, 8, 13, …; After six years, there are 13 total, and $13 - 8 = 5$ newly born.

3.   1, 1, 2, 3, 5, 8, 13, <u>21, 34, 55, 89, 144, 233, 377, 610, 987, 1597</u>

5.   1 parent, 2 grandparents, 3 great-grandparents, 5 great-great grandparents, 8 great-great-great grandparents

7.   Answers will vary.

9.

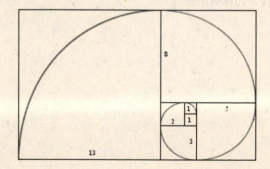

11.   a. Let $n = 30$.

$$\frac{1}{\sqrt{5}}\left[\left(\frac{1+\sqrt{5}}{2}\right)^{30} - \left(\frac{1-\sqrt{5}}{2}\right)^{30}\right] = 832,040$$

Let $n = 40$.

$$\frac{1}{\sqrt{5}}\left[\left(\frac{1+\sqrt{5}}{2}\right)^{40} - \left(\frac{1-\sqrt{5}}{2}\right)^{40}\right] = 102,334,155$$

b. Possible answer: It allows you to find a Fibonacci number without listing all the preceding numbers.

c. Possible answer: It is an unnecessarily lengthy formula when dealing with the first numbers, say the first 24 or so numbers.

13.   a. $\dfrac{a+b}{a} = \dfrac{a}{b}$ b. $ab \cdot \dfrac{a+b}{a} = ab \cdot \dfrac{a}{b}$ c. $ab + b^2 = a^2$

$$b(a+b) = a^2$$

d. $\dfrac{ab+b^2}{b^2} = \dfrac{a^2}{b^2}$ e. $\dfrac{a}{b} + 1 = \left(\dfrac{a}{b}\right)^2$ f. Let $\dfrac{a}{b} = x$

$\dfrac{ab}{b^2} + \dfrac{b^2}{b^2} = \dfrac{a^2}{b^2}$

$x + 1 = x^2$

$\dfrac{ab}{b^2} + 1 = \dfrac{a^2}{b^2}$

g. $x^2 - x - 1 = 0$ h. x is a ratio of lengths.

$x = \dfrac{-b \pm \sqrt{b^2 - 4ac}}{2a}$

$x = \dfrac{-(-1) \pm \sqrt{(-1)^2 - 4(1)(-1)}}{2(1)}$

$x = \dfrac{1 \pm \sqrt{5}}{2}$

$x = \dfrac{1 + \sqrt{5}}{2}$ is the golden ratio.

15.   Answers may vary.   17.   Answers may vary.

19. 1.61803398896 is an accurate approximation to the golden ratio up to nine decimal places.

21. Answers may vary. 23. Answers may vary.

Chapter 7 Review

1. $5372 = 5 \cdot 10^3 + 3 \cdot 10^2 + 7 \cdot 10^1 + 2 \cdot 10^0$ or $5 \cdot 10^3 + 3 \cdot 10^2 + 7 \cdot 10 + 2$

3. $325_8 = 3 \cdot 8^2 + 2 \cdot 8^1 + 5 \cdot 8^0$ or $3 \cdot 8^2 + 2 \cdot 8 + 5$

5. 390_8; does not exist because base eight only uses numerals 0 through 7.

7. $905_{16} = 9 \cdot 16^2 + 0 \cdot 16^1 + 5 \cdot 16^0$ or $9 \cdot 16^2 + 5$

9. $ABC_{16} = 10 \cdot 16^2 + 11 \cdot 16^1 + 12 \cdot 16^0$ or $10 \cdot 16^2 + 11 \cdot 16 + 12$

11. $11011001_2 = 1 \cdot 2^7 + 1 \cdot 2^6 + 0 \cdot 2^5 + 1 \cdot 2^4 + 1 \cdot 2^3 + 0 \cdot 2^2 + 0 \cdot 2^1 + 1 \cdot 2^0$

    or $2^7 + 2^6 + 2^4 + 2^3 + 1$

13. 101112_2 does not exist because base two only uses numerals 0 and 1.

15. $39,22,54_{60} = 39 \cdot 60^2 + 22 \cdot 60^1 + 54 \cdot 60^0$ or $39 \cdot 60^2 + 22 \cdot 60 + 54$

17. $25,44,34_{60} = 25 \cdot 60^2 + 44 \cdot 60^1 + 34 \cdot 60^0$ or $25 \cdot 60^2 + 44 \cdot 60 + 34$

19. $514_8 = 5 \cdot 8^2 + 1 \cdot 8^1 + 4 \cdot 8^0$

    $= 320 + 8 + 4 = 332$

21. $110101_2 = 1 \cdot 2^5 + 1 \cdot 2^4 + 0 \cdot 2^3 + 1 \cdot 2^2 + 0 \cdot 2^1 + 1 \cdot 2^0$

    $= 2^5 + 2^4 + 2^2 + 1 = 53$

23. $BC9A_{16} = 11 \cdot 16^3 + 12 \cdot 16^2 + 9 \cdot 16^1 + 10 \cdot 16^0$

    $= 45,056 + 3072 + 144 + 10 = 48,282$

25. $111000_2 = 1 \cdot 2^5 + 1 \cdot 2^4 + 1 \cdot 2^3 + 0 \cdot 2^2 + 0 \cdot 2^1 + 0 \cdot 2^0$

    $= 2^5 + 2^4 + 2^3 = 56$

27. $1254_9 = 1 \cdot 9^3 + 2 \cdot 9^2 + 5 \cdot 9^1 + 4 \cdot 9^0$

    $= 729 + 162 + 45 + 4 = 940$

29. $\begin{array}{r} 1 \\ 512{\overline{\smash{\big)}\,514}} \\ \underline{512} \\ 2 \end{array}$ $\qquad 514 = 1 \cdot 8^3 + 0 \cdot 8^2 + 0 \cdot 8^1 + 2 \cdot 8^0 = 1002_8$

31.
$$4096\overline{)8430}\qquad 16\overline{)238}$$

with quotients 2 and 14:

$$\frac{8192}{238}\qquad \frac{224}{14}$$

$$8{,}430 = 2\cdot16^3 + 0\cdot16^2 + 14\cdot16^1 + 14\cdot16^0 = 20EE_{16}$$

33.
$$3600\overline{)39287}\qquad 60\overline{)3287}$$

with quotients 10 and 54:

$$\frac{36000}{3287}\qquad \frac{3240}{47}$$

$$39{,}287 = 10\cdot60^2 + 54\cdot60^1 + 47\cdot60^0 = 10{,}54{,}47_{60}$$

35.
$$8192\overline{)12003}\quad 2048\overline{)3811}\quad 1024\overline{)1763}\quad 512\overline{)739}\quad 128\overline{)227}\quad 64\overline{)99}\quad 32\overline{)35}\quad 2\overline{)3}$$

with quotients 1, 1, 1, 1, 1, 1, 1, 1:

$$\frac{8192}{3811}\quad \frac{2048}{1763}\quad \frac{1024}{739}\quad \frac{512}{227}\quad \frac{128}{99}\quad \frac{64}{35}\quad \frac{32}{3}\quad \frac{2}{1}$$

$$12{,}003 = 1\cdot2^{13} + 0\cdot2^{12} + 1\cdot2^{11} + 1\cdot2^{10} + 1\cdot2^9 + 0\cdot2^8 + 1\cdot2^7 + 1\cdot2^6 + 1\cdot2^5$$
$$+ 0\cdot2^4 + 0\cdot2^3 + 0\cdot2^2 + 1\cdot2^1 + 1\cdot2^0$$
$$= 10111011100011_2$$

37.
$$16807\overline{)59375}\quad 2401\overline{)8954}\quad 343\overline{)1751}\quad 7\overline{)36}$$

with quotients 3, 3, 5, 5:

$$\frac{50421}{8954}\quad \frac{7203}{1751}\quad \frac{1715}{36}\quad \frac{35}{1}$$

$$59{,}375 = 3\cdot7^5 + 3\cdot7^4 + 5\cdot7^3 + 0\cdot7^2 + 5\cdot7^1 + 1\cdot7^0 = 335051_7$$

39.
$$543_x = 207$$

$$5\cdot x^2 + 4\cdot x^1 + 3\cdot x^0 = 207$$
$$5x^2 + 4x + 3 = 207$$
$$5x^2 + 4x - 204 = 0$$
$$(5x + 34)(x - 6) = 0$$
$$5x + 34 = 0 \qquad \text{or} \qquad x - 6 = 0$$
$$x = -\frac{34}{5} \quad \text{or} \qquad x = 6$$

The base is 6.

41.
$$421_x = 343$$

$$4\cdot x^2 + 2\cdot x^1 + 1\cdot x^0 = 343$$
$$4x^2 + 2x + 1 = 343$$
$$4x^2 + 2x - 342 = 0$$
$$2(2x^2 + x - 171) = 0$$
$$2(2x + 19)(x - 9) = 0$$
$$2x + 19 = 0 \qquad \text{or} \qquad x - 9 = 0$$
$$x = -\frac{19}{2} \quad \text{or} \qquad x = 9$$

The base is 9.

43.   a. 52_8

$\underline{+62_8}$

134_8

b. $52_8 = 5 \cdot 8^1 + 2 \cdot 8^0 = 42$

$62_8 = 6 \cdot 8^1 + 2 \cdot 8^0 = 50$

$134_8 = 1 \cdot 8^2 + 3 \cdot 8^1 + 4 \cdot 8^0 = 92$

Check: $42 + 50 = 92$

45.   a. 101_2

$\underline{+110_2}$

1011_2

b. $101_2 = 1 \cdot 2^2 + 0 \cdot 2^1 + 1 \cdot 2^0 = 5$

$110_2 = 1 \cdot 2^2 + 1 \cdot 2^1 + 0 \cdot 2^0 = 6$

$1011_2 = 1 \cdot 2^3 + 0 \cdot 2^2 + 1 \cdot 2^1 + 1 \cdot 2^0 = 11$

Check: $5 + 6 = 11$

47.   a. $A9_{16}$

$\underline{+9B_{16}}$

144_{16}

b. $A9_{16} = 10 \cdot 16^1 + 9 = 169$

$9B_{16} = 9 \cdot 16 + 11 = 155$

$144_{16} = 1 \cdot 16^2 + 4 \cdot 16 + 4 = 324$

Check: $169 + 155 = 324$

49.   a. $22,33,44_{60}$

$\underline{+55,44,33_{60}}$

$1,18,18,17_{60}$

b. $22,33,44_{60} = 22 \cdot 60^2 + 33 \cdot 60 + 44 = 81,224$

$55,44,33_{60} = 55 \cdot 60^2 + 44 \cdot 60 + 33 = 200,673$

$1,18,18,17_{60} = 1 \cdot 60^3 + 18 \cdot 60^2 + 18 \cdot 60 + 17 = 281,897$

Check: $81,224 + 200,673 = 281,897$

51.   a. $\overset{1}{\cancel{2}}\,\overset{\overset{7}{\cancel{1}}}{\cancel{0}}\,\overset{1}{1}_8$

$\underline{-1\ 1\ 2_8}$

$6\ 7_8$

b. $201_8 = 2 \cdot 8^2 + 0 \cdot 8^1 + 1 \cdot 8^0 = 129$

$112_8 = 1 \cdot 8^2 + 1 \cdot 8^1 + 2 \cdot 8^0 = 74$

$67_8 = 6 \cdot 8^1 + 7 \cdot 8^0 = 55$

Check: $129 - 74 = 55$

53.   a. $1\overset{\overset{0}{\cancel{1}}}{}01_2$

$\underline{-1010_2}$

0011_2 or 11_2

b. $1101_2 = 2^3 + 2^2 + 1 = 13$

$1010_2 = 2^3 + 2 = 10$

$11_2 = 2 + 1 = 3$

Check: $13 - 10 = 3$

55. a. $\overset{9\quad\overset{F}{\cancel{1}}\ 1}{\cancel{A}\ \cancel{0}\ 8_{16}}$

$-\quad 7B_{16}$

$\overline{\qquad 98D_{16}}$

b. $A08_{16} = 10\cdot16^2 + 0\cdot16^1 + 8\cdot16^0 = 2568$

$7B_{16} = 7\cdot16^1 + 11\cdot16^0 = 123$

$98D_{16} = 9\cdot16^2 + 8\cdot16^1 + 13\cdot16^0 = 2445$

Check: $2568 - 123 = 2445$

57. a. $\overset{21\quad 1}{\cancel{22},33,44_{60}}$

$-\ 10,44,33_{60}$

$\overline{\quad 11,49,11_{60}}$

b. $22,33,44_{60} = 22\cdot60^2 + 33\cdot60 + 44 = 81,224$

$10,44,33_{60} = 10\cdot60^2 + 44\cdot60 + 33 = 38,673$

$11,49,11_{60} = 11\cdot60^2 + 49\cdot60 + 11 = 42,551$

Check: $81,224 - 38,671 = 42,551$

59. a. $\overset{1}{2}2_8$

$\times 72_8$

$\overline{\qquad 44}$

$\underline{176\qquad}$

2024_8

b. $22_8 = 2\cdot8 + 2 = 18$

$72_8 = 7\cdot8 + 2 = 58$

$2024_8 = 2\cdot8^3 + 2\cdot8 + 4 = 1044$

Check: $18\times58 = 1044$

$2_8 \cdot 2_8 = 4_{10} = 0\cdot8^1 + 4\cdot8^0 = 4_8$

$2_8 \cdot 2_8 = 4_{10} = 0\cdot8^1 + 4\cdot8^0 = 4_8$

$2_8 \cdot 7_8 = 14_{10} = 1\cdot8^1 + 6\cdot8^0 = 16_8$

$\left(2_8 \cdot 7_8\right) + 1_8 = 15_{10} = 1\cdot8^1 + 7\cdot8^0 = 17_8$

61. a. 101_2

$\times\ 110_2$

$\overline{\qquad 000}$

$101\quad$

$\underline{101\qquad}$

$\overline{11110_2}$

b. $101_2 = 2^2 + 1 = 5$

$110_2 = 2^2 + 2 = 6$

$11110_2 = 2^4 + 2^3 + 2^2 + 2 = 30$

Check: $5\cdot6 = 30$

63.   a. B3_{16}

       $\times\ 92_{16}$

       166

       $\underline{64B}$

       6616_{16}

     $3_{16}\cdot 2_{16}=6_{10}=0\cdot16^1+6\cdot16^0=6_{16}$

     $B_{16}\cdot 2_{16}=22_{10}=1\cdot16^1+6\cdot16^0=16_{16}$

     $3_{16}\cdot 9_{16}=27_{10}=1\cdot16^1+B\cdot16^0=1B_{16}$

     $\left(B_{16}\cdot 9_{16}\right)+1_{16}=100_{10}=6\cdot16^1+4\cdot16^0=64_{16}$

   b. $B3_{16}=11\cdot16+3=179$

     $92_{16}=9\cdot16+2=146$

     $6616_{16}=6\cdot16^3+6\cdot16^2+1\cdot16+6=26,134$

     Check: $179\cdot146=26,134$

65.   a. 34_5

       $\times\ 41_5$

       34

       $\underline{301}$

       3044_5

     $4_5\cdot 1_5=4_{10}=0\cdot5^1+4\cdot5^0=4_5$

     $3_5\cdot 1_5=3_{10}=0\cdot5^1+3\cdot5^0=3_5$

     $4_5\cdot 4_5=16_{10}=3\cdot5^1+1\cdot5^0=31_5$

     $\left(3_5\cdot 4_5\right)+3_5=15_{10}=3\cdot5^1+0\cdot5^0=30_5$

   b. $34_5=3\cdot5+4=19$

     $41_5=4\cdot5+1=21$

     $3044_5=3\cdot5^3+4\cdot5+4=399$

     Check: $19\cdot21=399$

67. a.

$$12_8 \overline{)\,301_8} \quad \frac{23_8 \text{ Rem } 3_8}{}$$

$$\begin{array}{r} -24 \\ \hline 41 \\ -36 \\ \hline 3_8 \end{array}$$

b.

$$\begin{array}{r} 12_8 \\ \times\, 23_8 \\ \hline 36 \\ 24 \\ \hline 276_8 \end{array}$$

Check: $276_8 + 3_8 = 301_8$

71. a.

$$1B_{16} \overline{)\,A08_{16}} \quad 5F_{16} \quad 5F_{16} \text{ R } 3_{16}$$

$$\begin{array}{r} -87 \\ \hline 198 \\ 195 \\ \hline 3_{16} \end{array}$$

b.

$$\begin{array}{r} 1B_{16} \\ \times\; 5F_{16} \\ \hline 195 \\ 87 \\ \hline A05_{16} \end{array}$$

Check: $A05_{16} + 3_{16} = A08_{16}$

69. a.

$$101_2 \overline{)\,1101_2} \quad 10_2 \quad 10_2 \text{ R } 11_2$$

$$\begin{array}{r} -101 \\ \hline 0011_2 \end{array}$$

b.

$$\begin{array}{r} 101_2 \\ \times\, 10_2 \\ \hline 1010_2 \end{array}$$

Check: $1010_2 + 11_2 = 1101_2$

73. a.

$$10_2 \overline{)\,110110_2} \quad 11011_2$$

$$\begin{array}{r} -10 \\ \hline 10 \\ -10 \\ \hline 11 \\ -10 \\ \hline 10 \\ -10 \\ \hline 0_2 \end{array}$$

b.

$$\begin{array}{r} 11011_2 \\ \times\; 10_2 \\ \hline 110110_2 \end{array}$$

75. a. $33 = 3 \cdot 11$ b. composite

77. a. 41   b. prime

79. Since $\sqrt{59} = 7.6\ldots$, eliminate multiples of 2, then 3, then 4, etc., through 7.

50 51 52 53 54
55 56 57 58 59

The prime numbers are 53 and 59.

81. $2^n + 1 = 2^2 + 1 = 4 + 1 = 5$; prime $2^n + 1 = 2^3 + 1 = 8 + 1 = 9$; composite, $9 = 3 \cdot 3$

83. $2^n - 1 = 2^3 - 1 = 8 - 1 = 7$; prime because $7 = 1 \cdot 7$.

85.   Note: $\sqrt{70} = 8.3\ldots$

$\frac{70}{2} = 35 \Rightarrow 2$ and 35 are factors.

$\frac{70}{3} = 23.3\ldots$

$\frac{70}{4} = 17.5$

$\frac{70}{5} = 14 \Rightarrow 5$ and 14 are factors.

$\frac{70}{7} = 10 \Rightarrow 7$ and 10 are factors.

$\frac{70}{8} = 8.75$

The proper factors are 1, 2, 5, 7, 10, 14, 35.

87.   Note: $\sqrt{20} = 4.4\ldots$

$\frac{20}{2} = 10 \Rightarrow 2$ and 10 are factors.

$\frac{20}{3} = 6.6\ldots$

$\frac{20}{4} = 5 \Rightarrow 4$ and 5 are factors.

Since $1 + 2 + 4 + 5 + 10 = 22 > 20$, it is abundant.

The proper factors are 1, 2, 4, 5, 10.

89.   Note: $\sqrt{49} = 7$

$\frac{49}{2} = 24.5$

$\frac{49}{3} = 16.3\ldots$

$\frac{49}{5} = 9.8$

The proper factors are 1 and 7. Since $1 + 7 = 8 < 49$, it is deficient.

91.   a. Since $2^{17} - 1 = 2^n - 1$, $n = 17$.   b. 8.59×10^9

$\left(2^{n-1}\right)\left(2^n - 1\right) = \left(2^{17-1}\right)\left(2^{17} - 1\right)$

93.   1, 1, 2, 3, 5, 8, 13, 21, 34, 55, 89, 144, 233, 377, 610, 987, 1597, 2584, 4181, 6765

95.   1 parent, 2 grandparents, 3 great-grandparents, and 5 great-great grandparents

97.   The pattern is 1, 1, 2, 3, 5, …; After 4 years, there are 5 total, and $5 - 3 = 2$ newly born.

99.   a. Let $n = 25$.   b. $75{,}025 + 121{,}393 = 196{,}418$

$$\frac{1}{\sqrt{5}}\left[\left(\frac{1+\sqrt{5}}{2}\right)^{25} - \left(\frac{1-\sqrt{5}}{2}\right)^{25}\right] = 75{,}025$$

Let $n = 26$.

$$\frac{1}{\sqrt{5}}\left[\left(\frac{1+\sqrt{5}}{2}\right)^{26} - \left(\frac{1-\sqrt{5}}{2}\right)^{26}\right] = 121{,}393$$

101.   The rectangle's length is about 1.6 times its width.

103.   Answers may vary.

105.   Answers will vary.

8.1 Perimeter and Area

1. $A = \frac{1}{2}(9.2)(3.5) = 16.1$ square centimeters

3. $A = \frac{1}{2}(5)(5) = 12.5$ square inches

5. $A = (6.2)(3.5) = 21.7$ square feet

7. $A = \frac{1}{2}(13+19)(11) = 176$ square meters

9. a. $r = \frac{6.5}{2} = 3.25$ inches

   b. $C = 2\pi(3.25) = 20.4$ inches

   $A = \pi(3.25)^2 = 33.2$ square inches

11. Let $x = $ length of third side.

    Using the Pythagorean Theorem, $x^2 + 5^2 = 13^2$

    Solving, $x = 12$ meters

    a. $A = \frac{1}{2}(12)(5) = 30$ square meters

    b. $P = 5 + 13 + 12 = 30$ meters

13. a. Use Heron's Formula:

    $s = \frac{1}{2}(8+8+8)$

    $\quad = 12$

    $A = \sqrt{12(12-8)(12-8)(12-8)}$

    $\quad = \sqrt{768}$

    $\quad = 27.7\,\text{m}^2$

    b. $P = 8 + 8 + 8 = 24$ meters

15. Dividing the shape into two rectangles.

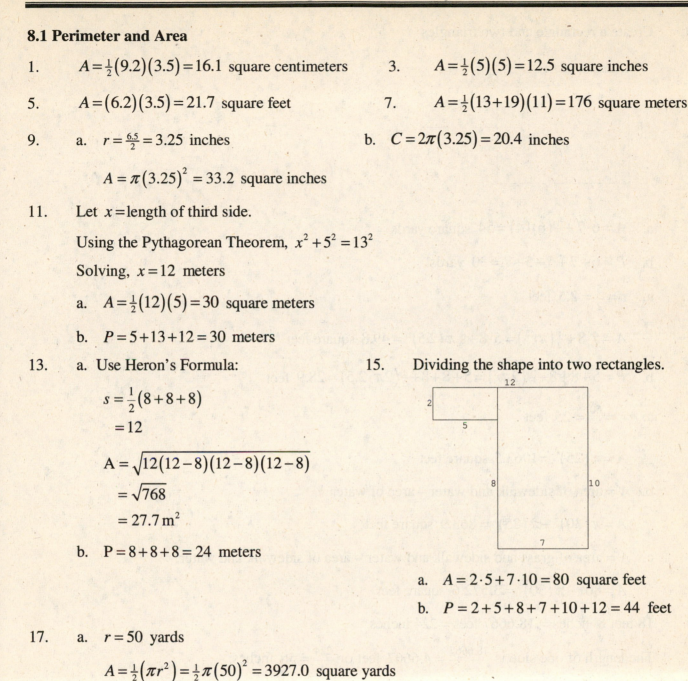

    a. $A = 2 \cdot 5 + 7 \cdot 10 = 80$ square feet

    b. $P = 2 + 5 + 8 + 7 + 10 + 12 = 44$ feet

17. a. $r = 50$ yards

    $A = \frac{1}{2}(\pi r^2) = \frac{1}{2}\pi(50)^2 = 3927.0$ square yards

    b. $P = 100 + \frac{1}{2}(2\pi r) = 100 + \frac{1}{2}(2\pi \cdot 50) = 257.1$ yards

19.   Create a rectangle and two triangles

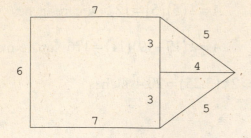

    a. $A = 6 \cdot 7 + \frac{1}{2}(6)(4) = 54$ square yards

    b. $P = 6 + 7 + 5 + 5 + 7 = 30$ yards

21.   a. $r = \frac{5}{2} = 2.5$ feet

        $A = 5 \cdot 8 + \frac{1}{2}(\pi r^2) = 5 \cdot 8 + \frac{1}{2}\pi(2.5)^2 = 49.8$ square feet

    b. $P = 5 + 8 + 8 + \frac{1}{2}(2\pi r) = 5 + 8 + 8 + \frac{1}{2}(2\pi \cdot 2.5) = 28.9$ feet

23.   a. $r = \frac{50}{2} = 25$ feet

        $A = \pi(25)^2 = 1963.5$ square feet

    b. $A =$ area of sidewalk and water $-$ area of water

        $A = \pi(30)^2 - \pi(25)^2 = 863.9$ square feet

    c. $A =$ area of grass and sidewalk and water $-$ area of sidewalk and water

        $A = 80^2 - \pi(30)^2 = 3{,}572.6$ square feet

25.   18 feet 8 inches = 18.6667 feet = 224 inches

The length of one side is $\frac{18.6667}{4} = 4.6667$ feet or $\frac{224}{4} = 56$ inches.

    a. $A = s^2 = (56)^2 = 3{,}136$ square inches

    b. $A = s^2 = (4.6667)^2 = 21.8$ square feet

27. $2\pi r = 10$

Solving, $r = 1.59$ feet

$A = \pi(1.59)^2 = 8.0$ square feet

29. $x^2 = (0.75)^2 + (1.5)^2$

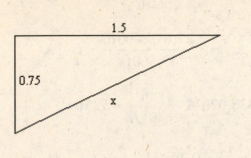

31. $x^2 + 6^2 = 10^2$

Solving, $x = 8$ feet

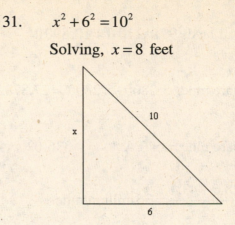

33. Together, the two semicircles have the same area as one circle and the same perimeter as one circle.

    a. $r = 20$ yards

       $A = 40 \cdot 100 + \pi(20)^2 = 5,256.6$ square yards

    b. $P = 100 + 100 + 2\pi(20) = 325.7$ yards

35. Using Heron's Formula, $s = \frac{1}{2}(100 + 140 + 180) = 210$

    $A = \sqrt{210(210-100)(210-140)(210-180)} = 6,964.9$ square feet

    You need 7 cans.

37. Small: $r = \frac{13}{2} = 6.5$ inches $\qquad A = \pi(6.5)^2 = 132.73$ square inches

    Price per square inch $= \frac{11.75}{132.73} = 0.0885$ about 8.9¢

    Large: $r = \frac{16}{2} = 8$ inches $\qquad A = \pi(8)^2 = 201.06$ square inches

    Price per square inch $= \frac{14.75}{201.06} = 0.0734$ about 7.3¢

    Super: $r = \frac{19}{2} = 9.5$ inches $\qquad A = \pi(9.5)^2 = 283.53$ square inches

    Price per square inch $= \frac{22.75}{283.53} = 0.0802$ about 8.0¢

    The large pizza is the best deal.

39. Area of land: $99\,\text{ft} \cdot 110\,\text{ft} = 10{,}890\,\text{ft}^2$ Acres: $\dfrac{10{,}890\,\text{ft}^2}{43{,}560\,\text{ft}^2} = 0.25\,\text{A}$

41. 1 square mile: $5{,}280\,\text{ft} \cdot 5{,}280\,\text{ft} = 27{,}878{,}400\,\text{ft}^2$ Acres: $\dfrac{27{,}878{,}400\,\text{ft}^2}{43{,}560\,\text{ft}^2} = 640\,\text{A}$

43. 1 square mile = 640 A $336{,}020\,\text{A} \cdot \dfrac{1\,\text{mi}^2}{640\,\text{A}} = 336{,}020\,\cancel{\text{A}} \cdot \dfrac{1\,\text{mi}^2}{640\,\cancel{\text{A}}} \approx 525\,\text{mi}^2$

45. a. $\dfrac{28.6 + 43.4}{2} = 36\,\text{million miles}$ b. $\dfrac{36}{93} = \dfrac{12}{31} = 0.39\,\text{AU}$

     c. The distance between Mercury and the earth is 0.39 times the distance between the earth and sun.

47. a. $\dfrac{4{,}080\,\text{trillion km}}{150\,\text{million km}} = 27{,}200{,}000\,\text{AU}$ b. $\dfrac{4{,}080\,\text{trillion km}}{9.46\,\text{trillion km}} = 431\,\text{ly}$

     c. $\dfrac{431\,\text{ly}}{3.26\,\text{ly}} = 132\,\text{pc}$

49. Area of shaded region = Area of polygon (square) RUST – Area of triangle TRY

     polygon (square) RUST: Let length = width = 4

        Area = length $\cdot$ width = $4 \cdot 4 = 16$

     Area of triangle TRY: Let base = 4, height $= \sqrt{4^2 - 2^2} = \sqrt{16 - 4} = \sqrt{12} = 2\sqrt{3}$

        Area $= \dfrac{1}{2} \cdot \text{base} \cdot \text{height} = \dfrac{1}{2} \cdot 4 \cdot 2\sqrt{3} = 4\sqrt{3}$

     Area of shaded region $= 16 - 4\sqrt{3}$, which is choice **c.**

51. Area of a square $= s^2$, find s if the area is 12.

     $s^2 = 12$
     $s = \sqrt{12} = 2\sqrt{3}$

     To find the radius of the circle, line KC, use the Pythagorean theorem.

     Let length of KE = length of EC = $s = 2\sqrt{3}$ (these are the legs of the right triangle). KC is the hypotenuse.

     $\text{KC}^2 = \left(2\sqrt{3}\right)^2 + \left(2\sqrt{3}\right)^2$
     $\text{KC}^2 = (4 \cdot 3) + (4 \cdot 3) = 12 + 12 = 24$

216

Area of a circle $= \pi \cdot \text{radius}^2 = \pi \cdot KC^2 = \pi \cdot 24 = 24\pi$

The correct choice is **c.**

53.   Use area of a circle $= \pi r^2$ and area $= 36\pi$ to find r.

$\pi r^2 = 36\pi$

$r^2 = 36$

$r = \pm\sqrt{36} = \pm 6$

The value -6 does not make sense in this problem, so use $r = 6$.

Circumference of a circle $= 2\pi r = 2\pi \cdot 6 = 12\pi$, which is choice **e.**

55.   A circle inscribed in a square, has a diameter equal to the length of the side of the square.

Area of a square $= s^2$, find s if the area is 8.

$s^2 = 8$

$s = \pm\sqrt{8} = \pm 2\sqrt{2}$

The value $-2\sqrt{2}$ does not make sense in this problem, so use $s = 2\sqrt{2}$.

Diameter $= 2\sqrt{2}$, so radius $= \dfrac{d}{2} = \dfrac{2\sqrt{2}}{2} = \sqrt{2}$

area of a circle $= \pi r^2 = \pi\left(\sqrt{2}\right)^2 = 2\pi$, which is choice **b.**

8.2 Volume and Surface Area

1.   a. $V = (5.2)(2.1)(3.5) = 38.22$ cubic meters

      b. $SA = 2(5.2)(2.1) + 2(5.2)(3.5) + 2(2.1)(3.5) = 72.94$ square meters

3.   a. $r = \dfrac{10}{2} = 5$ inches

      $A = \pi(5)^2 = 25\pi$ square inches

      $V = (25\pi)(10) = 785.40$ cubic inches

      b. $SA = 2\pi(5)^2 + 2\pi(5)(10) = 471.24$ square inches

5.   a. $r = \frac{1.75}{2} = 0.875$ inches

          $V = \frac{4}{3}\pi(0.875)^3$

             $= 2.81$ cubic inches

   b. $SA = 4\pi(0.875)^2$

             $= 9.62$ square inches

7. $r = \frac{4}{2} = 2$ feet

        $A_{\text{base}} = \pi(2)^2 = 4\pi$ square feet

        $V = \frac{1}{3}(4\pi)(4) = 16.76$ cubic feet

9. $A_{\text{base}} = (4)(4) = 16$ square feet

      $V = \frac{1}{3}(16)(4) = 21.33$ cubic feet

11. $V = A \cdot h$ where A is the area of the triangle.

     Using Heron's Formula, $s = \frac{1}{2}(3 + 4 + 5) = 6$

     $A = \sqrt{6(6-3)(6-4)(6-5)} = 6$ $V = (6)(6) = 36$ cubic feet

13. $V = A \cdot h$ where A is the area of the cross section (a washer).

     $A = \pi(2)^2 - \pi(1)^2 = 4\pi - \pi = 3\pi$ $V = (3\pi) \cdot 5 = 47.12$ cubic feet

15.   a. $h = 1.5,\ l = 16 - 2(1.5) = 13,\ w = 10 - 2(1.5) = 7$

          $V = (1.5)(13)(7) = 136.5$ cubic inches

   b. $SA = (13)(7) + 2(13)(1.5) + 2(7)(1.5) = 151$ square inches

17. $r = \frac{20}{2} = 10$ feet, $A = \pi(10)^2 = 100\pi$ square feet

     $V = A \cdot h + \frac{1}{2}\left(\frac{4}{3}\pi r^3\right) = (100\pi)(50) + \frac{1}{2} \cdot \frac{4}{3}\pi(10)^3 = 17{,}802.36$ cubic feet

19.   a.   Hardball: $2\pi r = 9$

          Solving, $r = 1.432394488$ inches

            $V = \frac{4}{3}\pi(1.432394488)^3 = 12.31$ cubic inches

      Softball: $2\pi r = 12$

          Solving, $r = 1.909859317$ inches

            $V = \frac{4}{3}\pi(1.909859317)^3 = 29.18$ cubic inches

     Volume of softball is 137% more than that of hardball.

b. Hardball: $SA = 4\pi(1.432394488)^2 = 25.78$ square inches

   Softball: $SA = 4\pi(1.909859317)^2 = 45.83$ square inches

   Surface area of softball is 78% more than that of hardball.

21.   Earth: $r = \dfrac{7920}{2} = 3,960$ miles

   $$V = \tfrac{4}{3}\pi(3960)^3 = 260,120,252,600 \text{ cubic miles}$$

   Moon: $r = \dfrac{2,160}{2} = 1,080$ miles

   $$V = \tfrac{4}{3}\pi(1080)^3 = 5,276,669,286 \text{ cubic miles}$$

   $$\dfrac{260,120,252,600}{5,276,669,286} = 49.30 \text{ about 49 moons}$$

23.   Earth: $r = \dfrac{7920}{2} = 3,960$ miles

   $$V = \tfrac{4}{3}\pi(3960)^3 = 260,120,252,600 \text{ cubic miles}$$

   Pluto: $r = \dfrac{1500}{2} = 750$ miles

   $$V = \tfrac{4}{3}\pi(750)^3 = 1,767,145,868 \text{ cubic miles}$$

   $$\dfrac{260,120,252,600}{1,767,145,868} = 147.20 \text{ about 147 Plutos}$$

25. $V = 4 \cdot 2 \cdot 10 = 80$ cubic feet

   $\dfrac{80}{128} = 0.625$ cords $(0.625)(190) = \$118.75$

   You did not get an honest deal. You should have paid \$118.75.

27. $r = \dfrac{2.5}{2} = 1.25$ inches $h = 3(2.5) = 7.5$ inches

   $A = \pi(1.25)^2 = 1.5625\pi$ square inches

   $V = (1.5625\pi)(7.5) = 36.82$ cubic inches

29.   36 feet = 12 yards   9 feet = 3 yards   6 inches = $\tfrac{1}{6}$ yard

   $V = (12)(3)(\tfrac{1}{6}) = 6$ cubic yards

   Cost $= (6)(4)(7.30) = \$175.20$

31. $A = (756)(756) = 571,536$ square feet

$$V = \frac{1}{3}(571,536)(480) = 91,445,760 \text{ cubic feet}$$

33.   Create two triangles to find the height.

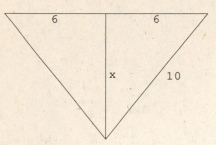

Using the Pythagorean Theorem, $x^2 + 6^2 = 10^2$

Solving, $x = 8$

$A = \pi(6^2) = 36\pi$ square feet

$V = \frac{1}{3}(36\pi)(8) = 301.59$ cubic feet

35. $1\,\text{ft} = 12\,\text{in.}$

$1\,\text{ft}^3 = (1\,\text{ft})(1\,\text{ft})(1\,\text{ft}) = (12\,\text{in.})(12\,\text{in.})(12\,\text{in.}) = 1,728\,\text{in}^3$, which is choice **e.**

37.   There are 12 edges on a cube and the sum of the length is 4, so $12x = 4$. The length of each edge

is $\frac{1}{3}$. Volume of a cube $= x^3 = \left(\frac{1}{3}\right)^3 = \frac{1}{27}$, which is choice **a.**

39.   Volume of a cube is $216\,\text{unit}^3$ and $V = x^3$.

$x^3 = 216$

$x = \sqrt[3]{216} = 6$

$SA = 6x^2 = 6 \cdot 6^2 = 6 \cdot 36 = 216\,\text{unit}^2$, which is choice **c.**

41.   Circumference is $C = 2\pi r$, so $r = \dfrac{C}{2\pi}$. Height of cylinder is 2 time circumference (2C).

$$V = \pi r^2 h = \pi \left(\frac{C}{2\pi}\right)^2 \cdot 2C = \pi \frac{C^2}{4\pi^2} \cdot 2C = \frac{C^3}{2\pi}, \text{ which is choice } \mathbf{b.}$$

8.3 Egyptian Geometry

1. $6^2 + 8^2 = 100 = 10^2$

This is a right triangle.

3. $3^2 + 5^2 = 34 \neq 6^2$

This is not a right triangle

.   a. $V = \frac{12}{13}\left(2^2 + 2 \cdot 15 + 15^2\right) = 1036$ cubic cubits or 1,554 khar

b. $A = (15)(15) = 225$ square cubits

$V = \frac{1}{3}(225)(18) = 1350$ cubic cubits or 2,025 khar

The regular pyramid holds more.

7.   36; 18; 9; 36; 18; 9; 63; 9; 3; 63; 189;189

9. $A_{\text{circle}} \approx (24)^2 - 2(8)^2 = 448 \approx 441 = A_{\text{square of side 21}}$

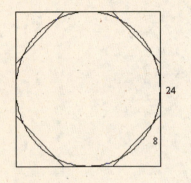

11. $r = \frac{24}{2} = 12$

$\pi(12)^2 = 441$

$\pi = \frac{49}{16} = 3.0625$

13.   a. $A = \frac{256}{81} \cdot (3)^2 = 28.4444$ square palms

b. $A = \pi(3)^2 = 28.2743$ square palms

c. $\frac{28.4444 - 28.2743}{28.2743} = 0.00623 \approx 0.6\%$

15.   a. $V = \frac{4}{3}\left(\frac{256}{81}\right)(3)^3 = 113.7778$ cubic palms

b. $V = \frac{4}{3}\pi(3)^3 = 113.0973$ cubic palms

c. $\frac{113.7778 - 113.0973}{113.0973} = 0.00602 \approx 0.6\%$

17.   2 cubits = 14 palms = 56 fingers, 5 palms = 20 fingers

First side = $56 + 20 + 3 = 79$ fingers

3 cubits = 21 palms = 84 fingers, 3 palms = 12 fingers

Second side = $84 + 12 + 2 = 98$ fingers

4 cubits = 28 palms = 112 fingers, 4 palms = 16 fingers

Third side = $112 + 16 + 3 = 131$ fingers

$P = 79 + 98 + 131 = 308$ fingers = 11 cubits

19.   1 cubic cubit $= (20.67)^3$ cubic inches $= 8,831.23$ cubic inches

1 khar $= \frac{2}{3}$ cubic cubit $= 5,887.49$ cubic inches

Since 1 cubic foot $= (12)^3$ cubic inches $= 1,728$ cubic inches

1 khar $= \frac{5887.49}{1728} = 3.41$ cubic feet

1 khar is larger.

21.   Using Heron's Formula: $s = \frac{1}{2}(150 + 200 + 250) = 300$

$A = \sqrt{300(300 - 150)(300 - 200)(300 - 250)} = 15,000$ square cubits

$A = \frac{15,000}{10,000} = 1.5$ setats

23. $s^2 = 4(10,000) = 40,000$ square cubits

$s = 200$ cubits

a. $\frac{200}{100} = 2$ khet   b.   200 cubits

25.   a. $A_{\text{circle of diameter 9}} = A_{\text{square of side 8}}$

$9^2 - 2(3)^2 = 63 \Rightarrow 64$

$V = (64)(10) = 640$ cubic cubits

$640 \div \frac{2}{3} = 960$ khar

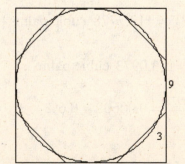

b. $r = \frac{9}{2} = 4.5$

$V = \pi(4.5)^2(10) = 636.17$ cubic cubits

$636.17 \div \frac{2}{3} = 954.2588$ khar

c. $\frac{960 - 954.2588}{954.2588} = 0.00602 \approx 0.6\%$

8.4 The Greeks

1. $\frac{x}{75} = \frac{6}{90}$ $\frac{y}{4} = \frac{90}{6}$ 3. $\frac{x}{3.8} = \frac{54.4}{3.2}$ $\frac{y}{42.5} = \frac{3.2}{54.4}$

    $90x = 450$ $6y = 360$ $3.2x = 206.72$ $54.4y = 136$

    $x = 5$ $y = 60$ $x = 64.6$ $y = 2.5$

5. $\frac{x}{6} = \frac{21}{3.5}$

    $3.5x = 126$

    $x = 36$ feet

7. The area of the big square is $(a+b)^2 = a^2 + 2ab + b^2$.

The area of the big square is also equal to the area of the smaller square plus the area of the four triangles.

Two small triangles could be put together to make a square with area ab.

The four triangles have total area $2(ab)$.

Thus, the area of the big square is also $c^2 + 2ab$.

Setting the two expressions equal: $a^2 + 2ab + b^2 = c^2 + 2ab$.

Simplifying: $a^2 + b^2 = c^2$

9. First find the length of the diagonal of the 3×4 rectangle, d.

$d^2 = 3^2 + 4^2 = 25$

$d = 5$

Use this to find the length of l:

$l^2 = 2^2 + 5^2 = 29$

$l = 5.3$ feet, (rounded down so the object will fit)

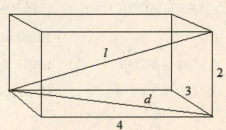

223

11.   The diameter of the circle is $2\left(2\tfrac{1}{4}\right) = 4.5$

$$l^2 = (4.5)^2 + 6^2 = 56.25$$

$$l = 7.5 \text{ inches}$$

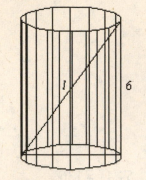

13.   1. $AD = CD$ — Given

       2. $AB = CB$ — Given

       3. $DB = DB$ — Anything equals itself.

       4. $\triangle DBA \cong \triangle DBC$ — SSS

       5. $\angle DBA = \angle DBC$ — Corresponding parts of congruent triangles are equal.

15.   1. $AD = BD$ — Given

       2. $\angle ADC = \angle BDC$ — Given

       3. $DC = DC$ — Anything is equal to itself.

       4. $\triangle ACD \cong \triangle BCD$ — SAS

       5. $AC = BC$ — Corresponding parts of congruent triangles are equal.

17.   1. $AE = CE$ — Given

       2. $AB = CB$ — Given

       3. $BE = BE$ — Anything is equal to itself.

       4. $\triangle ABE \cong \triangle CBE$ — SSS

       5. $\angle ABE = \angle CBE$ — Corresponding parts of congruent triangles are equal.

       6. $BD = BD$ — Anything is equal to itself.

       7. $\triangle ADB \cong \triangle CDB$ — SAS

       8. $\angle ADB = \angle CDB$ — Corresponding parts of congruent triangles are equal.

19.   a. $2\pi r \approx 8\sqrt{2-\sqrt{2}}\,r$

$\pi \approx 4\sqrt{2-\sqrt{2}} = 3.061467459$

b. $2\pi r \approx 8\cdot 2\left(\sqrt{2}-1\right)r$

$\pi \approx 8\left(\sqrt{2}-1\right) = 3.313708499$

21.   a. $2\pi r \approx 16\sqrt{2-\sqrt{2+\sqrt{2}}}\,r$

$\pi \approx 8\sqrt{2-\sqrt{2+\sqrt{2}}} = 3.121445152$

b. $2\pi r \approx 16\left(\dfrac{2\sqrt{2-\sqrt{2}}}{2+\sqrt{2+\sqrt{2}}}\right)r$

$\pi \approx 8\left(\dfrac{2\sqrt{2-\sqrt{2}}}{2+\sqrt{2+\sqrt{2}}}\right) = 3.182597878$

23.   Let r = radius of sphere. The volume of the sphere is $V = \frac{4}{3}\pi r^3$.

r is also the radius of the circle. The area of the circle is $A = \pi r^2$.

The height of the cylinder $= 2r$.

The volume of the cylinder is $V = \pi r^2\left(2r\right)$.

$$\frac{V_{\text{sphere}}}{V_{\text{cylinder}}} = \frac{\frac{4}{3}\pi^3}{\pi r^2\left(2r\right)} = \frac{\frac{4}{3}}{2} = \frac{2}{3}$$

8.5 Right Triangle Trigonometry

1. $\sin 30° = \dfrac{x}{6}$

$x = 6\cdot \sin 30° = 6\cdot \dfrac{1}{2} = 3$

$\cos 30° = \dfrac{y}{6}$

$y = 6\cdot \cos 30° = 6\cdot \dfrac{\sqrt{3}}{2} = 3\sqrt{3}$

$\cos\theta = \dfrac{3}{6} = \dfrac{1}{2}$

$\theta = 60°$

3. $\cos 60° = \dfrac{7}{x}$

$x = \dfrac{7}{\cos 60°} = \dfrac{7}{\frac{1}{2}} = 14$

$\sin 60° = \dfrac{y}{14}$

$y = 14\sin 60° = 14\cdot \dfrac{\sqrt{3}}{2} = 7\sqrt{3}$

$\sin\theta = \dfrac{7}{14} = \dfrac{1}{2}$

$\theta = 30°$

5. $\cos 45° = \dfrac{3}{y}$

$y = \dfrac{3}{\cos 45°} = \dfrac{3}{\frac{\sqrt{2}}{2}} = \dfrac{6}{\sqrt{2}} = \dfrac{6\sqrt{2}}{2} = 3\sqrt{2}$

$\sin 45° = \dfrac{x}{3\sqrt{2}}$

$x = 3\sqrt{2}\sin 45° = 3\sqrt{2} \cdot \dfrac{\sqrt{2}}{2} = 3$

$\tan\theta = \dfrac{3}{3} = 1$

$\theta = 45°$

7. $\sin 45° = \dfrac{x}{7}$

$x = 7\sin 45° = 7 \cdot \dfrac{\sqrt{2}}{2} = \dfrac{7\sqrt{2}}{2}$

$\cos 45° = \dfrac{y}{7}$

$y = 7\cos 45° = 7 \cdot \dfrac{\sqrt{2}}{2} = \dfrac{7\sqrt{2}}{2}$

$\sin\theta = \dfrac{\frac{7\sqrt{2}}{2}}{7} = \dfrac{\sqrt{2}}{2}$

$\theta = 45°$

9. $B = 180° - 37° - 90° = 53°$

$\sin 37° = \dfrac{12}{c}$

$c = \dfrac{12}{\sin 37°} = 19.9$

$\tan 53° = \dfrac{b}{12}$

$b = 12\tan 53° = 15.9$

11. $B = 180° - 90° - 54.3° = 35.7°$

$\cos 54.3° = \dfrac{5.6}{c}$

$c = \dfrac{5.6}{\cos 54.3°} = 9.6$

$\tan 54.3° = \dfrac{a}{5.6}$

$a = 5.6\tan 54.3° = 7.8$

13. $A = 180° - 90° - 49.9° = 40.1°$

$\cos 40.1° = \dfrac{b}{0.92}$

$b = 0.92\cos 40.1° = 0.70$

$\sin 40.1° = \dfrac{a}{0.92}$

$a = 0.92\sin 40.1° = 0.59$

15. $A = 180° - 90° - 9.15° = 80.85°$

$\sin 80.85° = \dfrac{1,546}{c}$

$c = \dfrac{1,546}{\sin 80.85°} = 1,566$

$\tan 9.15° = \dfrac{b}{1,546}$

$b = 1,546\tan 9.15° = 249$

226

17. $B = 180° - 90° - 53.125° = 36.875°$

$\sin 53.125° = \dfrac{a}{54.40}$

$a = 54.40 \sin 53.125° = 43.52$

$\cos 53.125° = \dfrac{b}{54.40}$

$b = 54.40 \cos 53.125° = 32.64$

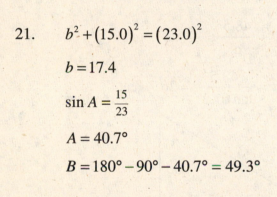

19. $A = 180° - 90° - 5° = 85.00°$

$\cos 85° = \dfrac{1.002}{c}$

$c = \dfrac{1.002}{\cos 85°} = 11.497$

$\tan 85° = \dfrac{a}{1.002}$

$a = 1.002 \tan 85° = 11.453$

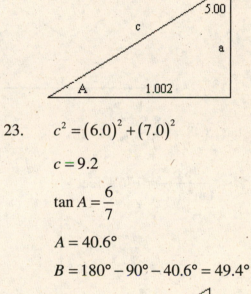

21. $b^2 + (15.0)^2 = (23.0)^2$

$b = 17.4$

$\sin A = \dfrac{15}{23}$

$A = 40.7°$

$B = 180° - 90° - 40.7° = 49.3°$

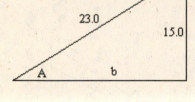

23. $c^2 = (6.0)^2 + (7.0)^2$

$c = 9.2$

$\tan A = \dfrac{6}{7}$

$A = 40.6°$

$B = 180° - 90° - 40.6° = 49.4°$

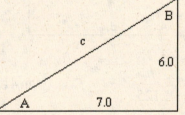

25. $a^2 + (0.123)^2 = (0.456)^2$

$a = 0.439$

$\cos A = \dfrac{0.123}{0.456}$

$A = 74.4°$

$B = 180° - 90° - 74.4° = 15.6°$

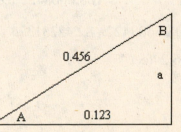

27.   a. $\tan 43.9° = \dfrac{h}{66.5}$ b. $l^2 = \left(66.5\right)^2 + \left(64.0\right)^2$

          $h = 66.5\tan 43.9° = 64.0$ feet $l = 92.3$ feet

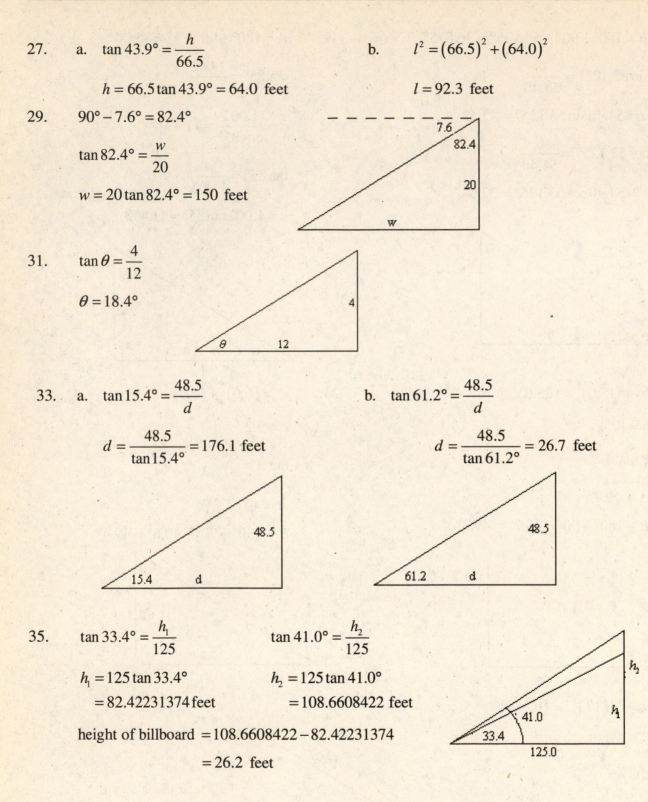

29. $90° - 7.6° = 82.4°$

      $\tan 82.4° = \dfrac{w}{20}$

      $w = 20\tan 82.4° = 150$ feet

31. $\tan\theta = \dfrac{4}{12}$

      $\theta = 18.4°$

33.   a. $\tan 15.4° = \dfrac{48.5}{d}$ b. $\tan 61.2° = \dfrac{48.5}{d}$

        $d = \dfrac{48.5}{\tan 15.4°} = 176.1$ feet $d = \dfrac{48.5}{\tan 61.2°} = 26.7$ feet

35. $\tan 33.4° = \dfrac{h_1}{125}$ $\tan 41.0° = \dfrac{h_2}{125}$

      $h_1 = 125\tan 33.4°$ $h_2 = 125\tan 41.0°$

         $= 82.42231374$ feet $= 108.6608422$ feet

    height of billboard $= 108.6608422 - 82.42231374$

                    $= 26.2$ feet

37.　　$\tan 61.0° = \dfrac{h}{x}$

$x = \dfrac{h}{\tan 61.0°}$

$\quad = \dfrac{h}{1.8084047755}$

$\tan 49.5° = \dfrac{h}{x + 50}$

$x + 50 = \dfrac{h}{\tan 49.5°} = \dfrac{h}{1.170849566}$

$x = \dfrac{h - 58.54247831}{1.170849566}$

So, $\dfrac{h}{1.804047755} = \dfrac{h - 58.54247831}{1.170849566}$

$1.170849566h = 1.804047755h - 105.6134266$

$0.633198189h = 105.6134266$

$h = 167 \ \text{feet}$

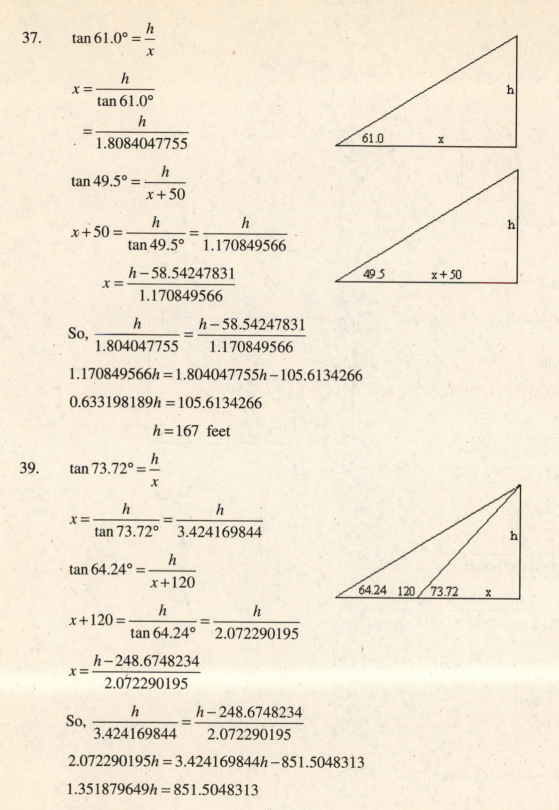

39.　　$\tan 73.72° = \dfrac{h}{x}$

$x = \dfrac{h}{\tan 73.72°} = \dfrac{h}{3.424169844}$

$\tan 64.24° = \dfrac{h}{x + 120}$

$x + 120 = \dfrac{h}{\tan 64.24°} = \dfrac{h}{2.072290195}$

$x = \dfrac{h - 248.6748234}{2.072290195}$

So, $\dfrac{h}{3.424169844} = \dfrac{h - 248.6748234}{2.072290195}$

$2.072290195h = 3.424169844h - 851.5048313$

$1.351879649h = 851.5048313$

$h = 630 \ \text{feet}$

41. $\tan 58.24° = \dfrac{h}{900}$

$h = 900 \tan 58.24° = 1,454 \text{ feet}$

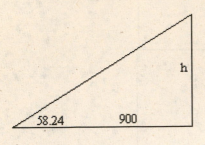

43. $\sin\left(12.5° + 28.3°\right) = \dfrac{h}{100}$

$h = 65.3420604$

$\cos\left(40.8°\right) = \dfrac{x}{100}$

$x = 100 \cos\left(40.8°\right) = 75.69950557$

$\tan 28.3° = \dfrac{h_1}{75.69950557}$

$h_1 = 40.7599842$

$h_2 = h - h_1 = 24.6 \text{ feet}$

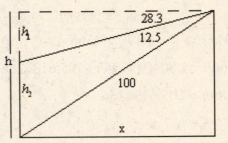

45.   Note: $1 \text{ pc} = \dfrac{1 \text{ AU}}{\tan\left(1 \text{ second}\right)}$

$\text{distance} = \dfrac{1 \text{ AU}}{\tan\left(2 \text{ seconds}\right)} \div \dfrac{1 \text{ AU}}{\tan\left(1 \text{ second}\right)}$

$= \dfrac{1 \text{ AU}}{\tan\left(\frac{2}{3600}\right)°} \cdot \dfrac{\tan\left(\frac{1}{3600}\right)°}{1 \text{ AU}}$

$= 0.5 \text{ pc}$

Alternative solution on the next page.

Alternative solution:

$$\tan\left(2 \text{ seconds}\right) = \frac{1 \text{ AU}}{x}$$

$$x = \frac{1 \text{ AU}}{\tan\left(\frac{2}{3600}\right)^\circ}$$

$$x = \frac{93,000,000 \text{ mi}}{\tan\left(\frac{2}{3600}\right)^\circ}$$

$$x = \frac{93}{\tan\left(\frac{2}{3600}\right)^\circ} \cdot \frac{1 \text{ ly}}{5,880,000}$$

$$x = 1.6311\ldots \text{ ly}$$

$$x = \frac{1.6311\ldots \text{ ly}}{3.26 \text{ ly}}$$

$$x = 0.5 \text{ pc}$$

47.   a.   From page 590, we know:

1 pc = 3.26(63,240 AU)

1 pc = 206,162.4 AU

0.9 pc = 0.9(206,162.4 AU)

= 185,546 AU

b. $\tan\left(\dfrac{x}{3600}\right)^\circ = \dfrac{1 \text{ AU}}{0.9 \text{ pc}}$

$$\tan\left(\frac{x}{3600}\right)^\circ = \frac{1 \text{ AU}}{185,546 \text{ AU}}$$

$$x = 3600 \cdot \tan^{-1}\left(\frac{1}{185,546}\right)$$

$$x = 1.11 \text{ seconds}$$

For 49, refer to the following figure.

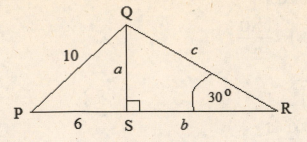

49.   The perimeter of $\Delta PQR = 10 + c + b + 6$. By finding the value of a, the relationship of the sides of a $30° - 60° - 90°$ triangle can be used to find b and c.

$$a^2 = 10^2 - 6^2$$
$$a^2 = 100 - 36$$
$$a^2 = 64$$
$$a = 8$$

Since a = 8, then $c = 2a = 2 \cdot 8 = 16$ and $b = x\sqrt{3} = 8\sqrt{3}$, $\Delta PQR = 10 + 16 + 8\sqrt{3} + 6 = 32 + 8\sqrt{3}$.

The correct answer is **e.**

51.   Use the following facts to solve for x: 1) the measure of a straight angle is $180°$ and 2) vertical angles are equal. The figure illustrates the values of the angles. The angles whose measure is $3y°$, $4y°$, and $2x°$ make a straight angle so $3y° + 4y° + 2x° = 180$, in addition the angles whose measure is $2x°, x°$, and $3y°$ make another straight angle, $2x° + x° + 3y° = 180$. Solve the system of equations, disregarding the degree symbol.

$$\begin{pmatrix} 3y + 4y + 2x = 180 \\ 2x + x + 3y = 180 \end{pmatrix} = \begin{pmatrix} 2x + 7y = 180 \\ 3x + 3y = 180 \end{pmatrix} = \begin{pmatrix} 2x + 7y = 180 \\ x + y = 60 \end{pmatrix}$$

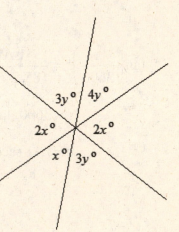

Solve the second equation for y.

$$x + y = 60$$
$$y = 60 - x$$

Substitute $60 - x$ for y in the first equation and solve for x.

$$2x + 7(60 - x) = 180$$
$$2x + 420 - 7x = 180$$
$$-5x = -240$$
$$x = 48$$

The correct answer is **d.**

8.6 Linear Perspective

1. a. one-point perspective
   b. above
   c. approximately $(17, 14)$
   d. $y = 14$

3. a. two-point perspective
   b. below
   c. approximately $(1, 2)$ and $(18, 2)$
   d. $y = 2$

5. Answers will vary. Possible drawing:

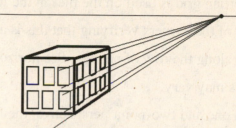

7. Anwers will vary.

9. Answers will vary.

11. Answers may vary.

   a. One-point perspective; the wall faces are parallel to the surface of the painting and there is one (main) vanishing point.

   b. The horizon line shares two vanishing points. The main vanishing point appears to be between the two central figures (Plato and Aristotle). Raphael might have chosen that point to put the focus on Plato and Aristotle. The second vanishing point is to the left, highlighting another discussion among scholars. The cabinet in front left has its own vanishing point. The outside arch has its own vanishing point below the main vanishing point. This makes the arch seem separate from the rest, in turn making the painting more 3-dimensional. See painting.

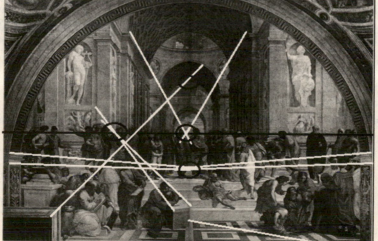

Raphael, *School of Athens*

233

c. Yes, in the front, a philosopher is leaning against a small table. It appears that two-point perspective was used.

d. Yes, see part b for description of vanishing points.

e. Yes, the tiles in the front.

f. Yes, in the foreground of the painting.

g. Albertian grid is used on the tiles in the foreground. An inserted diagonal line perfectly hits each of the corners verifying that this is an Albertian grid. See diagonal line on painting.

h. Yes, along the row of people. See horizon line on painting.

13. Answers may vary.

a. Both one and two-point perspective; the front column is parallel to the surface of the painting, whereas the wall faces are not all parallel to the surface of the painting, the building in the back right does not have a wall face parallel to the surface of the painting.

b. There appears to be three vanishing points at the bottom of the painting. Two are near the feet of St. James. The third is to the far left.

See painting. Mantegna might have chosen those points to put the viewer's focus on St. James, to convey the power of St. James' executioners.

c. Yes, the building in the background.

d. Yes, see part b for description of vanishing points.

e. Yes, the tiles in the ceiling.

f. No

g. Albertian grid is used on the tiles in the ceiling. See diagonal line on painting.

h. Yes, it is the floor.

Andrea Mantegna, *St. James on the Way to His Execution*

15. Answers may vary.

a. One-point perspective; many buildings, including the large building to the right, have walls parallel to the surface of the painting.

b. The vanishing point appears to be at the small grove of trees directly above the walking people. Pissaro might have chosen that point to convey the feeling of the distance that the travelers must walk or ride.

c. No   d. No, there is a central vanishing point. See painting.

e. No   f. No   g. No

h. Yes, horizon is along the tops of trees and buildings in the distance.

Pissaro, *Road to Louveciennes* 1872

17.   Answers may vary.

    a. One-point perspective; the wall faces are parallel to the surface of the painting.

    b. The vanishing point appears to be at knee level of the figures in front and the man sitting in the back approximately on the column. Della Francesca might have chosen that point to emphasize separation.

    c. No   d. No, there is a central vanishing point.

    e. Yes, the tiles in the ceiling and on the floor.   f. Yes, there are tiles on the floor.

    g. Albertian grid is used on the tiles in the ceiling. See diagonal line on painting.

    h. No

Piero della Francesca, *The Flagellation of Christ*

235

19. Answers may vary.

   a. Unclear as to one- or two-point perspective; the wall faces of the building in the middle are parallel to the surface of the painting, while the wall faces of the building on the right are not parallel to the surface of the painting.

   b. The vanishing point appears on the house. Wyeth might have chosen that point to convey the great distance that the woman was away from home/help.

   c. No   d. No, there is a central vanishing point. See painting.   e. No   f. No

   g. No   h. Yes, the horizon of the landscape.

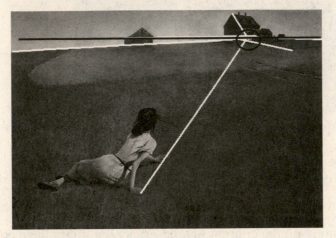

Andrew Wyeth, *Christina's World* 1948

21. Eakins drew center lines on the horizontal and vertical. Directly below that intersection, he placed the vanishing point. An Albertian grid was used to draw the boat. Eakins was so accurate that scholars have been able to determine the precise length of the boat and to pinpoint the exact time of day to 7:20 p.m.

   http://www.philamuseum.org/micro_sites/exhibitions/eakins/1872/main_frameset.html

23. Answers may vary.

   a. One-point perspective; the building at the end of the street is parallel to the surface of the painting.

   b. The vanishing point appears to be down the street on the window below the balcony. Sargent appears to be giving a sense of length to the street.

   c. No   d. No, there is a central vanishing point. See painting.

   e. Yes, on the street pavement and on the lower part of the wall on the left.

   f. Yes, on the street.   g. Albertian grid is used on the walls.

   h. No

236

John Singer Sargent, *A Street in Venice*

*25.   *The questions should be:*

   a. *What is three-point perspective?*

   b. *Under what general circumstances is three-point perspective appropriate?*

   c. *Why might Escher have used three-point perspective here?*

   d. *Where are the vanishing points?*

   e. *In this work, Escher uses perspective to trick the eye. What is the trick?*

   f. *Give a detailed description of how Escher accomplishes this trick.*

   Answers may vary.

   a. Three-point perspective has three vanishing points in which to portray three dimensions.

   b. When the object portrayed is viewed from an angle and from above or below.

   c. To make it easier for him to trick the eye into 'seeing' length, width, and height.

   d. The vanishing points are to the upper left, and upper right approximately on the same horizon,
      and the third vanishing point is below the bottom of the painting. See the painting.

*The first printing of the text did not include the correct questions for exercise 25.

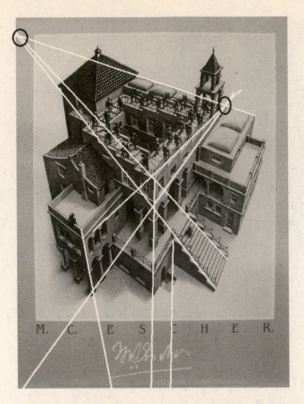

e. Look at the people going up the steps on top of the building. They keep going up and up but still they connect and form a loop, not a spiral.

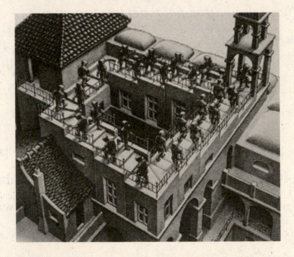

f. Features such as the varying heights of the walls, the use of shadowing on the steps, and the different sizes of the people, and the feet drawn to show walking up and stepping down are some of the techniques utilized to give the sense that the stairs are always going up.

8.7 Conic Sections and Analytic Geometry

1. $(x-0)^2 + (y-0)^2 = 1$

   Center $(0,0)$

   $r = 1$

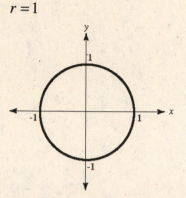

3. $(x^2 - 4x + 4) + (y-0)^2 = 5 + 4$

   $(x-2)^2 + (y-0)^2 = 9$

   Center $(2,0)$

   $r = 3$

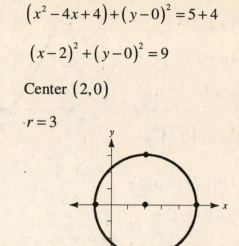

5. $(x^2 - 10x + 25) + (y^2 + 4y + 4) = -13 + 25 + 4$

   $(x-5)^2 + (y+2)^2 = 16$

   Center $(5,-2)$

   $r = 4$

7. $(x^2 - 10x + 25) + (y^2 - 10y + 25) = -25 + 25 + 25$

   $(x-5)^2 + (y-5)^2 = 25$

   Center $(5,5)$

   $r = 5$

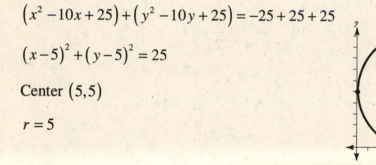

9. $4p = 1$

   $p = \frac{1}{4}$

   Focus $\left(0, \frac{1}{4}\right)$

239

11. $y = \frac{1}{2}x^2$

   $2y = x^2$

   $4p = 2$

   $p = \frac{1}{2}$

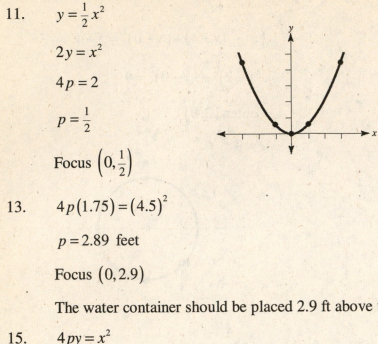

   Focus $\left(0, \frac{1}{2}\right)$

13. $4p(1.75) = (4.5)^2$

   $p = 2.89$ feet

   Focus $(0, 2.9)$

   The water container should be placed 2.9 ft above the bottom of the dish.

15. $4py = x^2$

   $4p(1) = (1.5)^2$

   $p = 0.5625$ inch

   Focus $(0, 0.5625)$

   The light bulb should be located $\frac{9}{16}$ inch above the bottom of the reflector.

17. $\frac{x^2}{9} + \frac{y^2}{4} = 1$ 19. $\frac{x^2}{4} + \frac{y^2}{25} = 1$

   $a = 3, b = 2$ $a = 2, b = 5$

   $c^2 = 3^2 - 2^2 = 5$ $c^2 = 5^2 - 2^2 = 21$

   $c = \pm\sqrt{5}$ $c = \pm\sqrt{21}$

   Foci $\left(-\sqrt{5}, 0\right)$ and $\left(\sqrt{5}, 0\right)$ Foci $\left(0, -\sqrt{21}\right)$ and $\left(0, \sqrt{21}\right)$

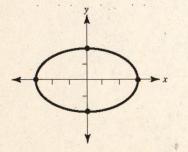

240

21. $\dfrac{x^2}{9} + \dfrac{y^2}{16} = 1$

    $a = 3, b = 4$

    $c^2 = 4^2 - 3^2 = 7$

    $c = \pm\sqrt{7}$

    Foci $\left(0, -\sqrt{7}\right)$ and $\left(0, \sqrt{7}\right)$

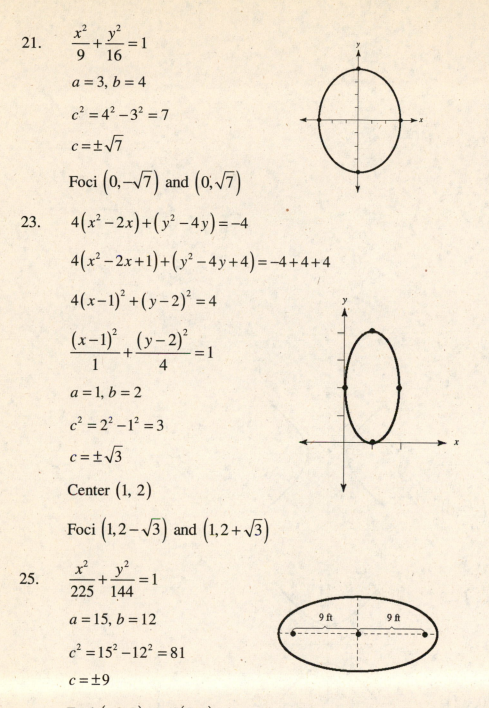

23. $4\left(x^2 - 2x\right) + \left(y^2 - 4y\right) = -4$

    $4\left(x^2 - 2x + 1\right) + \left(y^2 - 4y + 4\right) = -4 + 4 + 4$

    $4\left(x - 1\right)^2 + \left(y - 2\right)^2 = 4$

    $\dfrac{\left(x - 1\right)^2}{1} + \dfrac{\left(y - 2\right)^2}{4} = 1$

    $a = 1, b = 2$

    $c^2 = 2^2 - 1^2 = 3$

    $c = \pm\sqrt{3}$

    Center $(1, 2)$

    Foci $\left(1, 2 - \sqrt{3}\right)$ and $\left(1, 2 + \sqrt{3}\right)$

25. $\dfrac{x^2}{225} + \dfrac{y^2}{144} = 1$

    $a = 15, b = 12$

    $c^2 = 15^2 - 12^2 = 81$

    $c = \pm 9$

    Foci $(-9, 0)$ and $(9, 0)$

    They should stand at the foci, which are located 9 ft from the center, in the long direction.

241

27. $a + c = 94.51$

   $\underline{a - c = 91.40}$

   $2a = 185.91$

   $a = 92.955$

Substituting, $c = 1.555$

$c^2 = a^2 - b^2$

$(1.555)^2 = (92.955)^2 - b^2$

$b^2 = 8638.214$

$b = 92.942$

$$\frac{x^2}{92{,}955^2} + \frac{y^2}{92{,}942^2} = 1$$

29.   a. $\dfrac{x^2}{1} - \dfrac{y^2}{1} = 1$

   $a = 1,\ b = 1$

   $c^2 = 1^2 + 1^2 = 2$

   $c = \pm\sqrt{2}$

   Foci $\left(-\sqrt{2}, 0\right)$ and $\left(\sqrt{2}, 0\right)$

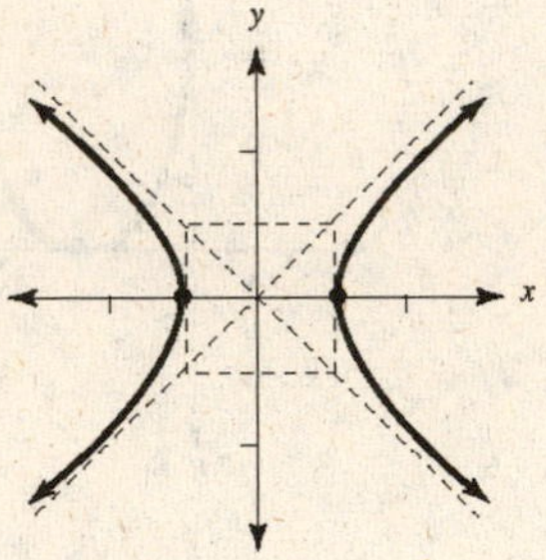

   b. $\dfrac{y^2}{1} - \dfrac{x^2}{1} = 1$

   $a = 1,\ b = 1$

   $c^2 = 1^2 + 1^2 = 2$

   $c = \pm\sqrt{2}$

   Foci $\left(0, -\sqrt{2}\right)$ and $\left(0, \sqrt{2}\right)$

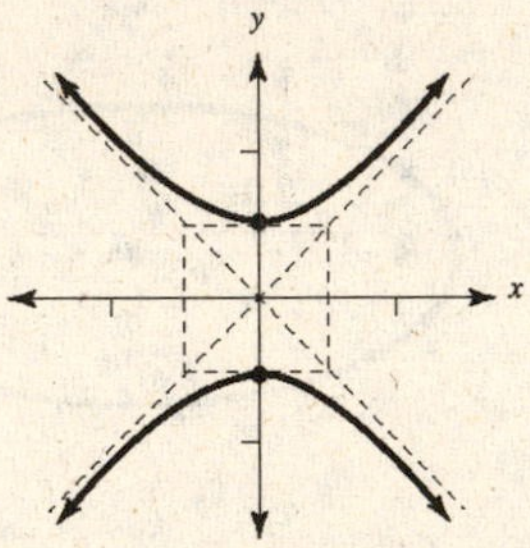

31. $\dfrac{x^2}{9} - \dfrac{y^2}{4} = 1$

    $a = 3,\ b = 2$

    $c^2 = 3^2 + 2^2 = 13$

    $c = \pm\sqrt{13}$

    Foci $\left(-\sqrt{13}, 0\right)$ and $\left(\sqrt{13}, 0\right)$

33. $\dfrac{y^2}{4} - \dfrac{x^2}{25} = 1$

    $b = 2,\ a = 5$

    $c^2 = 2^2 + 5^2 = 29$

    $c = \pm\sqrt{29}$

    Foci $\left(0, -\sqrt{29}\right)$ and $\left(0, \sqrt{29}\right)$

35. $\left(x^2 - 6x\right) - 4y^2 = -5$

    $\left(x^2 - 6x + 9\right) - 4y^2 = -5 + 9$

    $\left(x - 3\right)^2 - 4y^2 = 4$

    $\dfrac{\left(x - 3\right)^2}{4} - \dfrac{y^2}{1} = 1$

    $a = 2,\ b = 1$

    $c^2 = 2^2 + 1^2 = 5$

    $c = \pm\sqrt{5}$

    Center $\left(3, 0\right)$

    Foci $\left(3 - \sqrt{5}, 0\right)$ and $\left(3 + \sqrt{5}, 0\right)$

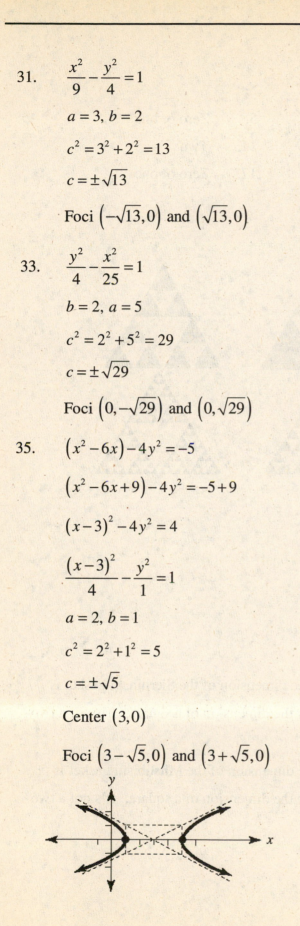

8.8 Non-Euclidean Geometry

1.	One	3.	Zero or one	5.	Zero or one
7.	None	9.	Zero, one, two, or three	11.	Two
13.	Infinitely many	15.	Zero or one	17.	Zero or one

8.9 Fractal Geometry

1.   a.    b.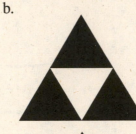

3.

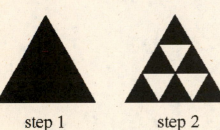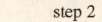

5.   step 1   step 2   step 3

7.   Using the formula $s^d = n$: solve $27^d = 3$. $d = \frac{1}{3}$

9.   Using the formula $s^d = n$: solve $2^d = 3$. $d \approx 1.6$ The dimension of the Sierpinski gasket is larger than the dimension of a circle, but smaller than the dimension of a square. It's not a two-dimensional as a regular triangle is.

11.   Using the formula $s^d = n$: solve $3^d = 6$. $d \approx 1.6$ The dimension of the Mitsubishi gasket is larger than the dimension of a circle, but smaller than the dimension of a square. It's not a two-dimensional as a regular triangle is.

13. Using the formula $s^d = n$: solve $4^d = 8$. $d = 1.5$ The dimension of the square snowflake is larger than the dimension of a circle, but smaller than the dimension of a square. It's not two-dimensional.

15. Using the formula $s^d = n$: solve $4^d = 64$. $d = 3$, the same dimension that we observed before.

17. Using the formula $s^d = n$: solve $2^d = 4$. $d = 2$. Also, using the formula $s^d = n$: solve $3^d = 9$. Again, $d = 2$.

8.10 The Perimeter and Area of a Fractal

1. a. In step 1, there is only one triangle. If each side is 1 ft. in length, then the perimeter is 3 ft.

    In step 2, the original triangle is modified by removing a center triangle. This results in three smaller triangles. Each side of the original triangle now has two triangles with sides of length $\frac{1}{2}$ ft. Thus, the perimeter of one of these smaller triangles is $\frac{1}{2} + \frac{1}{2} + \frac{1}{2} = 3 \cdot \frac{1}{2} = \frac{3}{2}$ ft. Since there are three such triangles, the total perimeter is $3 \cdot \frac{3}{2} = \frac{9}{2}$ ft.

    b.

Step	Number of Triangles	Length of each side (feet)	Perimeter of one triangle (feet)	Total perimeter of all triangles (feet)
1	1	1	3	3
2	3	$\frac{1}{2}$	$3 \cdot \frac{1}{2} = \frac{3}{2}$	$3 \cdot \frac{3}{2} = \frac{9}{2}$
3	9	$\frac{1}{4}$	$3 \cdot \frac{1}{4} = \frac{3}{4}$	$9 \cdot \frac{3}{4} = \frac{27}{4}$
4	27	$\frac{1}{8}$	$3 \cdot \frac{1}{8} = \frac{3}{8}$	$27 \cdot \frac{3}{8} = \frac{81}{8}$
5	81	$\frac{1}{16}$	$3 \cdot \frac{1}{16} = \frac{3}{16}$	$81 \cdot \frac{3}{16} = \frac{243}{16}$
6	243	$\frac{1}{32}$	$3 \cdot \frac{1}{32} = \frac{3}{32}$	$243 \cdot \frac{3}{32} = \frac{729}{32}$

    c. 3. Each triangle is divided into four smaller triangles, one of which is deleted.

    d. $\frac{1}{2}$. When the middle triangle is taken out, it creates two triangles along each side.

    e. $\frac{1}{2}$. Since each side is half as long, the perimeter is half as much. The perimeter of one triangle is decreasing.

    f. $\frac{3}{2}$. Three times the number of triangles, but each one has half the perimeter. The total perimeter of all triangles is increasing.

    g. $3\left(\frac{3}{2}\right)^{n-1}$

h. The perimeter is infinite. The numbers get larger and larger without bound.

3.   a. In step 1, there is only one square. If each side is 1 ft. in length, then the perimeter is 4 ft. In step 2, the original square is divided into nine smaller squares and modified by removing the center square. The removed square contributes to the perimeter. It has sides of length $\frac{1}{3}$ ft. Thus, the perimeter of this smaller square is $4 \cdot \frac{1}{3} = \frac{4}{3}$ ft. The total perimeter is the sum of the previous perimeter and the new contribution: $4 + \frac{4}{3}$ ft.

   b.

Step	Number of new squares	Length of each side (feet)	Perimeter of one new square (feet)	Total perimeter of all new squares (feet)	Total perimeter of all squares (feet)
1	1	1	$4 \cdot 1 = 4$	$1 \cdot 4 = 4$	4
2	1	$\frac{1}{3}$	$4 \cdot \frac{1}{3} = \frac{4}{3}$	$1 \cdot \frac{4}{3} = \frac{4}{3}$	$4 + \frac{4}{3}$
3	8	$\frac{1}{9}$	$4 \cdot \frac{1}{9} = \frac{4}{9}$	$8 \cdot \frac{4}{9} = \frac{32}{9}$	$4 + \frac{4}{3} + \frac{32}{9}$
4	64	$\frac{1}{27}$	$4 \cdot \frac{1}{27} = \frac{4}{27}$	$64 \cdot \frac{4}{27} = \frac{256}{27}$	$4 + \frac{4}{3} + \frac{32}{9} + \frac{256}{27}$
5	512	$\frac{1}{81}$	$4 \cdot \frac{1}{81} = \frac{4}{81}$	$512 \cdot \frac{4}{81} = \frac{2,048}{81}$	$4 + \frac{4}{3} + \frac{32}{9} + \frac{256}{27} + \frac{2048}{81}$
6	4,096	$\frac{1}{243}$	$4 \cdot \frac{1}{243} = \frac{4}{243}$	$4,096 \cdot \frac{4}{243} = \frac{16,384}{243}$	$4 + \frac{4}{3} + \frac{32}{9} + \frac{256}{27} + \frac{2048}{81} + \frac{16384}{243}$

   c. 8. Each step creates nine new squares, one of which is deleted.

   d. $\frac{1}{3}$. Each original side is divided into three new squares.

   e. $\frac{1}{3}$. Since each side is one third as long, the perimeter is one third as long.

   f. $\frac{8}{3}$. Eight new squares each of which has perimeter one third as much as before. The total perimeter of all new squares is increasing.

   g. $\dfrac{4 \cdot 8^{n-2}}{3^{n-1}}$ or $\dfrac{4}{3} \cdot \left(\dfrac{8}{3}\right)^{n-2}$, valid for $n > 1$

   h. The total perimeter is infinite. The numbers increase without bound.

   i. The total perimeter is infinite. It increases without bound.

5.   a. $3, 6, 12, 24, \ldots 3 \cdot 2^{n-1}$ ft. (If the original triangle has sides of length 1 ft.)

   b. Infinite ∞

   c. $\left(\dfrac{2}{3}\right)^{n-1} \cdot \dfrac{\sqrt{3}}{4}$ sq. ft. (If the original triangle has sides of length 1 ft.)

   d. 0

7. Inductive: noticing the pattern that the number of sides is increasing by a factor of 4, that the length of each side is decreasing by a factor of $\frac{1}{3}$, and that the perimeter is increasing by a factor of $\frac{4}{3}$.

Deductive: applying the general rule that the perimeter of a square is four times the length of a side, applying the general rule that the perimeter of a shape is the sum of all lengths of sides.

Chapter 8 Review

1. Area $= x(x-2)\ ft^2$

    $\qquad = x^2 - 2x$

    Perimeter $= x + (x-2) + x + (x-2)$ feet

    $\qquad\qquad = 4x - 4$ feet

3. $A = \frac{1}{2}(7+14)(7) = 73.5$ square inches

    $P = 7 + 9 + 14 + 11 = 41$ inches

5. Divide the shape into one triangle, one square and one rectangle.

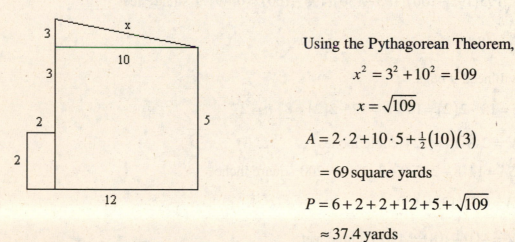

Using the Pythagorean Theorem,

$$x^2 = 3^2 + 10^2 = 109$$

$$x = \sqrt{109}$$

$$A = 2 \cdot 2 + 10 \cdot 5 + \tfrac{1}{2}(10)(3)$$

$$= 69\ \text{square yards}$$

$$P = 6 + 2 + 2 + 12 + 5 + \sqrt{109}$$

$$\approx 37.4\ \text{yards}$$

7. a. $\dfrac{239.7\ \text{trillion km}}{150\ \text{million km}} = 1{,}598{,}000\ \text{AU}$

    b. $\dfrac{239.7\ \text{trillion km}}{9.46\ \text{trillion km}} = 25.3\ \text{ly}$

    c. $\dfrac{25.3\ \text{ly}}{3.26\ \text{ly}} = 7.8\ \text{pc}$

9. $r_1 = \dfrac{2}{2} = 1$ centimeter

       $r_2 = \dfrac{4.2}{2} = 2.1$ centimeters

       $V = \pi r_2^2 h - \pi r_1^2 h = \pi(2.1)^2(6.8) - \pi(1)^2(6.8) = 72.8$ cubic centimeters

       $SA = 2\pi r_2 h + 2\pi r_1 h + 2\left(\pi r_2^2 - \pi r_1^2\right)$

           $= 2\pi(2.1)(6.8) + 2\pi(1)(6.8) + 2\pi\left[(2.1)^2 - (1)^2\right]$

           $= 153.9$ square centimeters

11. $V =$ volume of large box $-$ volume of small cube.

       $= (14)(8)(10) - (4)(4)(4) = 1,056$ cubic inches

     $A =$ bottom $+$ top $+$ left $+$ right $+$ back $+$ front $+$ small box faces

       $= (14)(8) + (14\cdot 8 - 4\cdot 4) + (8)(10) + (8\cdot 10 - 4\cdot 4) + (14)(10) + (14\cdot 10 - 4\cdot 4) + 3(4\cdot 4)$

       $= 112 + 96 + 80 + 64 + 140 + 124 + 48 = 664$ square inches

13.   Using Heron's Formula, $s = \dfrac{1}{2}(100 + 130 + 160) = 195$

     $A = \sqrt{195(195 - 100)(195 - 130)(195 - 160)} = 6,491.9$ square feet

     $\dfrac{6,491.9}{800} = 8.11$

     He will need 9 bags.

15.   a. $l = 18 - 2(2) = 14$, $w = 12 - 2(2) = 8$, $h = 2$

          $V = 2\cdot 8\cdot 14 = 224$ cubic inches

      b. $SA = 14\cdot 8 + 2\cdot 2\cdot 8 + 2\cdot 2\cdot 14 = 200$ square inches

17.   a. $A = (13)(13) = 169$

          $V = \frac{1}{3}(169)(8) = 450.7$ cubic feet

      b. $h^2 = 6.5^2 + 8^2 = 106.25$

        $h = \sqrt{106.25}$

        $SA = 4\left(\frac{1}{2}\right)(13)\left(\sqrt{106.25}\right) = 268.0$ square feet

19.   a. $V = \dfrac{4}{3} \cdot \dfrac{256}{81} (1)^3 = 4.2140$ cubic cubits

      b. $V = \dfrac{4}{3} \pi (1)^3 = 4.1888$ cubic cubits

      c. $\dfrac{4.2140 - 4.1888}{4.1888} = 0.00602 \approx 0.6\%$

21.   Combining the four triangles into two squares.

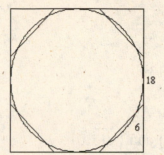

$A_{circle} \approx 18^2 + 2(6^2) = 252 \approx 256$

$A_{circle} \approx A_{square}$ with side 16.

23.   Using the Pythagorean Theorem, solve for x:

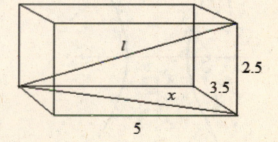

$x^2 = 3.5^2 + 5^2 = 37.25$

$x = \sqrt{37.25}$

Using the Pythagorean Theorem, solve for l:

$l^2 = 2.5^2 + \left(\sqrt{37.25}\right)^2 = 43.5$

$l = 6.5$ ft (rounded down so the object will fit)

25.   The first square has area c^2.

Rearranging the area creates two squares: one with area a^2 and another with area b^2.

Thus, $a^2 + b^2 = c^2$.

27.
1.	$\angle CBA = \angle DAB$	Given.
2.	$BC = AD$	Given.
3.	$AB = AB$	Anything is equal to itself.
4.	$\triangle CAB \cong \triangle DBA$	SAS
5.	$\angle CAB = \angle DBA$	Corresponding parts of congruent triangles are equal.
6.	$\angle CAD = \angle DBC$	Since $\angle CAD = \angle CAB - \angle DAB$

$$= \angle DBA - \angle CBA$$

$$= \angle DBC$$

29. $C \approx P = 12\left[\left(\sqrt{6}-\sqrt{2}\right)/2\right]r$

$2\pi r \approx 12\left[\left(\sqrt{6}-\sqrt{2}\right)/2\right]r$

$\pi \approx 6\left[\left(\sqrt{6}-\sqrt{2}\right)/2\right] = 3.105828541$

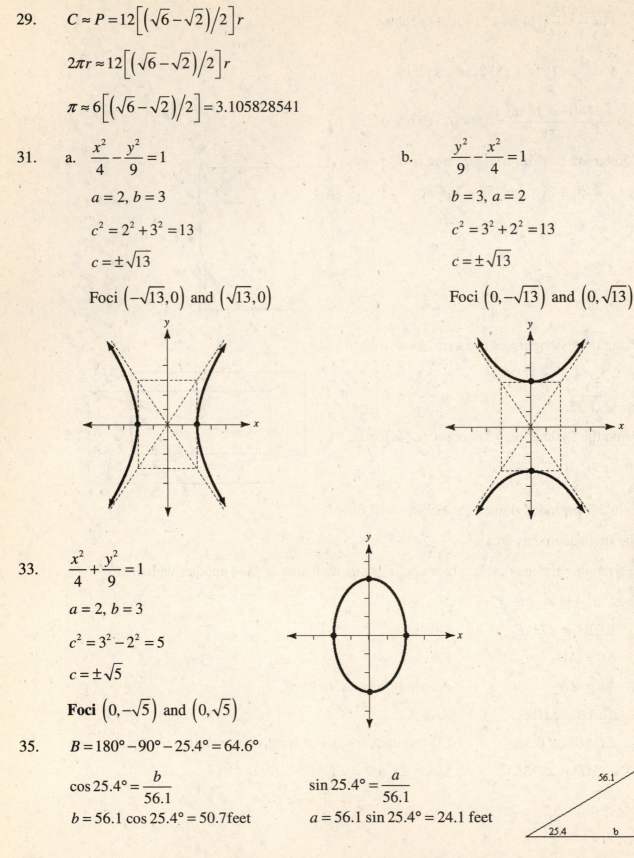

31.   a. $\dfrac{x^2}{4} - \dfrac{y^2}{9} = 1$ b. $\dfrac{y^2}{9} - \dfrac{x^2}{4} = 1$

$a = 2, \, b = 3$ $b = 3, \, a = 2$

$c^2 = 2^2 + 3^2 = 13$ $c^2 = 3^2 + 2^2 = 13$

$c = \pm\sqrt{13}$ $c = \pm\sqrt{13}$

Foci $\left(-\sqrt{13},0\right)$ and $\left(\sqrt{13},0\right)$ Foci $\left(0,-\sqrt{13}\right)$ and $\left(0,\sqrt{13}\right)$

33. $\dfrac{x^2}{4} + \dfrac{y^2}{9} = 1$

$a = 2, \, b = 3$

$c^2 = 3^2 - 2^2 = 5$

$c = \pm\sqrt{5}$

Foci $\left(0,-\sqrt{5}\right)$ and $\left(0,\sqrt{5}\right)$

35. $B = 180° - 90° - 25.4° = 64.6°$

$\cos 25.4° = \dfrac{b}{56.1}$ $\sin 25.4° = \dfrac{a}{56.1}$

$b = 56.1\cos 25.4° = 50.7\text{feet}$ $a = 56.1\sin 25.4° = 24.1\text{ feet}$

250

37. $\theta = 90° - 13.2° = 76.8°$

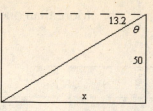

$\tan 76.8° = \dfrac{x}{50}$

$x = 50 \tan 76.8° = 213.2 \text{ feet}$

39.   Note: $1 \text{ pc} = \dfrac{1 \text{ AU}}{\tan(1 \text{ second})}$

distance $= \dfrac{1 \text{ AU}}{\tan(3 \text{ seconds})} \div \dfrac{1 \text{ AU}}{\tan(1 \text{ second})}$

$= \dfrac{1 \text{ AU}}{\tan\left(\frac{3}{3600}\right)°} \cdot \dfrac{\tan\left(\frac{1}{3600}\right)°}{1 \text{ AU}}$

$= 0.33 \text{ pc}$

Alternative solution:

$\tan(3 \text{ seconds}) = \dfrac{1 \text{ AU}}{x}$

$x = \dfrac{1 \text{ AU}}{\tan(3 \text{ seconds})}$

$x = \dfrac{93}{\tan\left(\frac{3}{3600}\right)°} \cdot \dfrac{1 \text{ ly}}{5,880,000}$

$x = 1.087\ldots \text{ ly}$

$x = \dfrac{1.087\ldots \text{ ly}}{3.26 \text{ ly}}$

$x = 0.33 \text{ pc}$

41.   a.   From page 590, we know:

$1 \text{ pc} = 3.26(63,240 \text{ AU})$

$1 \text{ pc} = 206,162.4 \text{ AU}$

$0.85 \text{ pc} = 0.85(206,162.4 \text{ AU})$

$= 175,238 \text{ AU}$

b. $\tan\left(\dfrac{x}{3600}\right)° = \dfrac{1 \text{ AU}}{0.85 \text{ pc}}$

$\tan\left(\dfrac{x}{3600}\right)° = \dfrac{1 \text{ AU}}{175,238 \text{ AU}}$

$x = 3600 \cdot \tan^{-1}\left(\dfrac{1}{175,238}\right)$

$x = 1.18 \text{ seconds}$

43. a. One   b. More than one   c. None

45. a. Step 1

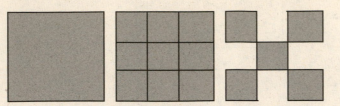

b.

Step	Number of Squares	Length of Side	Perimeter of Square	Total Perimeter
1	1	1	4	4
2	5	$\frac{1}{3}$	$\frac{4}{3}$	$\frac{20}{3}$
3	25	$\frac{1}{9}$	$\frac{4}{9}$	$\frac{100}{9}$
4	125	$\frac{1}{27}$	$\frac{4}{27}$	$\frac{500}{27}$
5	625	$\frac{1}{81}$	$\frac{4}{81}$	$\frac{2500}{81}$
n	5^{n-1}	$\dfrac{1}{3^{n-1}}$	$\dfrac{4}{3^{n-1}}$	$4 \cdot \dfrac{5^{n-1}}{3^{n-1}} = 4\left(\dfrac{5}{3}\right)^{n-1}$

c. $4\left(\dfrac{5}{3}\right)^{n-1}$ d. The perimeter is infinite.

e.

Step	Number of Squares	Length of Side	Area of Square	Total Area
1	1	1	1	1
2	5	$\frac{1}{3}$	$\frac{1}{9}$	$\frac{5}{9}$
3	25	$\frac{1}{9}$	$\frac{1}{81}$	$\frac{25}{81}$
4	125	$\frac{1}{27}$	$\frac{1}{729}$	$\frac{125}{729}$
5	625	$\frac{1}{81}$	$\frac{1}{6561}$	$\frac{625}{6561}$
n	5^{n-1}	$\dfrac{1}{3^{n-1}}$	$\left(\dfrac{1}{3^{n-1}}\right)^2$	$\left(\dfrac{5}{9}\right)^{n-1}$

f. 0   g. Using the formula $s^d = n$: solve $3^d = 5$. $d = 1.465$.

h. If we focus in on any smaller square it will look like the entire box fractal.

i. For any square that is not removed, the recursive rule is to remove the same four squares in the middle of the edges.

47. a. one-point perspective   b. below   c. approximately $(1, 2)$ d. $y = 2$

49. Answers will vary.

51. Answers may vary.

a. One-point perspective; the table front sides are all parallel to the surface of the painting.

b. The vanishing point appears to be on the girl playing. Vermeer might have chosen that point to draw the viewer's eye to the girl at the piano.

c. No

d. No, there is a central vanishing point. See painting.

e. Yes, on the tiles in the floor.

f. Yes, there are tiles on the floor.

g. Albertian grid is used on the tiles in the floor. See diagonal lines on painting.

h. Yes, horizon follows the line of the windows and piano.

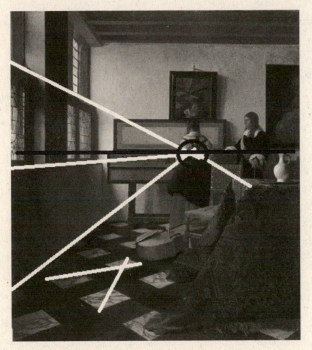

Jan Vermeer, *The Music Lesson*, 1665

9.1 A Walk Through Konigsberg

1. Yes. There are four vertices. If you eliminate one bridge, then two of the vertices have an odd number of edges, and the rest have an even number of edges. Start at one of the odd vertices and end at the other.

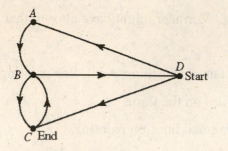

3. No; the closure of any one bridge is sufficient to create a bridge walk.

5. Yes. Start at one odd vertex and end at the other.

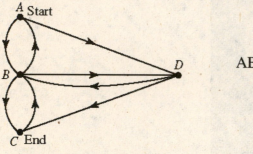

ABDBCBADC

7. 4 vertices, 4 edges, 0 loops

9. 5 vertices, 5 edges (one of which is a loop), 1 loop

11. a. 7 vertices, 6 edges, 0 loops   13.

    b. A family member is not its own child.

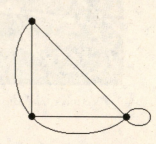

15. a. They are all graphs with 4 edges, 4 vertices, and 0 loops, and the edges in both graphs connect the same points.

    b.

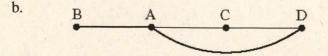

254

17.

Section 9.2 Graphs and Euler Trials

1.   a.   AB and AC   b.   A and B   c.   A-3, B-1, C-2, D-2

      d.   Yes; every pair of vertices is connected by a trail.

3.   a.   AB and BD   b.   A and B   c.   A-3, B-4, C-2, D-3

      d.   Yes; every pair of vertices is connected by a trail.

5.   a.   AB and AC   b.   A and B   c.   A-4, B-3, C-3, D-2

      d.   Yes; every pair of vertices is connected by a trail.

7.   a.   b.   c.

9.   a.   b.

11.   a.   b.

      c.   d.

13. a. There is an Euler trail because there are exactly two odd vertices.

b. Eulerize by adding edge between A and B. Euler circuit: BACDAB

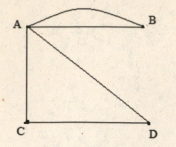

15. a. There is an Euler trail because there are exactly two odd vertices.

b. ACABDBD

17. a. There is an Euler trail because there are exactly two odd vertices.

b. Eulerize by adding an edge between B and C. Euler circuit: ABCDABCA

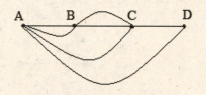

19. Answers will vary. Possible answer:

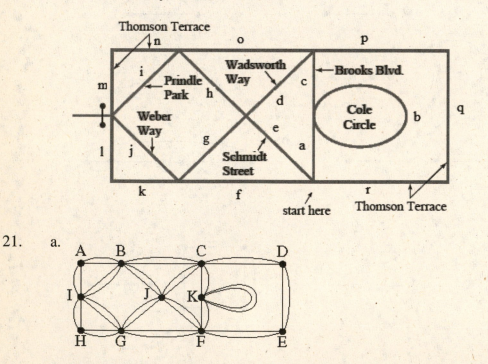

21. a.

Each edge is a double edge. Start in upper left-hand corner, go clockwise one

256

block, and detour to take Prindle, Weber, Wadsworth, Schmidt, Brooks Road,

Brooks Circle, Brooks Road, and Wadsworth and Schmidt back to where you detoured.

Continue along the outer border to where you started, and repeat for the other side.

b.
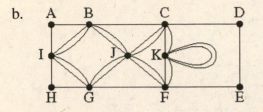

The outside border is single, and the inside edges are double. Start in upper left-hand corner

and go one block clockwise. Make a detour around the inside. (See part (a).) When you come

back, continue along the border clockwise to your return.

c.
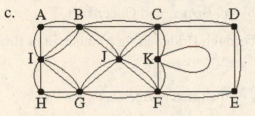

Everything is the same in part (a) except that when you repeat the inside circuit,

you skip Cole Circle the second time.

23.
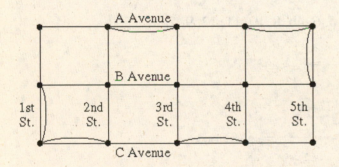

Eulerize by adding six extra edges, Avenue A from 2nd to 3rd and from 4th to 5th,

Avenue C from 1st to 2nd and from 3rd to 4th, First Street from B to C, and Fourth

Street from A to B.

Start at A on 1st. Go over to 5th, down to C, over to 1st, up to B, over to

5th, back up to A, back to 4th, down to C, over to 3rd, up to A, over to 2nd, down

to C, over to 1st, up to B and return back to A.

25.

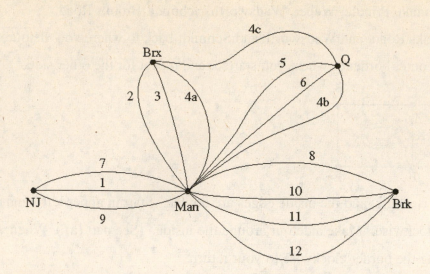

a. Manhattan has degree 13, NJ has degree 3, Brooklyn, Bronx, and Queens all

have degree 4. Label the three edges of the Triborough Bridge as 4a Manhattan to Bronx, 4b

Manhattan to Queens, and 4c Bronx to Queens.

b. No, there are two odds. To Eulerize, revisit any of the three bridges between

NJ and Manhattan. Start in Manhattan: 7, 1, 9, 1(or 7 or 9), 2, 3, 4a, 4c, 5, 6,

4b, 8, 10, 11, 12, ending in Manhattan.

c. There are exactly two vertices with odd degrees. Start in NJ and end in M.

Here is the route NJ, 1, 2, 3, 4a, 4c, 5, 6, 4b, 8, 10, 11, 12, 9, 7, M.

27.   a.

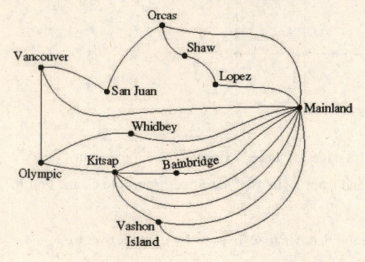

b. No, there are four odd vertices. To Eulerize, revisit a ferry from Vancouver to

Orcas, and a ferry from Vashon Island to Olympic Peninsula. Now all the

vertices could be considered even, and you can start and end at the same point.

c. No, there are four odd vertices. To Eulerize, revisit a ferry connecting two of the odd vertices. The other pair of odd vertices are the starting and stopping points.

29. a.

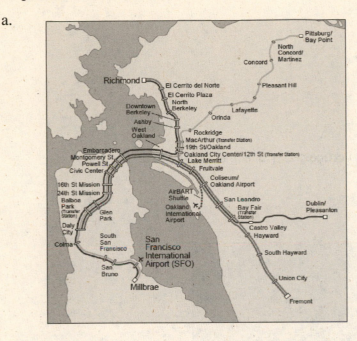

b. No, since there are only two stations with odd-degree: Pittsburgh and Dublin, both with 1. Millbrae has 2 (the red and the yellow); Daly City has 6 (2 reds, 2 yellows, 1 blue, and 1 green). To Eulerize, take the yellow from Pittsburg to Millbrae, then the red from Millbrae to Richmond, the orange from Richmond to Fremont, the green from Fremont to Daly City, the blue from Daly City to Dublin, and then the blue from Dublin to West Oakland and the yellow back to Pittsburg.

c. Yes, it is possible to start and end at different points since there are only two odd-degrees stations. Therefore, follow the route described in part (b) except end at Dublin. Do not proceed from Dublin back to Pittsburg.

d. They can patrol all stations efficiently.

e. Part (b) is more useful so they can start and end their day at the same place.

31. a. Graph of Figure 9.15:

San Diego Padres	Colorado Rockies	San Francisco Giants	Los Angeles Dodgers	Arizona Diamondbacks

Number of edges in the graph: 11

Sum of the degrees of all of the vertices: $2 + 5 + 6 + 6 + 3 = 22$

Graph of Figure 9.17:

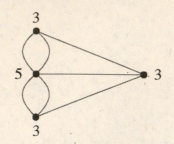

Number of edges in the graph: 7

Sum of the degrees of all of the vertices: $3 + 3 + 5 + 3 = 14$

Graph of Figure 9.20:

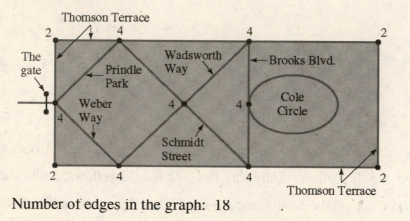

Number of edges in the graph: 18

Sum of the degrees of all of the vertices:

$$2 + 4 + 4 + 2 + 4 + 4 + 4 + 2 + 4 + 4 + 2 = 36$$

b. Sum of degrees of vertices = $2\times$ number of edges   *or*

   number of edges $= \dfrac{\text{sum of degrees of vertices}}{2}$

c. 2

d. 1 to each

e. An edge connects two vertices.

Section 9.3 Hamilton Circuits

Note: When a tie occurs, you have to look at all the cases. This means that there may be more sequences than it appears for the repetitive nearest neighbor algorithm and cheapest edge algorithm.

This might not be practical if there are too many choices. If you simply pick one choice at random, you might not get the best sequence. In some cases, answers may differ because of a tie.

1. a. $(4-1)! = 3 \cdot 2 \cdot 1 = 6$

    b. A→B: $104

    B→D: $310

    D→P: $156

    P→A: $357

    A→B→D→P→A: $927

    c. A→B: $104

    B→P: $444

    P→D: $156

    D→A: $144

    A→B→P→D→A: $848

    d. A→B→P→D→A: $848

    e. 3! = 6 routes; A→B→P→D→A: $848

3. a. 4! = 24

    b. A → B : $104

    B → Po : $444

    Po → Px : $104

    Px → D : $122

    D → A : $144

    A → B → Po → Px → D → A : $685

    c. A → B → Po → Px → D → A: $685

    d. A → B : $104

    B → Px : $444

    Px → Po : $104

    Po → D : $156

    D → A : $144

    A → B → Px → Po → D → A : $918

    e. There are too many possibilities.

5. F → I : 10 min

    I → K : 8 min

    K → B : 15 min

    B → C : 18 min

    C → L : 37 min

    L → F : 35 min

    F → I → K → B → C → L → F : 123 min

7. F → K : 14 min

    K → I : 8 min

    I → C : 14 min

    C → B : 18 min

    B → L : 19 min

    L → F : 35 min

    F → K → I → C → B → L → F : 108 min

9. F → I : 10 min

    I → C : 14 min

    C → R : 11 min

    R → B : 29 min

    B → L : 19 min

    L → F : 35 min

    F → I → C → R → B → L → F : 118 min

11. F → I : 10 min

    I → L : 15 min

    L → B : 19 min

    B → C : 18 min

    C → R : 11 min

    R → F : 11 min

    F → I → L → B → C → R → F : 84 min

13. a. $(4-0)+(1-0)=4+1=5$ mm   b. $(4-3)+(7-1)=1+6=7$ mm

     c. $(3-0)+(7-0)=3+7=10$ mm   d. $5+7+10=22$ mm

15. $(0,0)\rightarrow(2,2):2+2=4$ mm

     $(2,2)\rightarrow(3,1):1+1=2$ mm

     $(3,1)\rightarrow(4,5):1+4=5$ mm

     $(4,5)\rightarrow(6,4):2+1=3$ mm

     $(6,4)\rightarrow(7,5):1+1=2$ mm

     $(7,5)\rightarrow(0,0):7+5=12$ mm

     $(0,0)\rightarrow(2,2)\rightarrow(3,1)\rightarrow(4,5)\rightarrow(6,4)\rightarrow(7,5)\rightarrow(0,0):28$ mm

     Or

     $(0,0)\rightarrow(3,1):3+1=4$ mm

     $(3,1)\rightarrow(2,2):1+1=2$ mm

     $(2,2)\rightarrow(4,5):2+3=5$ mm

     $(4,5)\rightarrow(6,4):2+1=3$ mm

     $(6,4)\rightarrow(7,5):1+1=2$ mm

     $(7,5)\rightarrow(0,0):7+5=12$ mm

     $(0,0)\rightarrow(3,1)\rightarrow(2,2)\rightarrow(4,5)\rightarrow(6,4)\rightarrow(7,5)\rightarrow(0,0):28$ mm

7. $(0,0)\rightarrow(2,2):2+2=4$ mm

     $(2,2)\rightarrow(3,1):1+1=2$ mm

     $(3,1)\rightarrow(4,5):1+4=5$ mm

     $(4,5)\rightarrow(6,4):2+1=3$ mm

     $(6,4)\rightarrow(7,5):1+1=2$ mm

     $(7,5)\rightarrow(0,0):7+5=12$ mm

     $(0,0)\rightarrow(2,2)\rightarrow(3,1)\rightarrow(4,5)\rightarrow(6,4)\rightarrow(7,5)\rightarrow(0,0):28$ mm

19. $(0, 0) \rightarrow (5, 1): 5 + 1 = 6 \, \text{mm}$

    $(5, 1) \rightarrow (4, 3): 1 + 2 = 3 \, \text{mm}$

    $(4, 3) \rightarrow (3, 4): 1 + 1 = 2 \, \text{mm}$

    $(3, 4) \rightarrow (2, 7): 1 + 3 = 4 \, \text{mm}$

    $(2, 7) \rightarrow (1, 8): 1 + 1 = 2 \, \text{mm}$

    $(1, 8) \rightarrow (0, 0): 1 + 8 = 9 \, \text{mm}$

    $(0, 0) \rightarrow (5, 1) \rightarrow (4, 3) \rightarrow (3, 4) \rightarrow (2, 7) \rightarrow (1, 8) \rightarrow (0, 0): 26 \, \text{mm}$

21. $(0, 0) \rightarrow (1, 8): 1 + 8 = 9 \, \text{mm}$

    $(1, 8) \rightarrow (2, 7): 1 + 1 = 2 \, \text{mm}$

    $(2, 7) \rightarrow (3, 4): 1 + 3 = 4 \, \text{mm}$

    $(3, 4) \rightarrow (4, 3): 1 + 1 = 2 \, \text{mm}$

    $(4, 3) \rightarrow (5, 1): 1 + 2 = 3 \, \text{mm}$

    $(5, 1) \rightarrow (0, 0): 5 + 1 = 6 \, \text{mm}$

    $(0, 0) \rightarrow (1, 8) \rightarrow (2, 7) \rightarrow (3, 4) \rightarrow (4, 3) \rightarrow (5, 1) \rightarrow (0, 0): 26 \, \text{mm}$

23. In some cases answers may differ because of a tie.

    ATL → WASH : $74

    WASH → BOS : $54

    BOS → SFO : $132

    SFO → PORT : $84

    PORT → PHX : $104

    PHX → DEN : $156

    DEN → ATL : $144

    ATL → WASH → BOS → SFO → PORT → PHX → DEN → ATL: $748

    The cheapest route is $748.

25. In some cases, answers may differ because of a tie.

    ATL → WASH : $74

    WASH → BOS : $54

    BOS → SFO : $132

    SFO → PORT : $84

    PORT → PHX : $104

    PHX → DEN : $156

    DEN → ATL : $144

    ATL → WASH → BOS → SFO → PORT → PHX → DEN → ATL: $748

    Or

    ATL → DEN : $144

    DEN → PHX : $156

    PHX → PORT : $104

    PORT → SFO : $84

    SFO → BOS : $134

    BOS → WASH : $52

    WASH → ATL : $74

    ATL → DEN → PHX → PORT → SFO → BOS → WASH → ATL: $748

27. In some cases, answers may differ because of a tie.

    ATL → DEN : $144

    DEN → SFO : $192

    SFO → PORT : $84

    PORT → PHX : $104

    PHX → BOS : $446

    BOS → WASH : $52

    WASH → ATL : $74

    ATL → DEN → SFO → PORT → PHX → BOS → WASH → ATL: $1,096

29. Exercises 23 and 25 both yielded $748.

31.   In some cases, answers may differ because of a tie.

NYC → SEA : $132

SEA → CHI : $179

CHI → MIA : $124

MIA → HOU : $214

HOU → LAX : $284

LAX → NYC : $199

NYC → SEA → CHI → MIA → HOU → LAX → NYC: $1,132

Section 9.4 Networks

1.   Yes; it is a connected graph that has no circuits.

3.   No; it is not connected.   5.   No; it has a circuit.

7.   No; it has a circuit.   9.   Yes; it is a connected graph that has no circuits.

11.   No; it is not connected.   13.   Answers will vary. Possible answer: a, d, e

15.   Answers will vary. Possible answer: a, c, d, e, f, g, h, i, j, k, l, m

17.   Answers will vary. Possible answer: a, b, d, e, g, h

19. $10 + 20 + 35 = 65$ 21. $20 + 25 + 30 + 40 = 115$

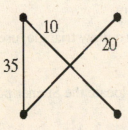

23. $120 + 25 + 35 + 22 + 40 + 22 + 47 + 36 + 29 + 37 + 35 + 36 + 10 + 37 = 531$

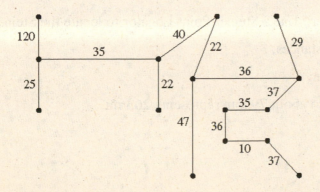

25. $N \to B : 187$ mi

B $\to$ C : 851 mi

C $\to$ K : 407 mi

K $\to$ F : 1,500 mi

$N \to B \to C \to K \to F$: 2,945 miles

27. $N \to Ph : 86$ mi

Ph $\to$ A : 658 mi

A $\to$ D : 1,213 mi

D $\to$ S : 947 mi

S $\to$ Po : 538 mi

$N \to Ph \to A \to D \to S \to Po$: 3,442 miles

29.   S $\to$ Po : 538 mi

Po $\to$ K : 1,493 mi

K $\to$ C : 407 mi

C $\to$ A : 585 mi

A $\to$ Ph : 658 mi

Ph $\to$ N : 86 mi

N $\to$ B : 187 mi

$S \to Po \to K \to C \to A \to Ph \to$

$N \to B$: 3,954 miles

31. $8 \times 7 + 9 \times 5 + 10 \times 3 = 131$

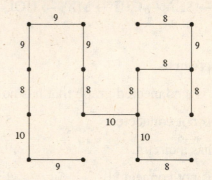

33.   B because the sum of the distances is the shortest and it is given that one of them is a Steiner point.

35.   Answers will vary. Possible answer: Use a ruler to measure the distances.

4.8 in. + 2.4 in. + 3.3 in. = 10.5 in. This represents a length of 10.5 km.

37.   Answers will vary. Possible answer: 4 mi + 4 mi + 6 mi = 14 mi   (To draw triangle, used scale: 1 mile = 1/8 inch)

39.   Answers may vary. Possible answer: Use a Steiner Point Locator to locate the Steiner point.

Then use a ruler to measure the distances.

64 mm + 46 mm + 42 mm = 152 mm

About 28 mm represents 100 mi, so about 152 mm represents 543 mi.

41.   Answers will vary. Possible answer: Use a Steiner Point Locator to locate the Steiner point.

Then use a ruler to measure the distances.

14 mm + 13 mm + 52 mm = 79 mm

About 30 mm represents 100 mi, so about 79 mm represents 263 mi.

43. a. $\tan\theta = \dfrac{opp}{adj}$

        $\tan(20° + 20°) = \dfrac{x+y}{100}$

           $\tan 40° = \dfrac{x+y}{100}$

            $x + y = 100\tan 40°$

    b. $\tan\theta = \dfrac{opp}{adj}$

        $\tan 20° = \dfrac{x}{100}$

        $x = 100\tan 20°$

    c. Substitute.

        $x + y = 100\tan 40°$

    $100\tan 20° + y = 100\tan 40°$

            $y = 100\tan 40° - 100\tan 20°$

            $y = 100(\tan 40° - \tan 20°)$

45. a. $40° + 90° + a = 180°$

        $a = 50°$

    b. $20° + 90° + b = 180°$

        $b = 70°$

    c. $c = 180° - 70°$

        $c = 110°$

47. a. $\tan\theta = \dfrac{opp}{adj}$

        $\tan(20° + 25°) = \dfrac{x+y}{100}$

           $\tan 45° = \dfrac{x+y}{100}$

           $x + y = 100\tan 45°$

           $x + y = 100 \cdot 1$

           $x + y = 100$

    b. $\tan\theta = \dfrac{opp}{adj}$

        $\tan 25° = \dfrac{x}{100}$

        $x = 100\tan 25°$

    c. Substitute.

        $x + y = 100$

    $100\tan 25° + y = 100$

            $y = 100 - 100\tan 25°$

49. a. $45° + 90° + a = 180°$

        $a = 45°$

    b. $25° + 90° + b = 180°$

        $b = 65°$

    c. $c = 180° - b$

        $c = 180° - 65°$

        $c = 115°$

51. Since the three points form a triangle with angles that are less than $120°$, find a Steiner point inside the triangle. Use the results of Exercises 43 – 48.

    Let $y = $ perpendicular edge length and $h = $ length of each other edge.

$$y = 50\left(\tan 40^\circ - \tan 30^\circ\right) = 50\left(\tan 40^\circ - \frac{\sqrt{3}}{3}\right) = 50\left(\tan 40^\circ\right) - 50 \cdot \frac{\sqrt{3}}{3}$$

$$\sin 60^\circ = \frac{50}{h}, \; h = \frac{50}{\sin 60^\circ} = \frac{50}{\frac{\sqrt{3}}{2}} = \frac{100}{\sqrt{3}} = \frac{100\sqrt{3}}{3}$$

$$y + 2h = 50\left(\tan 40^\circ\right) - 50 \cdot \frac{\sqrt{3}}{3} + 200 \cdot \frac{\sqrt{3}}{3}$$

$$= 50\tan 40^\circ + 150 \cdot \frac{\sqrt{3}}{3} = 50\tan 40^\circ + 50\sqrt{3} = 50\left(\tan 40^\circ + \sqrt{3}\right)$$

53.   Since the three points form a triangle with an angle that is 120° or more (Durham, 130°), then

   the shortest network consists of the sum of the two shortest sides of the triangle.

   Let h = length of one short side.

$$\sin 65^\circ = \frac{280}{h}$$

$$h = \frac{280}{\sin 65^\circ}$$

$$2h = \frac{560}{\sin 65^\circ}$$

55.   a. $2 \times 500 + 600 = 1,600$ b. $2 \times 600 + 500 = 1,700$

   c. $500\sqrt{3} + 600$ d. $600\sqrt{3} + 500$

   Therefore, c has the shortest network.

57.   Use two Steiner points to form triangles with the shorter sides: $600\sqrt{3} + 700$

59.   a. $3 \times 500 = 1,500$ b. $2\sqrt{500^2 + 500^2} = 2\sqrt{500,000}$

   c. $500\sqrt{3} + 500$

   Therefore, c has the shortest network.

Section 9.5 Scheduling

1.

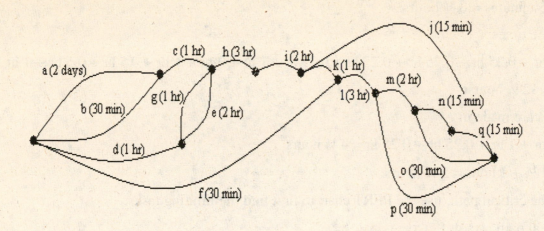

To find the critical path, use the PERT chart to first find the limiting tasks. The critical path is a, c, h, i, k, l, m, n, q.

3. From the PERT chart, it will take 4 workers.

5. Create the Gantt chart from the PERT chart.

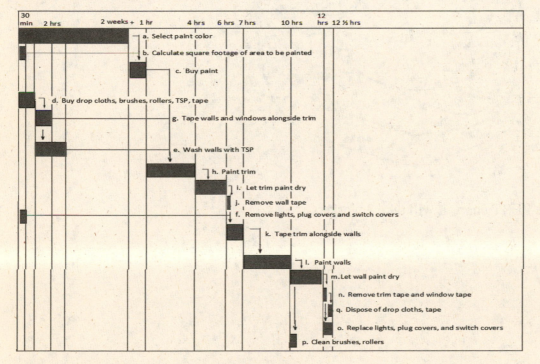

7. Let wall paint dry is m. Add times through m.

a + c + h + i + k + l + m = 2 days 12 hours.

9.   3 hours + 1 hour + 15 minutes + 15 minutes + 1 hour + 1 hour + 30 minutes + 30 minutes + 1 hour + 15 minutes = 8 ¾ hours

11.   Add all the times:

3 hr + 1 hr + 0.25 hr + 0.25 hr + 0.25 hr + 0.25 hr + 1 hr + 3 hr + 1 hr + 0.5 hr + 0.5 hr + 1 hr + 0.25 hr = 12 ¼ hours

13.   Purchases are made in *g*.

Start: 3 hr + 1 hr + 0.25 hr + 0.25 hr = 4 ½ hours

Finish: 4 ½ + 1 hour = 5 ½ hours

15.   To find the critical path, use the PERT chart to first find the limiting tasks.

The critical path is c, d, f, n, o.

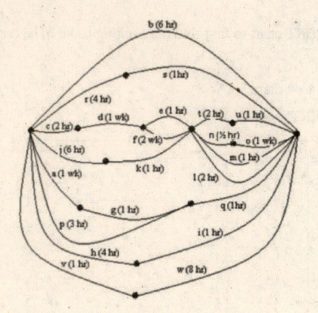

17.   From the PERT chart, it will take 3 workers.

19.   Create the Gantt chart from the PERT chart.

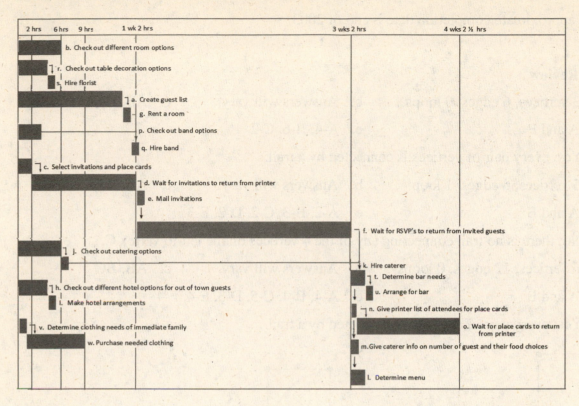

21.   Purchase needed clothing is w, so start after v. Start after 1 hour into the schedule and finish before the wedding starts.

23.   4 wk + 4 wk + 6 wk + 4 wk + 3 wk + 3 wk + 4 wk + 6 wk + 1 wk + 2 wk = 37 wk

25.   The roof is built at g. So, start right before g and end after g.

        Start: 4 + 4 + 6 + 4 + 3 + 3 = 24 weeks

        Finish: 24 + 4 = 28 weeks

27.   The inspection takes place at p. So start right before p, 35 weeks, and finish right after p, 37 weeks.

29.   It was dependent on running closed ESS in lab. The task's approximate length is 1 year. The task's completion date is end of 1999.

31.   It was dependent on install solar array, perform targeted lightweighting, improve reliability of motors, complete environmental control system installation, upgrade PMRF facilities, solar cell procurement, and Helios prototype functional test. The task's approximate length is 1 year. The task's completion date is end of 2001.

33.   2003 − 1994 = 9 years

The approximate completion date is end of 2003.

Chapter 9 Review

1.   a.   3 vertices, 6 edges, 0 loops   b.   Answers will vary.   c.   AB, BC

d.   A and B   e.   A-4, B-6, C-2

f.   Yes; every pair of vertices is connected by a trail.

3.   a.   5 vertices, 6 edges, 1 loop   b.   Answers will vary.   c.   AB, BE

d.   A and B   e.   A-2, B-3, C-2, D-2, E-3

f.   No; there is no trail connecting any of the 4 vertices on the left to vertex C.

5.   a.   6 vertices, 12 edges, 0 loops   b.   Answers will vary.   c.   AB, BC

d.   A and B   e.   A-4, B-4, C-5, D-3, E-4, F-4

f.   Yes; every pair of vertices is connected by a trail.

7.   9.

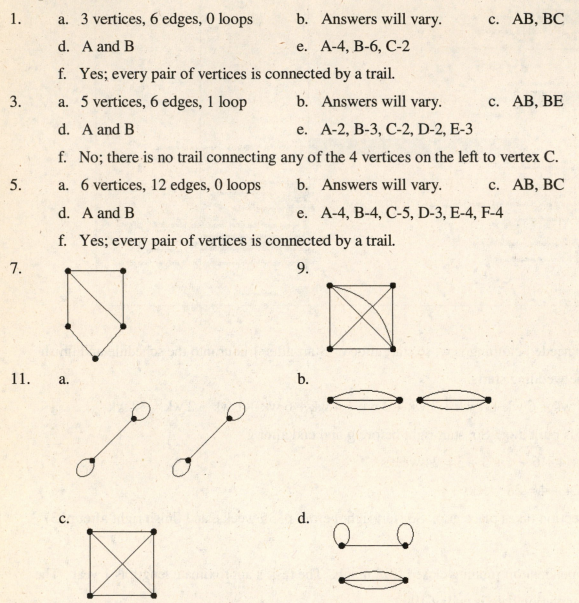

11.   a.   b.

c.   d.

13.   a.   All even vertices means that there is an Euler circuit.

b.   Euler circuit: BABABCB

15.   a.   None because the graph is not connected.

b.   Euler circuit: ABCCEBDEA

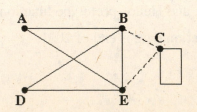

17.   a.   There are exactly two odd vertices. There is an Euler trail starting at C and
          ending at D.

     b.   Eulerize by adding an edge connecting any two adjacent vertices, for instance, from C to D.
          Euler circuit: CAFCEBFEDCBADC

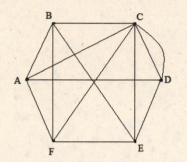

19.   a.   There are more than two odd vertices. There is no Euler trail or circuit. Eulerize.

     b.   Eulerize by adding two edges, one connecting the two top vertices and the other connecting
          the two bottom vertices. Euler circuit: ACACDCDECECBCBA

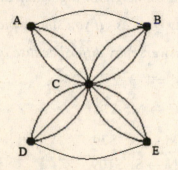

21.   a.   No. There are more than two odd vertices.

     b.

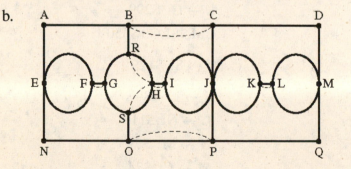

          Eulerize with seven extra edges. B-C, F-G, R-H, H-S, H-I, K-L, O-P. The
          route is E-F-G-R-B-C-J-P-O-S-H-I-J-K-L-M-L-K-J-I-H-R-H-S-G-F-E-A-B-C-
          D-M-Q-P-O-N-E.

23.   Eulerize Avenue A from 2nd to 3rd, Avenue D from 2nd to 3rd, 4th from B to C, 1st from B to C. Start on A and 1st, go right on A one block, down to D, over to 3rd, back to A, loop to 2nd and back, over to 4th, down to D, over to 1st, up to C, over to 4th, up to B over to 1st, loop to C and back to A1.

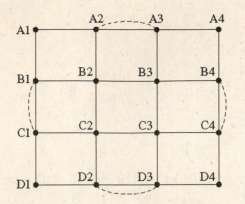

For Problems 25 – 33, note that when a tie occurs, you have to look at all the cases. This means that there might be more choices than it appears for the repetitive nearest neighbor algorithm and cheapest edge algorithm. This might not be practical if there are too many choices. If you simply pick one choice at random, you might not get the best answer. In some cases, answers may differ because of a tie.

25.   a. $3! = 3 \cdot 2 \cdot 1 = 6$

       b. C → M: \$74

           M → N: \$79

           N → L: \$119

           L → C: \$299

           CMNLC: \$571

       c. C → N: \$243

           N → L: \$119

           L → M: \$124

           M → C: \$74

           CNLMC: \$560

       d. CMNLC: \$571

       e. CNLMC: \$560

27.   a. $4! = 4 \cdot 3 \cdot 2 \cdot 1 = 24$

       b. C → M: \$74

           M → N: \$79

           N → S: \$109

           S → L: \$89

           L → C: \$299

           CMNSLC: \$650

       c. C → S: \$129

           S → L: \$89

           L → N: \$119

           N → M: \$79

           M → C: \$74

           CSLNMC: \$490

       d. CSLNMC: \$490

       e. Too many possibilities.

274

29. $C \rightarrow M$: $74

   $M \rightarrow N$: $79
   $N \rightarrow S$: $109
   $S \rightarrow L$: $89
   $L \rightarrow H$: $154
   $H \rightarrow C$: $133
   CMNSLHC: $638

31. $C \rightarrow M$: $74

   $M \rightarrow N$: $79
   $N \rightarrow S$: $109
   $S \rightarrow L$: $89
   $L \rightarrow H$: $154
   $H \rightarrow C$: $133
   CMNSLHC: $638

33. $(0,0) \rightarrow (2,4)$: $2+4 = 6$ mm

   $(2,4) \rightarrow (5,7)$: $3+3 = 6$ mm
   $(5,7) \rightarrow (6,3)$: $1+4 = 5$ mm
   $(6,3) \rightarrow (7,2)$: $1+1 = 2$ mm
   $(7,2) \rightarrow (1,1)$: $6+1 = 7$ mm
   $(1,1) \rightarrow (0,0)$: $1+1 = 2$ mm
   $(0,0) \rightarrow (2,4) \rightarrow (5,7) \rightarrow (6,3) \rightarrow (7,2) \rightarrow (1,1) \rightarrow (0,0)$: 28 mm

35. Answers will vary. Possible answer:

37. Answers will vary. Possible answer:

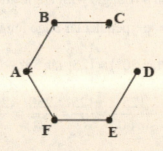

39. Answers will vary. Possible answer:

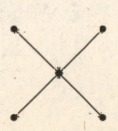

41. Answers will vary. Possible answer:

43. $25 + 25 + 10 = 60$

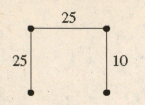

45. $45 + 50 + 50 + 55 = 200$

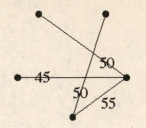

47. $50 + 300 + 100 + 200 + 100 + 250 = 1,000$

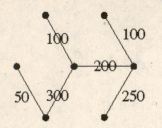

49.   B → N: 187 mi

N → C: 714 mi

C → K: 407 mi

K → D: 558 mi

BNCKD: 1,866 mi

51.   B → N: 187 mi

N → Ph: 86 mi

Ph → C: 664 mi

C → K: 407 mi

K → D: 558 mi

D → S: 947 mi

BNPhCKDS: 2,849 mi

53.   Since the three points form a triangle with angles that are less than $120°$, find a Steiner point inside the triangle. Let y = perpendicular edge length and h = length of each other edge.

$$y = 125\left(\tan 35° - \tan 30°\right) = 125\left(\tan 35° - \frac{\sqrt{3}}{3}\right) = 125\left(\tan 35°\right) - 125 \cdot \frac{\sqrt{3}}{3}$$

$$\sin 60° = \frac{125}{h},\ h = \frac{125}{\sin 60°} = \frac{125}{\frac{\sqrt{3}}{2}} = \frac{250}{\sqrt{3}} = \frac{250\sqrt{3}}{3}$$

$$y + 2h = 125\left(\tan 35°\right) - 125 \cdot \frac{\sqrt{3}}{3} + 500 \cdot \frac{\sqrt{3}}{3}$$

$$= 125\tan 35° + 375 \cdot \frac{\sqrt{3}}{3}$$

$$= 125\left(\tan 35° + \sqrt{3}\right)$$

55.   Since the three points form a triangle with an angle that is $120°$ or more (Pleasant Hill, $124°$), then the shortest network consists of the sum of the two shortest sides of the triangle. Let h = length of one short side.

$$\sin 62° = \frac{150}{h}$$

$$h = \frac{150}{\sin 62°}$$

$$2h = \frac{300}{\sin 62°} \text{ or } \frac{300}{\cos 28°}$$

57.   Place two Steiner points, so the shorter sides are bases of isosceles triangles. $700\sqrt{3} + 900$

59.

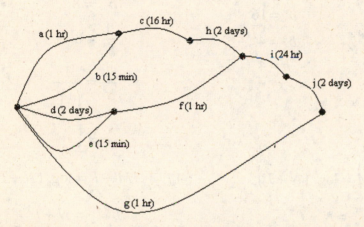

To find the critical path, use the PERT chart to first find the limiting tasks. The critical path is a, c, h, i, j.

61.   From the PERT chart, it takes 2 workers.

63.   Create the Gantt chart from the PERT chart.

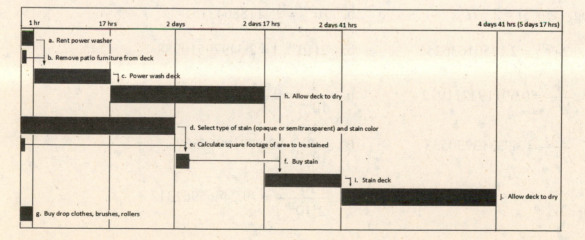

10.0A Review of Exponentials and Logarithms

1. $v = \log_2 4$

      $2^v = 4$
      $2^v = 2^2$
       $v = 2$

3. $v = \log_2\left(\frac{1}{16}\right)$

      $2^v = \frac{1}{16}$
      $2^v = \frac{1}{2^4}$
      $2^v = 2^{-4}$
      $v = -4$

5. $\log_5 u = 2$

      $5^2 = u$
      $25 = u$

7. $\log_3 u = 0$

      $3^0 = u$
       $1 = u$

9. $\log_b 16 = 2$

      $b^2 = 16$
      $b^2 = 4^2$
       $b = 4$

11. $\log_b 8 = -3$

      $b^{-3} = 8$
      $\frac{1}{b^3} = 8$
      $\frac{1}{b^3} = \frac{1}{\left(\frac{1}{2}\right)^3}$
       $b = \frac{1}{2}$

13. $P = \log_b Q$

      $b^P = Q$

15. $M = \log_b(N+T)$

      $b^M = N+T$

17. $M + R = \log_b(N+T)$

      $b^{M+R} = N+T$

19. $b^F = G$

      $\log_b G = F$

21. $b^{F+2} = G$

      $\log_b G = F+2$

23. $b^{CD} = E - F$

      $\log_b(E-F) = CD$

25. $b^{2-H} = Z + 3$

      $\log_b(Z+3) = 2 - H$

27.   a. $e^{1.4} = 4.05519996684$

      b. $10^{1.4} = 25.1188643151$

29.   a. $2e^{0.07} = 2.14501636251$ b. $2\left(10^{0.07}\right) = 2.34979510988$

31.   a. $\frac{1}{e^{1.2}} = 0.301194211912$ b. $\frac{1}{10^{1.2}} = 0.063095734448$

33.   a. $\frac{5}{e^{0.24}} = 3.93313930533$ b. $\frac{5}{10^{0.24}} = 2.87719968669$

35.   a. $\frac{e^{5.6}}{2e^{3.4}} = 4.51250674972$ b. $\frac{10^{5.6}}{2\left(10^{3.4}\right)} = 79.2446596231$

37. a. $\dfrac{2e^{5.6}}{e^{3.4}} = 18.0500269989$ b. $\dfrac{2(10^{5.6})}{10^{3.4}} = 316.978638492$

39. a. $\dfrac{e^{4.5}}{10^{1.2}} = 5.67969701231$ b. $\dfrac{10^{1.2}}{e^{4.5}} = 0.176065729885$

41. a. $\ln 2.67 = 0.982078472412$ b. $\log 2.67 = 0.426511261365$

43. a. $\ln 0.85 = -0.162518929498$ b. $\log 0.85 = -0.070581074286$

45. a. $\ln(e^{4.1}) = 4.1$ b. $\log(10^{4.1}) = 4.1$

47. a. $\ln(2e^{4.1}) = 4.79314718056$ b. $\log\left[2(10^{4.1})\right] = 4.40102999566$

49. a. $e^{\ln 2.3} = 2.3$ b. $10^{\log 2.3} = 2.3$

51. a. $e^{3\ln 2} = 8$ b. $10^{3\log 2} = 8$

53. a. $\ln 10 = 2.30258509299$ b. $\ln(10^2) = 4.60517018599$

  c. $\ln(10^3) = 6.90775527898$

55. a. $\log e = 0.434294481903$ b. $\log(e^2) = 0.868588963807$

  c. $\log(e^3) = 1.30288344571$

57. $e^{2.9} = 18.17414537$

  $\ln 2.9 = 1.064710737$

  $10^{2.9} = 794.3282347$

  $\log 2.9 = 0.4623979979$

  $\log 2.9,\ \ln 2.9,\ e^{2.9},\ 10^{2.9}$

10.0B Review of Properties of Logarithms

1. $\log(10^{6x}) = \log_{10}(10^{6x}) = 6x$

3. $\ln(e^{-0.036x}) = \log_e(e^{-0.036x}) = -0.036x$

5. $10^{\log(2x+5)} = 10^{\log_{10}(2x+5)} = 2x + 5$

7. $e^{\ln(1-x)} = e^{\log_e(1-x)} = 1 - x$

9. a. $e^{2\ln x} = e^{\ln x^2} = x^2$

11. a. $e^{2\ln(3x)} = e^{\ln(3x)^2} = (3x)^2 = 9x^2$

  b. $10^{2\log x} = 10^{\log x^2} = x^2$

  b. $10^{2\log(3x)} = 10^{\log(3x)^2} = (3x)^2 = 9x^2$

13. $\log\left(\dfrac{x}{4}\right) = \log x - \log 4$

15. $\ln(1.8x) = \ln 1.8 + \ln x$

17. $\log\left(1.225^x\right) = x\log 1.225$

19. $\ln\left(3x^4\right) = \ln 3 + \ln x^4$

$$= \ln 3 + 4\ln x$$

21. $\log\left(\dfrac{5x^2}{7}\right) = \log\left(5x^2\right) - \log 7$

$$= \log 5 + \log x^2 - \log 7$$
$$= \log 5 + 2\log x - \log 7$$

23. $\ln\left(4x\right) + \ln 5 = \ln\left(4x\cdot 5\right)$

$$= \ln\left(20x\right)$$

25. $\log\left(6x\right) - \log 2 = \log\left(\dfrac{6x}{2}\right) = \log\left(3x\right)$

27. $\ln x - \ln 3 + \ln 6 = \ln\left(\dfrac{x}{3}\right) + \ln 6$

$$= \ln\left(\dfrac{x}{3}\cdot 6\right)$$
$$= \ln\left(2x\right)$$

29. $3\log\left(2x\right) - \log 8 = \log\left(2x\right)^3 - \log 8$

$$= \log\left(\dfrac{\left(2x\right)^3}{8}\right)$$
$$= \log\left(\dfrac{8x^3}{8}\right)$$
$$= \log x^3$$

31. $\ln\left(9x\right) + \ln\left(4x\right) - 2\ln\left(6x\right) = \ln\left(9x\cdot 4x\right) - 2\ln\left(6x\right)$

$$= \ln\left(36x^2\right) - \ln\left(6x\right)^2$$
$$= \ln\left(36x^2\right) - \ln\left(36x^2\right)$$
$$= \ln\left(\dfrac{36x^2}{36x^2}\right)$$
$$= \ln 1$$
$$= 0$$

33. a. $e^x = 0.35$

$$\ln\left(e^x\right) = \ln 0.35$$
$$x = \ln 0.35$$

b. $10^x = 0.35$

$$\log\left(10^x\right) = \log 0.35$$

$$x = \log 0.35$$

35.   a. $145e^{0.024x} = 290$ b. $145(10)^{0.024x} = 290$

$$e^{0.024x} = \frac{290}{145} \qquad\qquad\qquad 10^{0.024x} = \frac{290}{145}$$

$$e^{0.024x} = 2 \qquad\qquad\qquad\qquad 10^{0.024x} = 2$$

$$\ln\left(e^{0.024x}\right) = \ln 2 \qquad\qquad \log\left(10^{0.024x}\right) = \log 2$$

$$0.024x = \ln 2 \qquad\qquad\qquad 0.024x = \log 2$$

$$x = \frac{\ln 2}{0.024} \qquad\qquad\qquad\quad x = \frac{\log 2}{0.024}$$

37.   a. $2000e^{0.004x} = 8500$ b. $2000(10)^{0.004x} = 8500$

$$e^{0.004x} = \frac{8500}{2000} \qquad\qquad\qquad 10^{0.004x} = \frac{8500}{2000}$$

$$e^{0.004x} = \frac{17}{4} \qquad\qquad\qquad\qquad 10^{0.004x} = \frac{17}{4}$$

$$\ln\left(e^{0.004x}\right) = \ln\left(\frac{17}{4}\right) \qquad\qquad \log\left(10^{0.004x}\right) = \log\left(\frac{17}{4}\right)$$

$$0.004x = \ln\left(\frac{17}{4}\right) \qquad\qquad\quad 0.004x = \log\left(\frac{17}{4}\right)$$

$$x = \frac{1}{0.004}\ln\left(\frac{17}{4}\right) \qquad\qquad x = \frac{1}{0.004}\log\left(\frac{17}{4}\right)$$

39.   a. $50e^{-0.035x} = 25$ b. $50(10)^{-0.035x} = 25$

$$e^{-0.035x} = \frac{25}{50} \qquad\qquad\qquad 10^{-0.035x} = \frac{25}{50}$$

$$e^{-0.035x} = \frac{1}{2} \qquad\qquad\qquad\qquad 10^{-0.035x} = \frac{1}{2}$$

$$e^{0.035x} = 2 \qquad\qquad\qquad\qquad 10^{0.035x} = 2$$

$$\ln\left(e^{0.035x}\right) = \ln 2 \qquad\qquad \log\left(10^{0.035x}\right) = \log 2$$

$$0.035x = \ln 2 \qquad\qquad\qquad 0.035x = \log 2$$

$$x = \frac{1}{0.035}\ln 2 \qquad\qquad\qquad x = \frac{1}{0.035}\log 2$$

41. a. $80e^{-0.0073x} = 65$

$$e^{-0.0073x} = \frac{65}{80}$$

$$e^{-0.0073x} = \frac{13}{16}$$

$$e^{0.0073x} = \frac{16}{13}$$

$$\ln\left(e^{0.0073x}\right) = \ln\left(\frac{16}{13}\right)$$

$$0.0073x = \ln\left(\frac{16}{13}\right)$$

$$x = \frac{1}{0.0073}\ln\left(\frac{16}{13}\right)$$

b. $80(10)^{-0.0073x} = 65$

$$10^{-0.0073x} = \frac{65}{80}$$

$$10^{-0.0073x} = \frac{13}{16}$$

$$10^{0.0073x} = \frac{16}{13}$$

$$\log\left(10^{0.0073x}\right) = \log\left(\frac{16}{13}\right)$$

$$0.0073x = \log\left(\frac{16}{13}\right)$$

$$x = \frac{1}{0.0073}\log\left(\frac{16}{13}\right)$$

43. a. $\ln x = 0.66$

$$e^{\ln x} = e^{0.66}$$

$$x = e^{0.66}$$

b. $\log x = 0.66$

$$10^{\log x} = 10^{0.66}$$

$$x = 10^{0.66}$$

45. a. $\ln x = 3.66$

$$e^{\ln x} = e^{3.66}$$

$$x = e^{3.66}$$

b. $\log x = 3.66$

$$10^{\log x} = 10^{3.66}$$

$$x = 10^{3.66}$$

47. a. $\ln x + \ln 6 = 2$

$$\ln(6x) = 2$$

$$e^{\ln(6x)} = e^2$$

$$6x = e^2$$

$$x = \frac{e^2}{6}$$

b. $\log x + \log 6 = 2$

$$\log(6x) = 2$$

$$10^{\log(6x)} = 10^2$$

$$6x = 100$$

$$x = \frac{100}{6}$$

$$x = \frac{50}{3}$$

49. a. $\ln x - \ln 6 = 2$

$$\ln\left(\frac{x}{6}\right) = 2$$

$$e^{\ln\left(\frac{x}{6}\right)} = e^2$$

$$\frac{x}{6} = e^2$$

$$x = 6e^2$$

b. $\log x - \log 6 = 2$

$$\log\left(\frac{x}{6}\right) = 2$$

$$10^{\log\left(\frac{x}{6}\right)} = 10^2$$

$$\frac{x}{6} = 100$$

$$x = 6 \cdot 100$$

$$x = 600$$

282

51. a.
$$\ln x = 4.8 + \ln 6.9$$

$$\ln x - \ln 6.9 = 4.8$$

$$\ln\left(\frac{x}{6.9}\right) = 4.8$$

$$e^{\ln\left(\frac{x}{6.9}\right)} = e^{4.8}$$

$$\frac{x}{6.9} = e^{4.8}$$

$$x = 6.9e^{4.8}$$

b.
$$\log x = 4.8 - \log 6.9$$

$$\log x + \log 6.9 = 4.8$$

$$\log(6.9x) = 4.8$$

$$10^{\log(6.9x)} = 10^{4.8}$$

$$x = \frac{10^{4.8}}{6.9}$$

53. a.
$$\ln 0.9 = 3.1 - \ln(4x)$$

$$\ln 0.9 + \ln(4x) = 3.1$$

$$\ln(0.9 \cdot 4x) = 3.1$$

$$\ln(3.6x) = 3.1$$

$$e^{\ln(3.6x)} = e^{3.1}$$

$$3.6x = e^{3.1}$$

$$x = \frac{1}{3.6} \cdot e^{3.1}$$

b.
$$\log 0.9 = 3.1 - \log(4x)$$

$$\log 0.9 + \log(4x) = 3.1$$

$$\log(0.9 \cdot 4x) = 3.1$$

$$\log(3.6x) = 3.1$$

$$10^{\log(3.6x)} = 10^{3.1}$$

$$3.6x = 10^{3.1}$$

$$x = \frac{1}{3.6} \cdot 10^{3.1}$$

55. Let $a = \ln A$, $b = \ln B$, then $A = e^a$, $B = e^b$

$$\ln\left(\frac{A}{B}\right) = \ln\left(\frac{e^a}{e^b}\right) = \ln\left(e^{(a-b)}\right) = a - b = \ln A - \ln B$$

57. Let $a = \log A$, $b = \log B$, then $A = 10^a$, $B = 10^b$

$$\log(A \cdot B) = \log\left(10^a \cdot 10^b\right) = \log\left(10^{a+b}\right) = a + b = \log A + \log B$$

59. $pH = -\log\left[H^+\right]$ $pH < 7$ acid; $pH > 7$ base

$$pH = -\log\left(3 \times 10^{-4}\right) = -\log 3 - \log 10^{-4} = -\log 3 + 4 = -0.477 + 4 \approx 3.5 < 7 \ \text{acid}$$

61. $$pH = -\log\left(3.7 \times 10^{-8}\right) = -\left(\log 3.7 + \log 10^{-8}\right) = -\log 3.7 + 8 = 7.4 > 7 \ \text{base}$$

63. $$pH = -\log\left(1.3 \times 10^{-5}\right) = -\left(\log 1.3 + \log 10^{-5}\right) = -\log 1.3 + 5 = 4.9 < 7 \ \text{acid}$$

65. $$-\log x = 7$$

$$\log x = -7$$

$$x = 10^{-7} \ \text{mole per liter}$$

67. a. $pH = -\log\left(3.5 \times 10^{-7}\right)$ b. $pH = -\log\left(3.5 \times 10^{-4}\right)$

$= -\left(\log 3.5 + \log 10^{-7}\right)$ $= -\left(\log 3.5 + \log 10^{-4}\right)$

$= -\log 3.5 + 7 \approx 6.5$ $= -\log 3.5 + 4 \approx 3.5$

Yes; plant because   No; do not plant because

pH = 6.5 is acceptable.   pH = 3.5 is not acceptable.

69. $-\log x = 7.0$ $-\log x = 8.5$

$\log x = -7$ $\log x = -8.5$

$x = 10^{-7}$ $x = 10^{-8.5}$

Paprika prefers soil that has hydrogen ion concentration from 3.16×10^{-9} to 10^{-7} mole per liter.

71. a. $(1 + 0.05)^t = 2$ b. $\left(1 + \dfrac{0.05}{4}\right)^{4t} = 2$

$\log(1 + 0.05)^t = \log 2$ $\log\left(1 + \dfrac{0.05}{4}\right)^{4t} = \log 2$

$t \log(1 + 0.05) = \log 2$ $4t \log\left(1 + \dfrac{0.05}{4}\right) = \log 2$

$t = \dfrac{\log 2}{\log(1 + 0.05)}$

$4t = \dfrac{\log 2}{\log\left(1 + \dfrac{0.05}{4}\right)}$

$t \approx 14.21$ years; $n \approx 14.21$ periods

$4t \approx 55.80$ periods (n)

$t \approx 13.95$ years

c. $\left(1 + \dfrac{0.05}{12}\right)^{12t} = 2$ d. $\left(1 + \dfrac{0.05}{365}\right)^{365t} = 2$

$\log\left(1 + \dfrac{0.05}{12}\right)^{12t} = \log 2$ $\log\left(1 + \dfrac{0.05}{365}\right)^{365t} = \log 2$

$12t \log\left(1 + \dfrac{0.05}{12}\right) = \log 2$ $365t \log\left(1 + \dfrac{0.05}{365}\right) = \log 2$

$12t = \dfrac{\log 2}{\log\left(1 + \dfrac{0.05}{12}\right)}$

$365t = \dfrac{\log 2}{\log\left(1 + \dfrac{0.05}{365}\right)}$

$12t \approx 166.70$ periods (n)

$365t \approx 5060.32$ periods (n)

$t \approx 13.89$ years

$t \approx 13.86$ years

e. The larger n is (the shorter the compounding period), the shorter the doubling time.

73. To accumulate $15,000:

$$10,000\left(1+\frac{0.08125}{365}\right)^{365t}=15,000$$

$$\left(1+\frac{0.08125}{365}\right)^{365t}=1.5$$

$$\log\left(1+\frac{0.08125}{365}\right)^{365t}=\log 1.5$$

$$365t\log\left(1+\frac{0.08125}{365}\right)=\log 1.5$$

$$365t=\frac{\log 1.5}{\log\left(1+\frac{0.08125}{365}\right)}$$

$$365t\approx 1821.68 \text{ periods}(n)$$

$$t\approx 4.99 \text{ years}$$

To accumulate $100,000:

$$10,000\left(1+\frac{0.08125}{365}\right)^{365t}=100,000$$

$$\left(1+\frac{0.08125}{365}\right)^{365t}=10$$

$$\log\left(1+\frac{0.08125}{365}\right)^{365t}=\log 10$$

$$365t\log\left(1+\frac{0.08125}{365}\right)=\log 10$$

$$365t=\frac{\log 10}{\log\left(1+\frac{0.08125}{365}\right)}$$

$$365t\approx 10,345.07 \text{ periods}(n)$$

$$t\approx 28.34 \text{ years}$$

75. To accumulate $30,000:

$$20,000\left(1+\frac{0.0625}{365}\right)^{365t}=30,000$$

$$\left(1+\frac{0.0625}{365}\right)^{365t}=1.5$$

$$\log\left(1+\frac{0.0625}{365}\right)^{365t}=\log 1.5$$

$$365t\log\left(1+\frac{0.0625}{365}\right)=\log 1.5$$

$$365t=\frac{\log 1.5}{\log\left(1+\frac{0.0625}{365}\right)}$$

$$365t\approx 2368.12 \text{ periods}(n)$$

$$t\approx 6.49 \text{ years}$$

To accumulate $100,000:

$$20,000\left(1+\frac{0.0625}{365}\right)^{365t}=100,000$$

$$\left(1+\frac{0.0625}{365}\right)^{365t}=5$$

$$\log\left(1+\frac{0.0625}{365}\right)^{365t}=\log 5$$

$$365t\log\left(1+\frac{0.0625}{365}\right)=\log 5$$

$$365t=\frac{\log 5}{\log\left(1+\frac{0.0625}{365}\right)}$$

$$365t\approx 9399.92 \text{ periods}(n)$$

$$t\approx 25.75 \text{ years}$$

77. a. $2^d=3$; $d=\frac{\log 3}{\log 2}$ b. $d=\frac{\log 3}{\log 2}=1.584962501\approx 1.58$

79. a. $3^d=6$; $d=\frac{\log 6}{\log 3}$ b. $d=\frac{\log 6}{\log 3}=1.630929754\approx 1.63$

81. a. $4^d=8$; $d=\frac{\log 8}{\log 4}$ b. $d=\frac{\log 8}{\log 4}=1.50$

10.1 Exponential Growth

1. $t=4$ $p=30e^{0.0198026273(4)}=32.473$

Population is 32,473.

285

3. $140e^{0.0676586485t} = 250$

   $e^{0.0676586485t} = 1.785714286$

   $0.0676586485t = \ln(1.785714286)$

   $t = 8.57$ years (8 years 7 months)

   July 2012

5. a. $(0, 9392)$ and $(4, 9786)$

   b. $\Delta t = 4 - 0 = 4$ years

   c. $\Delta p = 9,786 - 9,392$

      $= 394$ thousand people

   d. $\dfrac{\Delta p}{\Delta t} = \dfrac{394}{4}$

      $= 98.5$ thousand people/year

   e. $\dfrac{\left(\frac{\Delta p}{\Delta t}\right)}{p} = \dfrac{98.5}{9392} \approx 0.01 = 1\%$ per year

7. a. $(0, 18731)$ and $(4, 19007)$

   b. $\Delta t = 4 - 0 = 4$ years

   c. $\Delta p = 19,007 - 18,731$

      $= 276$ thousand people

   d. $\dfrac{\Delta p}{\Delta t} = \dfrac{276}{4} = 69$ thousand people/year

   e. $\dfrac{\left(\frac{\Delta p}{\Delta t}\right)}{p} = \dfrac{69}{18,731} \approx 0.0037$

      $= 0.37\%$ /year

9. a. $p = ae^{bt}$

      $(0, 9392) \Rightarrow a = 9,392$

      $p(t) = 9,392e^{bt}$

      $(4, 9786) \Rightarrow 9,786 = 9,392e^{4b}$

      $e^{4b} = \dfrac{9,786}{9,392} = 1.041950596$

      $4b = \ln(1.041950596) = 0.0410945298$

      $b = 0.0102736324$

      $p(t) = 9,392e^{0.0102736324t}$

   b. $t = 2012 - 2004 = 8$ years

      $p(8) = 9,392e^{0.0102736324(8)} \approx 10,196.529$ thousands $= 10,196,529$

   c. $t = 2017 - 2004 = 13$ years

      $p(13) = 9,392e^{0.0102736324(13)} \approx 10,733.992$ thousands $= 10,733,992$

286

d. $2(9,392) = 9,392e^{0.0102736324t}$

$2 = e^{0.0102736324t}$

$\ln 2 = 0.0102736324t$

$67.5 \approx t$

67.5 years from 2004

11. a. $(0, 18731) \Rightarrow p(t) = 18,731e^{bt}$

$(4, 19007) \Rightarrow 19,007 = 18,731e^{4b}$

$e^{4b} = 1.014734931$

$4b = \ln(1.014734931) \approx 0.0146274271$

$b = 0.0036568568$

$p(t) = 18,731e^{0.0036568568t}$

b. $t = 2010 - 2004 = 6$ years

$p(6) = 18,731e^{0.0036568568(6)} \approx 19,146.521 \text{ thousands} = 19,146,521$

c. $t = 2017 - 2004 = 13$ years

$p(13) = 18,731e^{0.0036568568(13)} \approx 19,642.961 \text{ thousands} = 19,642,961$

d. $1.5(18,731) = 18,731e^{0.0036568568t}$

$1.5 = e^{0.0036568568t}$

$\ln 1.5 = 0.0036568568t$

$110.9 \approx t$

110.9 years from 2004

13. a. $p(t) = ae^{bt}$

$(0, 2510) \Rightarrow a = 2,510 \qquad p(t) = 2,510e^{bt}$

$(3, 5380) \Rightarrow 5,380 = p(3) = 2,510e^{3b}$

$e^{3b} = \frac{5380}{2510} = 2.14$

$3b = \ln(2.14)$

$b = \frac{1}{3}\ln(2.14) \Rightarrow b = 0.2541352 \qquad p(t) = 2,510e^{0.2541352t}$

b. $p(7) = 2{,}510e^{0.2541352(7)} = 14{,}868$

c. $5020 = 2510e^{0.2541352t}$

$2 = e^{0.2541352t}$

$\ln 2 = 0.2541352t$

$t = \dfrac{\ln 2}{0.2541352} = 2.7 \text{ days}$

15. a. $t = 6$ months $\qquad p(t) = ae^{bt}$

$(0, 230000) \Rightarrow a = 230{,}000$

$p(t) = 230{,}000e^{bt}$

$310{,}000 = p(6) = 230{,}000e^{6b}$

$\dfrac{310{,}000}{230{,}000} = e^{6b}$

$6b = \ln\left(\dfrac{310{,}000}{230{,}000}\right) = 0.2984929886$

$b = 0.0497488314$

$p(t) = 230{,}000e^{0.0497488314t}$

c. $p(12) = 230{,}000e^{0.0497488314(12)}$

$= \$417{,}826$

b. $460{,}000 = 230{,}000e^{0.0497488314t}$

$2 = e^{0.0497488314t}$

$\ln 2 = 0.0497488314t$

$t = \dfrac{\ln 2}{0.0497488314} \approx 13.9 \approx 14 \text{ months}$

August 2009 + 14 months = October 2010

d. $\dfrac{\Delta v}{\Delta t} = \dfrac{417{,}826 - 230{,}000}{12}$

$= \$15{,}652 \text{ per month}$

17. a. $(0, 1.6) \Rightarrow p(t) = 1.6e^{bt}$

$(18, 5.0) \Rightarrow 5.0 = 1.6e^{18b}$

$e^{18b} = \dfrac{5.0}{1.6}$

$b = \dfrac{1}{18}\ln\left(\dfrac{5.0}{1.6}\right) = 0.0633019046$

$p(t) = 1.6e^{0.0633019046t}$

b. $p(t) = 1.6e^{0.0633019046t}$

$10.0 = 1.6e^{0.0633019046t}$

$e^{0.0633019046t} = \dfrac{10.0}{1.6}$

$t = \dfrac{1}{0.0633019046}\ln\dfrac{10.0}{1.6}$

$t \approx 28.94986297 = 29 \text{ months}$

There will be 10.1 billion tweets approximately by September 2010.

19. a. Let $t =$ the time after 2000.

$(0, 109478) \Rightarrow p(t) = 109,478e^{bt}$

$(2, 140766) \Rightarrow 140,766 = 109,478e^{2b}$

$e^{2b} = \dfrac{140,766}{109,478}$

$b = \dfrac{1}{2}\ln\left(\dfrac{140,766}{109,478}\right) = 0.1256876607$

$p(t) = 109,478e^{0.1256876607t}$

b. $t = 2004 - 2000 = 4$ years

$p(4) = 109,478e^{0.1256876607 \cdot 4} \approx 180,996$ thousands

c. $t = 2006 - 2000 = 6$ years

$p(6) = 109,478e^{0.1256876607 \cdot 6} \approx 232,723$ thousands

21. a. $(0, 3929214)$ $p(t) = 3,929,214e^{bt}$

$(10, 5308483)$ $3,929,214e^{10b} = p(10) = 5,308,483$

$e^{10b} = \dfrac{5,308,483}{3,929,214} = 1.351029239$

$b = \dfrac{1}{10}\ln(1.351029239) = 0.0300866701$

$p(t) = 3,929,214e^{0.0300866701t}$

b. 1810. The exponential growth can't continue forever.

c. 1810: $t = 20$ $p(20) = 7,171,916$

2000: $t = 210$ $p(210) = 2,179,040,956$

23. Africa:

$(0, 643)$ $p(t) = 643e^{bt}$

$(10, 794)$ $643e^{10b} = p(10) = 794$

$e^{10b} = \dfrac{794}{643}$

$b = \dfrac{1}{10}\ln\left(\dfrac{794}{643}\right)$

$= 0.0210938737$

$p(t) = 643e^{0.0210938737t}$

North America:

$(0, 277)$ $p(t) = 277e^{bt}$

$(10, 314)$ $277e^{10b} = p(10) = 314$

$e^{10b} = \dfrac{314}{277}$

$b = \dfrac{1}{10}\ln\left(\dfrac{314}{277}\right)$

$= 0.012537548$

$p(t) = 277e^{0.012537548t}$

289

Doubling Time: $e^{0.0210938737t} = 2$

$0.0210938737t = \ln 2$

$t \approx 32.9$ years

Doubling Time: $e^{0.012537548t} = 2$

$t = \dfrac{\ln 2}{0.012537548}$

$t \approx 55.3$ years

Europe:

$(0, 509) \qquad p(t) = 509e^{bt}$

$(10, 727) \qquad 509e^{10b} = p(10) = 727$

$e^{10b} = \dfrac{727}{509}$

$b = \dfrac{1}{10}\ln\left(\dfrac{727}{509}\right)$

$\quad = 0.0356478461$

$p(t) = 509e^{0.0356478461t}$

Doubling Time: $e^{0.0356478461t} = 2$

$t = \dfrac{\ln 2}{0.0356478461}$

$t \approx 19.4$ years

25. $\quad 10.566 = 5.283e^{0.014083947107t}$

$\quad 2 = e^{0.014083947107t}$

$\ln 2 = 0.014083947107t$

49 years $\approx t$

27. a. Each year the house is worth 10% more than its current value (not the original value)

b. $p(t) = 125,000e^{bt}$

(10% increase)

$125,000e^{b} = p(1) = 137,500$

$e^{b} = 1.1$

$b = \ln(1.1) = 0.0953101798$

$p(t) = 125,000e^{0.0953101798t}$

(50% increase)

$125,000e^{0.0953101798t} = 187,500$

$e^{0.0953101798t} = 1.5$

$0.0953101798t = \ln(1.5)$

$t \approx 4.25$ years

29. $$p = 18,323(1+i)^t$$

$$18,323(1+i)^4 = 18,710$$

$$(1+i)^4 = 1.021120995$$

$$1+i = 1.005238935$$

$$p = 18,323(1.005238935)^t$$

31. $$p = 9098(1+i)^t$$

$$9098(1+i)^4 = 9392$$

$$(1+i)^4 = 1.032314794$$

$$1+i = 1.007982606$$

$$p = 9098(1.007982606)^t$$

33. $e^{0.007950913484} = 1.007982606 = (1+i)$ so the equations are mathematically interchangeable.

35. $$2P = P\left(1+\frac{0.05}{365}\right)^n$$

$$2 = (1.000136986)^n$$

$$\ln 2 = n \cdot \ln(1.000136986)$$

$$n = 5060.3 \text{ days or } 13.86 \text{ years}$$

37. $$200 \cdot \frac{\left(1+\frac{0.05}{12}\right)^{12 \cdot t} - 1}{\frac{0.05}{12}} = 500,000$$

$$\left(1+\frac{0.05}{12}\right)^{12 \cdot t} - 1 = 10.41666667$$

$$\ln\left(\left(1+\frac{0.05}{12}\right)^{12 \cdot t}\right) = \ln(11.41666667)$$

$$12t \cdot \ln\left(1+\frac{0.05}{12}\right) = \ln(11.41666667)$$

$$12t = 586 \text{ months}$$

$$t = 48.8\overline{3} \text{ or } 48 \text{ years and } 10 \text{ months}$$

10.2 Exponential Decay

1. 3 weeks = 21 days

$$Q(21) = 20e^{-0.086643397(21)} = 3.2 \text{ g}$$

3. $Q(2) = 8.2e^{-0.053319013(2)} = 7.4$ g

5.   half-life = 2.6 hours

   a. $Q(t) = 50e^{bt}$

   $(2.6, 25) \Rightarrow Q(2.6) = 50e^{2.6b} = 25$

   $e^{2.6b} = \frac{1}{2}$

   $2.6b = \ln\left(\frac{1}{2}\right)$

   $b = -0.266595069$

   $Q(t) = 50e^{-0.266595069t}$

   b. $Q(1) = 50e^{-0.2665950694(1)} = 38.3$ mg

   c. $Q(24) = 50e^{-0.2665950694(24)} = 0.08$ mg

   d. $\dfrac{\Delta Q}{\Delta t} = \dfrac{38.3 - 50}{1} = -11.7$ mg per hour

   e. $\left(\dfrac{\Delta Q}{\Delta t}\right) \Big/ Q = \dfrac{-11.7 \text{ mg per hour}}{50 \text{ mg}} = -0.234$ per hour or -23.4% per hour.

   f. $\dfrac{\Delta Q}{\Delta t} = \dfrac{0.1 - 50}{24} = -2.1$ mg per hour

   g. $\left(\dfrac{\Delta Q}{\Delta t}\right) \Big/ Q = \dfrac{-2.1 \text{ mg per hour}}{50 \text{ mg}} = -0.042$ per hour or -4.2% per hour.

   h.   Radioactive substances decay faster when there is more substance present. The rate of decay is proportional to the amount present.

7.   half-life = 21.0 hours

   a. 21.0 hours   b. 42.0 hours   c. 63.0 hours

9.   half-life = 13 years

   $Q(t) = 500e^{bt}$ $Q(t) = 500e^{-0.0533190139t}$

   $(13, 250) \Rightarrow Q(13) = 500e^{13b} = 250$ $500e^{-0.0533190139t} = 100$

   $e^{13b} = \frac{1}{2}$ $e^{-0.0533190139t} = 0.2$

   $13b = \ln\left(\frac{1}{2}\right)$ $-0.0533190139t = \ln(0.2)$

   $b \approx -0.0533190139$ $t \approx 30.2$ years

11. half-life = 86 years

    $Q(t) = 30e^{bt}$

    $(86, 15) \Rightarrow Q(86) = 30e^{86b} = 15$

    $e^{86b} = \frac{1}{2}$

    $86b = \ln\left(\frac{1}{2}\right)$

    $b \approx -0.0080598509$

    $Q(t) = 30e^{-0.0080598509t}$

    $30e^{-0.0080598509t} = 20$

    $e^{-0.0080598509t} = \frac{2}{3}$

    $-0.0080598509t = \ln\left(\frac{2}{3}\right)$

    $t \approx 50.3$ years

13. half-life = 24,400 years

    $Q(t) = ae^{bt}$

    $\left(24400, \frac{1}{2}a\right) \Rightarrow Q(24400) = ae^{24400b} = \frac{1}{2}a$

    $e^{24400b} = \frac{1}{2}$

    $24400b = \ln\left(\frac{1}{2}\right)$

    $b \approx -0.0000284077$

    $Q(t) = ae^{-0.0000284077t}$

    If it loses 90%, the amount will be .1a

    $ae^{-0.0000284077t} = 0.1a$

    $e^{-0.0000284077t} = 0.1$

    $-0.0000284077t = \ln(0.1)$

    $t \approx 81,055$ years

15. half-life = 10 seconds

    $Q(t) = ae^{bt}$

    $\left(10, \frac{1}{2}a\right) \Rightarrow Q(10) = ae^{10b} = \frac{1}{2}a$

    $e^{10b} = \frac{1}{2}$

    $10b = \ln\left(\frac{1}{2}\right)$

    $b \approx -0.0693147181$

    $Q(t) = ae^{-0.0693147181t}$

    If it loses 99.9%, the amount will be $0.001a$.

    $ae^{-0.0693147181t} = 0.001a$

    $e^{-0.0693147181t} = 0.001$

    $-0.0693147181t = \ln(0.001)$

    $t \approx 99.7$ seconds

17. half-life = 5,730 years

$Q(t) = ae^{bt}$

$\left(5730, \frac{1}{2}a\right) \Rightarrow Q(5730) = ae^{5730b} = \frac{1}{2}a$

$e^{5730b} = \frac{1}{2}$

$5730b = \ln\left(\frac{1}{2}\right)$

$b \approx -0.0001209681$

$Q(t) = ae^{-0.0001209681t}$

$Q(5250) = ae^{-0.0001209681(5250)} = 0.53a$

53% of the original amount expected in

a living organism

19. $Q(t) = 58e^{bt}$

$(10, 52) \Rightarrow Q(10) = 58e^{10b} = 52$

$e^{10b} = 0.8965517241$

$10b = \ln(0.8965517241)$

$b = -0.0109199292$

$Q(t) = 58e^{-0.0109199292t}$

$58e^{-0.0109199292t} = 29$

$e^{-0.0109199292t} = 0.5$

$-0.0109199292t = \ln(0.5)$

21. $Q(t) = ae^{-0.0001209681t} = 0.84a$

$e^{-0.0001209681t} = 0.84$

$-0.0001209681t = \ln(0.84)$

$t \approx 1,441$ years old

23. If 83% is lost, 17% remains.

$Q(t) = ae^{-0.0001209681t} = 0.17a$

$e^{-0.0001209681t} = 0.17$

$-0.0001209681t = \ln(0.17)$

$t \approx 14,648$ years old

25. $Q(t) = ae^{-0.0001209681t} = 0.7a$

$e^{-0.0001209681t} = 0.7$

$-0.0001209681t = \ln(0.7)$

$t \approx 2,949$ years old

27. $Q(t) = ae^{-0.0001209681t}$

$t = 1988 - 1350 = 638$

$Q(638) = ae^{-0.0001209681(638)} \approx 0.926a$

92.6% of the original amount expected in a living organism

$t = 1988 - 33 = 1955$

$Q(1955) = ae^{-0.0001209681(1955)} \approx 0.789a$

78.9% of the original amount expected in a living organism

29. $Q(t) = ae^{-0.0001209681t}$

    $Q(5730) = ae^{-0.0001209681(5730)} \approx 0.5a$

    50% of the original amount expected in a living organism

31. $Q(t) = ae^{-0.0001209681t}$

    $Q(5000) = ae^{-0.0001209681(5000)} \approx 0.546a$

    A 5,000 year old mummy should have

    about 55% of the original amount.

    62% remaining carbon-14 would

    indicate approximately 3,950 years.

    The museum's claim is not justified.

33. If it lost 63.5%, then 36.5% remains.

    $Q(t) = ae^{-0.0001209681t} = 0.365a$

    $e^{-0.0001209681t} = 0.365$

    $-0.0001209681t = \ln(0.365)$

    $t \approx 8,300$ years old

35. a. $Q = ae^{-0.1151407277t}$

    $t = 2$ hr; $Q = ae^{-0.1151407277(2)} \approx 0.794a$

    $t = 3$ hr; $Q = ae^{-0.1151407277(3)} \approx 0.708a$

    $t = 4$ hr; $Q = ae^{-0.1151407277(4)} \approx 0.631a$

    $t = 5$ hr; $Q = ae^{-0.1151407277(5)} \approx 0.562a$

t hours after injection (hr)	1	2	3	4	5
Q portion remaining (%)	0.891	0.794	0.708	0.631	0.562

b.

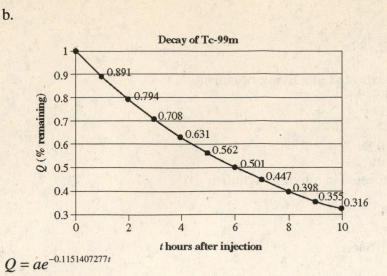

37. $Q = ae^{-0.1151407277t}$

$0.005 = e^{-0.1151407277t}$

$t = \dfrac{\ln 0.005}{-0.1151407277}$

$= 46.0 \text{ hours}$

39. a. From the article on pg. 784, there are 3.7×10^7 atomic disintegrations/second.

b. $10 \times \left(3.7 \times 10^7\right) = 3.7 \times 10^8$ atomic disintegrations/second

c. $2 \times \left(3.7 \times 10^8\right) = 7.4 \times 10^8$ atomic disintegrations/second

10.3 Logarithmic Scales

1. $M = \log A - \log A_0 = \log\left(3.9 \times 10^4\right) - (-3.0) \approx 7.6$

3. $M = \log A - \log A_0 = \log 250 - (-2.8) \approx 5.2$

Since the measurement is independent of distance, the other measurements will yield the same M value.

296

5.   a. $1989: M_2 = 7.1, \quad 1906: M_1 = 8.3$

$$M_1 - M_2 = \log\left(\frac{A_1}{A_2}\right)$$

$$8.3 - 7.1 = \log\left(\frac{A_1}{A_2}\right)$$

$$1.2 = \log\left(\frac{A_1}{A_2}\right)$$

$$10^{1.2} = \frac{A_1}{A_2}$$

$$A_1 \approx 15.8 A_2$$

The 1906 earthquake's amplitude was almost 16 times that of the 1989 quake.

b.   1906: $\log E_1 \approx 11.8 + 1.45M = 11.8 + 1.45(8.3) = 23.835$

$$E_1 \approx 10^{23.835}$$

1989: $\log E_2 \approx 11.8 + 1.45M = 11.8 + 1.45(7.1) = 22.095$

$$E_2 \approx 10^{22.095}$$

$$\frac{E_1}{E_2} = \frac{10^{23.835}}{10^{22.095}} = 10^{1.74} \approx 55$$

$$E_1 \approx 55 E_2$$

The 1906 quake released about 55 times as much energy as the 1989 quake.

7.   a.   SF: $M_1 = 8.3$,   LA: $M_2 = 6.8$

$$M_1 - M_2 = \log\left(\frac{A_1}{A_2}\right)$$

$$8.3 - 6.8 = \log\left(\frac{A_1}{A_2}\right)$$

$$1.5 = \log\left(\frac{A_1}{A_2}\right)$$

$$10^{1.5} = \frac{A_1}{A_2}$$

$$A_1 \approx 31.6A_2$$

The San Francisco earthquake's amplitude was about 32 times that of the LA quake.

b. SF: $\log E_1 \approx 11.8 + 1.45M = 11.8 + 1.45(8.3) = 23.835$

$$E_1 \approx 10^{23.835}$$

LA: $\log E_2 \approx 11.8 + 1.45M = 11.8 + 1.45(6.8) = 21.66$

$$E_2 \approx 10^{21.66}$$

$$\frac{E_1}{E_2} = \frac{10^{23.835}}{10^{21.66}} = 10^{2.175} \approx 149.6$$

$$E_1 \approx 150E_2$$

The San Francisco quake released about 150 times as much energy as the LA quake.

9.   a. Indian Ocean: $M_1 = 9.2$, San Francisco (1989): $M_2 = 7.1$

$$M_1 - M_2 = \log\left(\frac{A_1}{A_2}\right)$$

$$9.2 - 7.1 = \log\left(\frac{A_1}{A_2}\right)$$

$$2.1 = \log\left(\frac{A_1}{A_2}\right)$$

$$10^{2.1} = \frac{A_1}{A_2}$$

$$A_1 \approx 10^{2.1}, \quad A_2 \approx 126 A_2$$

The Indian Ocean earthquake's amplitude was about 126 times that of the 1989 San Francisco quake.

b. Indian Ocean: $\log E_1 \approx 11.8 + 1.45M = 11.8 + 1.45(9.2) = 25.14$

$$E_1 \approx 10^{25.14}$$

298

San Francisco (1989): $\log E_2 \approx 11.8 + 1.45M = 11.8 + 1.45(7.1) = 22.095$

$$E_2 \approx 10^{22.095}$$

$$\frac{E_1}{E_2} = \frac{10^{25.14}}{10^{22.095}} = 10^{3.045} \approx 1109$$

$$E_1 \approx 10^{3.045} E_2 \approx 1,109 E_2$$

The Indian Ocean quake released about 1,109 times as much energy as the 1989 San Francisco quake.

11.   a.   New Madrid: $M_1 = 8.7$,   Coalinga: $M_2 = 6.5$

$$M_1 - M_2 = \log\left(\frac{A_1}{A_2}\right)$$

$$8.7 - 6.5 = \log\left(\frac{A_1}{A_2}\right)$$

$$2.2 = \log\left(\frac{A_1}{A_2}\right)$$

$$10^{2.2} = \frac{A_1}{A_2}$$

$$A_1 \approx 158 A_2$$

The New Madrid earthquake's amplitude was more than 158 times that of the Coalinga quake.

   b.   New Madrid: $\log E_1 \approx 11.8 + 1.45M = 11.8 + 1.45(8.7) = 24.415$

$$E_1 \approx 10^{24.415}$$

LA: $\log E_2 \approx 11.8 + 1.45M = 11.8 + 1.45(6.5) = 21.225$

$$E_2 \approx 10^{21.225}$$

$$\frac{E_1}{E_2} = \frac{10^{24.415}}{10^{21.225}} = 10^{3.19} \approx 1,548.8$$

$$E_1 \approx 1,549 E_2$$

The New Madrid quake released about 1,549 times as much energy as the Coalinga quake.

13. a. $M_1 - M_2 = \log\left(\dfrac{A_1}{A_2}\right)$

b. $\log E_1 \approx 11.8 + 1.45M = 11.8 + 1.45(7.1) = 22.095$

$7.1 - 7.0 = \log\left(\dfrac{A_1}{A_2}\right)$

$E_1 \approx 10^{22.095}$

$0.1 = \log\left(\dfrac{A_1}{A_2}\right)$

$\log E_2 \approx 11.8 + 1.45M = 11.8 + 1.45(7.0) = 21.95$

$10^{0.1} = \dfrac{A_1}{A_2}$

$E_2 \approx 10^{21.95}$

$A_1 \approx 1.26A_2$

$\dfrac{E_1}{E_2} = \dfrac{10^{22.095}}{10^{21.95}} = 10^{0.145} \approx 1.4$

About a 26% increase.

$E_1 \approx 1.4E_2$

About a 40% increase.

15. $M_1 = M_2 + 1$

$\dfrac{E_1}{E_2} = \dfrac{10^{11.8}10^{1.45}10^{1.45M_2}}{10^{11.8}10^{1.45M_2}} = 10^{1.45} \approx 28.2$

$\log E_1 \approx 11.8 + 1.45M_1 = 11.8 + 1.45(M_2 + 1)$

$E_2 \approx 10^{11.8 + 1.45(M_2)} = 10^{11.8}10^{1.45M_2}$

$E_1 \approx 10^{11.8 + 1.45(M_2 + 1)} = 10^{11.8}10^{1.45}10^{1.45M_2}$

$E_1 \approx 28E_2$

$\log E_2 \approx 11.8 + 1.45(M_2)$

Energy released is magnified by a factor of 28.

17. $D = 10 \cdot \log\left(\dfrac{10^{-9}}{10^{-16}}\right) = 10 \cdot \log(10^7) = 70\,\text{dB}$

19. $D = 10 \cdot \log\left(\dfrac{2.5 \times 10^{-9}}{10^{-16}}\right) = 10 \cdot \log(25,000,000) \approx 74\,\text{dB}$

21. $D_1 - D_2 = 10 \cdot \log\left(\dfrac{10^{-13}}{10^{-14}}\right) = 10\,\text{dB gain}$

23. $D_1 - D_2 = 10 \cdot \log\left(\dfrac{3.2 \times 10^{-8}}{3.9 \times 10^{-9}}\right) = 9.1\,\text{dB gain}$

25. $D_1 - D_2 = 10 \cdot \log\left(\dfrac{I_1}{I_2}\right)$

      $81 - 74 = 10 \cdot \log\left(\dfrac{I_1}{I_2}\right)$

          $0.7 = \log\left(\dfrac{I_1}{I_2}\right)$

            $\dfrac{I_1}{I_2} = 10^{0.7} \approx 5$

              $I_1 \approx 5I_2$

It requires 5 singers to reach the higher dB level. Thus, 4 singers joined him.

27. $77 - 74 = 10 \cdot \log\left(\dfrac{I_1}{I_2}\right)$

         $0.3 = \log\left(\dfrac{I_1}{I_2}\right)$

           $\dfrac{I_1}{I_2} = 10^{0.3} \approx 2$

            $I_1 \approx 2I_2$

It requires 2 singers to reach the higher dB level. Thus, 1 singer joined her.

29. $85.8 - 78 = 10 \cdot \log\left(\dfrac{I_1}{I_2}\right)$

         $0.78 = \log\left(\dfrac{I_1}{I_2}\right)$

           $\dfrac{I_1}{I_2} = 10^{0.78} \approx 6$

            $I_1 \approx 6I_2$

It requires 6 players to reach the higher dB level. Thus, 5 players joined in.

31. $I_1 = 2I_2$

$$D_1 - D_2 = 10 \cdot \log\left(\dfrac{I_1}{I_2}\right) = 10 \cdot \log\left(\dfrac{2I_2}{I_2}\right) = 10 \cdot \log 2 \approx 3 \text{ dB gain}$$

33.　a.　$130 - 100 = 10 \cdot \log\left(\dfrac{I_1}{I_2}\right)$

$$3.0 = \log\left(\dfrac{I_1}{I_2}\right)$$

$$\dfrac{I_1}{I_2} = 10^{3.0} \approx 1{,}000$$

$I_1 = 1{,}000 I_2$　American iPod's impact is about 1,000 times that of the European iPod.

　b.　American: less than 2 minutes, European: 2 hours

35.　$180 - 215 = 10 \cdot \log\left(\dfrac{I_1}{I_2}\right)$

$$-3.5 = \log\left(\dfrac{I_1}{I_2}\right)$$

$$I_1 = 10^{-3.5}\, I_2$$

Chapter 10 Review

1.　$x = 4$

3.　$x = 5^5 = 3{,}125$

5.　$\ln\left(e^x\right) = x$ and $e^{\ln x} = x$

7.　$\ln\dfrac{A}{B} = \ln A - \ln B$

9.　$\ln\left(A^n\right) = n \cdot \ln A$

11.　$\ln\left(A \cdot B\right) = \ln A + \ln B$

13.　$\log(x + 2)$

15.　$e^{0.03x} = \dfrac{730}{520} = 1.403846154$

$$0.03x = \ln(1.403846154)$$

$$x = 11.30719075$$

17.　$\log\left(\dfrac{5x \cdot x^2}{x}\right) = 12$

$$\log\left(5x^2\right) = 12$$

$$5x^2 = 10^{12}$$

$$x^2 = 200{,}000{,}000{,}000$$

$$x = 447{,}213.6$$

19. a. half-life = 5.3 years

$$p(t) = 300e^{bt}$$

$$(5.3, 150) \Rightarrow p(5.3) = 300e^{5.3b} = 150$$

$$e^{5.3b} = 0.5$$

$$5.3b = \ln 0.5$$

$$b = -0.130782486$$

$$p(t) = 300e^{-0.130782486t}$$

b. 2 weeks = 0.0384615385 years

$$p(0.0384615385) = 300e^{-0.1307824869(0.0384615385)} = 298.49 \text{ g}$$

They lose about 1.5 g

c. $p(1) = 300e^{-0.1307824869(1)} = 263.2 \text{ g}$

d. If 90% decays, 10% is left.

$$300e^{-0.1307824869t} = 30$$

$$e^{-0.1307824869t} = 0.1$$

$$-0.1307824869t = \ln 0.1$$

$$t = 17.6 \text{ years}$$

21. $M = \log A - \log A_0 = \log(25) - (-1.7) = 3.1$ on the Richter scale

23. $D = 10 \cdot \log\left(\dfrac{1.6 \times 10^{-6}}{10^{-16}}\right) \approx 102 \, dB$

25. $84 - 78 = 10 \cdot \log\left(\dfrac{I_1}{I_2}\right)$

$$0.6 = \log\left(\dfrac{I_1}{I_2}\right)$$

$$\dfrac{I_1}{I_2} = 10^{0.6} \approx 4$$

$$I_1 \approx 4I_2$$

It requires 4 players to reach the highest dB level.   Thus, 3 trumpet players have joined in.

303

11.0 Review of Matrices

1.   a. 3×2 3.   a. 2×1 5.   a. 1×2

     b. none of these   b.   column matrix   b.   row matrix

7.   a. 3×3 9.   a. 3×1

     b. square matrix   b.   column matrix

11. $a_{21} = 22$ 13. $c_{21} = 41$ 15. $e_{11} = 3$

16. $f_{22} = -3$ 17. $g_{12} = -11$ 19. $j_{21} = -3$

21.   a. $AC = \begin{bmatrix} 5 \cdot 23 + 0 \cdot 41 \\ 22 \cdot 23 - 3 \cdot 41 \\ 18 \cdot 23 + 9 \cdot 41 \end{bmatrix} = \begin{bmatrix} 115 \\ 383 \\ 783 \end{bmatrix}$ b. CA does not exist.

23.   a. AD does not exist.   b. DA does not exist.

25.   a. CG does not exist.   b. GC does not exist.

27.   a. JB does not exist.   b. $BJ = \begin{bmatrix} 2243 \\ 1056 \\ 52 \end{bmatrix}$

29.   a. $AF = \begin{bmatrix} -10 & 50 \\ -56 & 229 \\ 0 & 153 \end{bmatrix}$ b. FA does not exist.

31. $\begin{bmatrix} 5 & 3 \end{bmatrix} \begin{bmatrix} 14 & 19 \\ 20 & 25 \end{bmatrix} = \begin{bmatrix} 5 \cdot 14 + 3 \cdot 20 & 5 \cdot 19 + 3 \cdot 25 \end{bmatrix} = \begin{bmatrix} 130 & 170 \end{bmatrix}$

    Sale: \$130, Regular: \$170

33. $\begin{bmatrix} 2 & 1 \end{bmatrix} \begin{bmatrix} 1.25 & 1.30 \\ 0.95 & 1.10 \end{bmatrix} = \begin{bmatrix} 3.45 & 3.70 \end{bmatrix}$

    Blondie: \$3.45, Slice Man: \$3.70

35.   In matrix form:

$$\begin{bmatrix} 4 & 3 & 3 & 2 \\ 2 & 4 & 4 & 2 \\ 3 & 4 & 3 & 4 \end{bmatrix} \begin{bmatrix} 4 \\ 4 \\ 3 \\ 3 \end{bmatrix} = \begin{bmatrix} 43 \\ 42 \\ 49 \end{bmatrix}$$

304

$$\frac{1}{14}\begin{bmatrix} 43 \\ 42 \\ 49 \end{bmatrix} = \begin{bmatrix} 3.07 \\ 3.00 \\ 3.50 \end{bmatrix}$$

Jim: 3.07, Eloise: 3.00, Sylvie: 3.50

37.   a. $\begin{bmatrix} 105.8 & 105.5 \\ 61.9 & 66.3 \end{bmatrix}\begin{bmatrix} -1 \\ 1 \end{bmatrix} = \begin{bmatrix} 105.8(-1)+105.5(1) \\ 61.9(-1)+66.3(1) \end{bmatrix} = \begin{bmatrix} -0.3 \\ 4.4 \end{bmatrix}$

     b. This represents the change in sales from 2002 to 2003 for hotels and restaurants.

39.   a. $\begin{bmatrix} 6.25 & 8.97 & 4.97 & 24.85 & 6.98 & 3.88 \\ 6.10 & 8.75 & 5.25 & 22.12 & 6.98 & 3.75 \end{bmatrix}$ $\begin{bmatrix} 2 \\ 1 \\ 2 \\ 1 \\ 9 \\ 4 \end{bmatrix}$

     b. $\begin{bmatrix} 6.25(2)+8.97(1)+4.97(2)+24.85(1)+6.98(9)+3.88(4) \\ 6.10(2)+8.75(1)+5.25(2)+22.12(1)+6.98(9)+3.75(4) \end{bmatrix} = \begin{bmatrix} 134.60 \\ 131.39 \end{bmatrix}$

      The cost is \$134.60 at Piedmont Lumber and \$131.39 at Truitt and White.

41.   a. $\begin{bmatrix} 53,594 & 64,393 & 100,237 & 63,198 \end{bmatrix}$

     b. $\begin{bmatrix} 0.9885 & 0.0015 & 0.0076 & 0.0024 \\ 0.0011 & 0.9901 & 0.0057 & 0.0032 \\ 0.0018 & 0.0042 & 0.9897 & 0.0043 \\ 0.0017 & 0.0035 & 0.0077 & 0.9870 \end{bmatrix}$

     c. $\begin{bmatrix} 53,594 & 64,393 & 100,237 & 63,198 \end{bmatrix}\begin{bmatrix} 0.9885 & 0.0015 & 0.0076 & 0.0024 \\ 0.0011 & 0.9901 & 0.0057 & 0.0032 \\ 0.0018 & 0.0042 & 0.9897 & 0.0043 \\ 0.0017 & 0.0035 & 0.0077 & 0.9870 \end{bmatrix}$

      $= \begin{bmatrix} 53,336 & 64,478 & 100,466 & 63,142 \end{bmatrix}$ This is the new population in 2001.

43. $BC = \begin{bmatrix} 1 \\ 22 \end{bmatrix}$, so $A(BC) = \begin{bmatrix} 106 \\ 68 \end{bmatrix}$ $AB = \begin{bmatrix} 17 & 20 & -6 \\ -3 & 12 & -8 \end{bmatrix}$, so $(AB)C = \begin{bmatrix} 106 \\ 68 \end{bmatrix}$

45. $\begin{bmatrix} 3 & -2 \\ 4 & 0 \end{bmatrix}$ 47.   Does not exist

49.
$$\begin{bmatrix} 19 & 7 & 34 \\ 74 & 0 & -11 \\ 13 & -2 & 44 \end{bmatrix}$$

63.
$$BA = \begin{bmatrix} 62 & 32 \\ -40 & 56 \end{bmatrix}$$

65. a.
$$EF = \begin{bmatrix} -1178 & -2101 & -2378 \\ 5970 & -3091 & 498 \\ 5580 & -660 & 2340 \end{bmatrix}$$

   b.
$$FE = \begin{bmatrix} 1892 & -2520 \\ 10,723 & -3821 \end{bmatrix}$$

67. a.
$$(BF)C = \begin{bmatrix} 109,348 & 23,813 & -16,663 \\ -23,840 & 1688 & -7720 \end{bmatrix}$$

   b.
$$B(FC) = \begin{bmatrix} 109,348 & 23,813 & -16,663 \\ -23,840 & 1688 & -7720 \end{bmatrix}$$

69. a.
$$C^2 = \begin{bmatrix} 376 & 64 & 281 \\ -932 & -1203 & 614 \\ 952 & -808 & 293 \end{bmatrix}$$

   b.
$$C^5 = \begin{bmatrix} 4,599,688 & -2,586,492 & 9,810,101 \\ -1,887,820 & -24,086,567 & 2,293,357 \\ 42,244,296 & -7,140,820 & -4,471,911 \end{bmatrix}$$

11.1 Markov Chains

1. a. 0.9   b. 0.2 = 20%   e.

   c. $PT = \begin{bmatrix} 0.8 & 0.2 \end{bmatrix}\begin{bmatrix} 0.1 & 0.9 \\ 0.7 & 0.3 \end{bmatrix} = \begin{bmatrix} 0.22 & 0.78 \end{bmatrix}$

   d. $PT^2 = PT \cdot T$

   $= \begin{bmatrix} 0.22 & 0.78 \end{bmatrix}\begin{bmatrix} 0.1 & 0.9 \\ 0.7 & 0.3 \end{bmatrix} = \begin{bmatrix} 0.568 & 0.432 \end{bmatrix}$

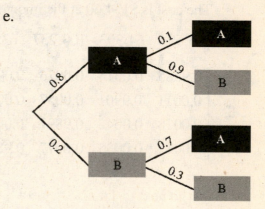

3. a. 0.8   b. 0.4 = 40%   e

   c. $PT = \begin{bmatrix} 0.6 & 0.4 \end{bmatrix}\begin{bmatrix} 0.2 & 0.8 \\ 0 & 1 \end{bmatrix} = \begin{bmatrix} 0.12 & 0.88 \end{bmatrix}$

   d. $PT^2 = PT \cdot T$

   $= \begin{bmatrix} 0.12 & 0.88 \end{bmatrix}\begin{bmatrix} 0.2 & 0.8 \\ 0 & 1 \end{bmatrix} = \begin{bmatrix} 0.024 & 0.976 \end{bmatrix}$

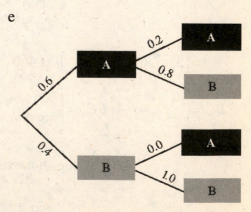

5.   a. $p(\text{a cola drinker chooses KickKola}) = 0.14$

$p(\text{a cola drinker doesn't choose KickKola}) = 1 - 0.14 = 0.86$

b. $\begin{bmatrix} 0.14 & 0.86 \end{bmatrix}$

7.   a. $p(\text{health club user in Metropolis uses Silver's Gym}) = 0.48$

$p(\text{health club user in Metropolis uses Fitness Lab}) = 0.37$

$p(\text{health club user in Metropolis uses ThinNFit}) = 1 - 0.48 - 0.37 = 0.15$

b. $\begin{bmatrix} 0.48 & 0.37 & 0.15 \end{bmatrix}$

9.   a. $p(\text{next purchases is KickKola}\,|\,\text{current purchase is not KickKola}) = 0.12$

$p(\text{next purchase is not KickKola}\,|\,\text{current purchase is not KickKola}) = 0.88$

$p(\text{next purchase is KickKola}\,|\,\text{current purchase is KickKola}) = 0.63$

$p(\text{next purchase is not KickKola}\,|\,\text{current purchase is KickKola}) = 0.37$

b.   The transition matrix is given by:

$$\begin{array}{cc} K & K' \\ \begin{bmatrix} 0.63 & 0.37 \\ 0.12 & 0.88 \end{bmatrix} & \begin{array}{c} K \\ K' \end{array} \end{array}$$

11.   a. $p(\text{go next to Silvers}\,|\,\text{go now to Silvers}) = 0.71$

$p(\text{go next to Fitness Lab}\,|\,\text{go now to Silvers}) = 0.12$

$p(\text{go next to ThinNFit}\,|\,\text{go now to Silvers}) = 1 - 0.71 - 0.12 = 0.17$

$p(\text{go next to Silvers}\,|\,\text{go now to Fitness Lab}) = 0.32$

$p(\text{go next to ThinNFit}\,|\,\text{go now to Fitness Lab}) = 0.34$

$p(\text{go next to Fitness lab}\,|\,\text{go now to Fitness Lab}) = 1 - 0.32 - 0.34 = 0.34$

$p(\text{go next to ThinNFit}\,|\,\text{go now to ThinNFit}) = 0.96$

$p(\text{go next to Silvers}\,|\,\text{go now to ThinNFit}) = 0.02$

$p\left(\text{go next to Fitness Lab}\,\middle|\,\text{go now to ThinNFit}\right) = 0.02$

b. The transition matrix is given by:

$$\begin{array}{ccc} S & F & T \end{array}$$
$$\begin{bmatrix} 0.71 & 0.12 & 0.17 \\ 0.32 & 0.34 & 0.34 \\ 0.02 & 0.02 & 0.96 \end{bmatrix} \begin{array}{c} S \\ F \\ T \end{array}$$

13. a.

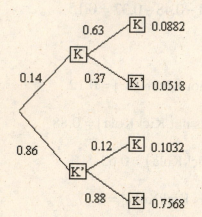

Market share: $0.0882 + 0.1032 = 0.1914 \approx 19\%$

b.

Market share:

$0.055566 + 0.006216 + 0.065016 + 0.090816$

$= 0.217614 \approx 22\%$

c. The first following purchase:

$$\begin{bmatrix} 0.14 & 0.86 \end{bmatrix} \begin{bmatrix} 0.63 & 0.37 \\ 0.12 & 0.88 \end{bmatrix} = \begin{bmatrix} 0.1914 & 0.8086 \end{bmatrix}$$
    Market share: $0.1914 \approx 19\%$

d. The second following purchase:

$$\begin{bmatrix} 0.14 & 0.86 \end{bmatrix} \begin{bmatrix} 0.63 & 0.37 \\ 0.12 & 0.88 \end{bmatrix}^2 = \begin{bmatrix} 0.217614 & 0.782386 \end{bmatrix}$$ Market share: $0.217614 \approx 22\%$

e. $[0.14 \quad 0.86]\begin{bmatrix} 0.63 & 0.37 \\ 0.12 & 0.88 \end{bmatrix}^6 = [0.2431 \quad 0.7569]$ Market share: $0.2431 \approx 24\%$

f. Trees don't require any knowledge of matrices and are very visual. The matrices are quicker and don't require drawings.

15. a. In one year:

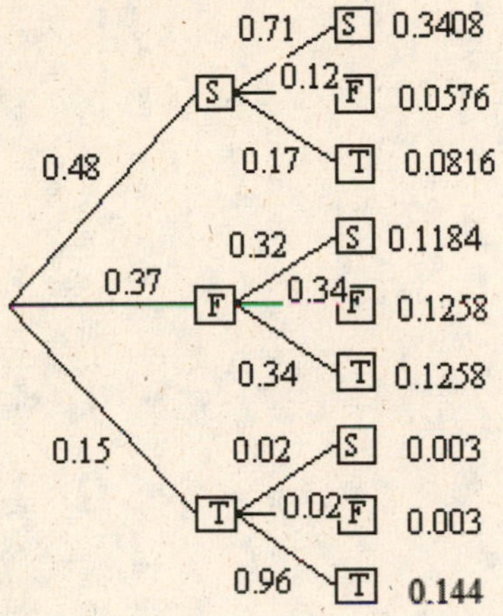

Market share:

Silver's $= 0.3408 + 0.1184 + 0.003 = 0.4622 \approx 46\%$

Fitness Lab $= 0.0576 + 0.1258 + 0.003 = 0.1864 \approx 19\%$

ThinNFit $= 0.0816 + 0.1258 + 0.144 = 0.3514 \approx 35\%$

b. In one year:

$$[0.48 \quad 0.37 \quad 0.15]\begin{array}{c} \begin{matrix} S & F & T \end{matrix} \\ \begin{bmatrix} 0.71 & 0.12 & 0.17 \\ 0.32 & 0.34 & 0.34 \\ 0.02 & 0.02 & 0.96 \end{bmatrix} \end{array}\begin{matrix} S \\ F \\ T \end{matrix} = [0.4622 \quad 0.1864 \quad 0.3514]$$

Market share: Silver's $= 0.4622 \approx 46\%$, Fitness Lab $= 0.1864 \approx 19\%$, ThinNFit $= 0.3514 \approx 35\%$

c. In two years:

$$[0.48 \quad 0.37 \quad 0.15]\begin{bmatrix} 0.71 & 0.12 & 0.17 \\ 0.32 & 0.34 & 0.34 \\ 0.02 & 0.02 & 0.96 \end{bmatrix}^2 = [0.3948 \quad 0.1259 \quad 0.4793]$$

Market share: Silver's $= 0.3948 \approx 39\%$, Fitness Lab $= 0.1259 \approx 13\%$, ThinNFit $= 0.4793 \approx 48\%$

d. In three years:

$$[0.48 \quad 0.37 \quad 0.15]\begin{bmatrix} 0.71 & 0.12 & 0.17 \\ 0.32 & 0.34 & 0.34 \\ 0.02 & 0.02 & 0.96 \end{bmatrix}^3 = [0.3302 \quad 0.0998 \quad 0.5700]$$

Market share: Silver's $= 0.3302 \approx 33\%$, Fitness Lab $= 0.0998 \approx 10\%$, ThinNFit $= 0.5700 \approx 57\%$

e. $[0.48 \quad 0.37 \quad 0.15] \begin{bmatrix} 0.71 & 0.12 & 0.17 \\ 0.32 & 0.34 & 0.34 \\ 0.02 & 0.02 & 0.96 \end{bmatrix}^5 = [0.2371 \quad 0.0750 \quad 0.6879]$

Market share: Silver's $= 0.2371 \approx 24\%$, Fitness Lab $= 0.0750 \approx 8\%$, ThinNFit $= 0.6879 \approx 69\%$

(More than 100% due to rounding.)

17.   a. $[0.32 \quad 0.68] \begin{bmatrix} 0.97 & 0.03 \\ 0.12 & 0.88 \end{bmatrix} = [0.39 \quad 0.61]$

         Homeowners: 39%; Renters: 61%

      b. $[0.32 \quad 0.68] \begin{bmatrix} 0.97 & 0.03 \\ 0.12 & 0.88 \end{bmatrix}^2 = [0.45 \quad 0.55]$

         Homeowners: 45%; Renters: 55%

      c. $[0.32 \quad 0.68] \begin{bmatrix} 0.97 & 0.03 \\ 0.12 & 0.88 \end{bmatrix}^3 = [0.51 \quad 0.49]$

         Homeowners: 51%; Renters: 49%.

19.   In four years:

$[0.41 \quad 0.59] \begin{bmatrix} 0.12 & 0.88 \\ 0.31 & 0.69 \end{bmatrix}^2 = [0.2659 \quad 0.7341]$ Market share: $0.2659 \approx 27\%$

11.2 Systems of Linear Equations

1.   Solution   3.   Not a solution   5.   Not a solution

7.   a. $2y = -5x + 4$ 9.   a. $3y = -4x + 12$

          $y = -\frac{5}{2}x + 2$ $y = \frac{-4}{3}x + 4$

          slope $= -\frac{5}{2}$ $y-\text{int} = 2$ slope $= -\frac{4}{3}$ $y-\text{int} = 4$

          $-19y = -6x + 72$ $6y = -8x + 24$

          $y = \frac{6}{19}x - \frac{72}{19}$ $y = \frac{-8}{6}x + 4$

          slope $= \frac{6}{19}$ $y-\text{int} = -\frac{72}{19}$ slope $= \frac{-8}{6} = -\frac{4}{3}$ $y-\text{int} = 4$

      b.   One solution   b.   Infinite number of solutions

310

11.   a. $y = -x + 7$

  slope $= -1$ $y-\text{int} = 7$

  $2y = -3x + 8$

  $y = -\frac{3}{2}x + 4$

  slope $= -\frac{3}{2}$ $y-\text{int} = 4$

  $2y = -2x + 14$

  $y = -x + 7$

  slope $= -1$ $y-\text{int} = 7$

  b.   One solution

13.   This system could have a single solution since it has 3 equations and 3 unknowns.

15.   This system could not have a single solution because the first and second equations are equivalent.

17.   This system could not have a single solution because there are fewer equations than unknowns.

11.3 Long-Range Predictions with Markov Chains

1.   a. $L = \begin{bmatrix} x & y \end{bmatrix}$, where $x + y = 1$.   b. $LT = \begin{bmatrix} \frac{2}{11} & \frac{9}{11} \end{bmatrix} \begin{bmatrix} 0.1 & 0.9 \\ 0.2 & 0.8 \end{bmatrix} = \begin{bmatrix} \frac{2}{11} & \frac{9}{11} \end{bmatrix}$

  $\begin{bmatrix} x & y \end{bmatrix} \begin{bmatrix} 0.1 & 0.9 \\ 0.2 & 0.8 \end{bmatrix} = \begin{bmatrix} x & y \end{bmatrix}$

  $\begin{aligned} 0.1x + 0.2y &= x \\ 0.9x + 0.8y &= y \end{aligned}$ becomes $\begin{aligned} -0.9x + 0.2y &= 0 \\ 0.9x - 0.2y &= 0 \end{aligned}$

  Replacing one equation with $x + y = 1$:

  $\begin{aligned} -0.9x + 0.2y &= 0 \\ x + y &= 1 \end{aligned}$

  Multiplying the second equation by .9:

  $\begin{aligned} -0.9x + 0.2y &= 0 \\ \underline{0.9x + 0.9y} &= \underline{0.9} \\ 1.1y &= 0.9 \\ y &= \tfrac{9}{11} \end{aligned}$

Substituting, $x = 1 - y = \frac{2}{11}$

$L = \begin{bmatrix} \frac{2}{11} & \frac{9}{11} \end{bmatrix} = \begin{bmatrix} 0.1818 & 0.8182 \end{bmatrix}$

3. a. $L = \begin{bmatrix} x & y \end{bmatrix}$, where $x + y = 1$. b. $LT = \begin{bmatrix} \frac{8}{15} & \frac{7}{15} \end{bmatrix} \begin{bmatrix} 0.3 & 0.7 \\ 0.8 & 0.2 \end{bmatrix} = \begin{bmatrix} \frac{8}{15} & \frac{7}{15} \end{bmatrix}$

$\begin{bmatrix} x & y \end{bmatrix} \begin{bmatrix} 0.3 & 0.7 \\ 0.8 & 0.2 \end{bmatrix} = \begin{bmatrix} x & y \end{bmatrix}$

$\begin{aligned} 0.3x + 0.8y &= x \\ 0.7x + 0.2y &= y \end{aligned}$ becomes $\begin{aligned} -0.7x + 0.8y &= 0 \\ 0.7x - 0.8y &= 0 \end{aligned}$

Replacing one equation with $x + y = 1$: Multiplying the second equation by 0.7:

$\begin{aligned} -0.7x + 0.8y &= 0 \\ x + y &= 1 \end{aligned}$ $\begin{aligned} -0.7x + 0.8y &= 0 \\ \underline{0.7x + 0.7y} &= \underline{0.7} \end{aligned}$

$\begin{aligned} 1.5y &= 0.7 \\ y &= \frac{7}{15} \end{aligned}$

Substituting, $x = 1 - y = \frac{8}{15}$

$L = \begin{bmatrix} \frac{8}{15} & \frac{7}{15} \end{bmatrix} = \begin{bmatrix} 0.5333 & 0.4667 \end{bmatrix}$

5. $L = \begin{bmatrix} x & y \end{bmatrix}$, where $x + y = 1$.

$\begin{bmatrix} x & y \end{bmatrix} \begin{bmatrix} 0.63 & 0.37 \\ 0.12 & 0.88 \end{bmatrix} = \begin{bmatrix} x & y \end{bmatrix}$

$\begin{aligned} 0.63x + 0.12y &= x \\ 0.37x + 0.88y &= y \end{aligned}$ becomes $\begin{aligned} -0.37x + 0.12y &= 0 \\ 0.37x - 0.12y &= 0 \end{aligned}$

Replacing one equation with $x + y = 1$: Multiplying the second equation by 0.37:

$\begin{aligned} -0.37x + 0.12y &= 0 \\ x + y &= 1 \end{aligned}$ $\begin{aligned} -0.37x + 0.12y &= 0 \\ \underline{0.37x + 0.37y} &= \underline{0.37} \end{aligned}$

$\begin{aligned} 0.49y &= 0.37 \\ y &= \frac{0.37}{0.49} = \frac{37}{49} \end{aligned}$

Substituting, $x = \frac{12}{49}$

$L = \begin{bmatrix} \frac{12}{49} & \frac{37}{49} \end{bmatrix} = \begin{bmatrix} 0.24 & 0.76 \end{bmatrix}$ Market share: 24%

7. $L = \begin{bmatrix} x & y \end{bmatrix}$, where $x + y = 1$.

$$\begin{bmatrix} x & y \end{bmatrix} \begin{bmatrix} 0.88 & 0.12 \\ 0.03 & 0.97 \end{bmatrix} = \begin{bmatrix} x & y \end{bmatrix}$$

$$\begin{array}{ll} 0.88x + 0.03y & = & x \\ 0.12x + 0.97y & = & y \end{array} \quad \text{becomes} \quad \begin{array}{ll} -0.12x + 0.03y & = & 0 \\ 0.12x - 0.03y & = & 0 \end{array}$$

Replacing one equation with $x + y = 1$:

$$\begin{array}{rl} -0.12x + 0.03y & = & 0 \\ x + y & = & 1 \end{array}$$

Multiplying the second equation by 0.12:

$$\begin{array}{rl} -0.12x + 0.03y & = & 0 \\ 0.12x + 0.12y & = & 0.12 \end{array}$$

$$\begin{array}{rl} 0.15y & = & 0.12 \\ y & = & \frac{0.12}{0.15} = \frac{4}{5} \end{array}$$

Substituting: $x = \frac{1}{5}$ $L = \begin{bmatrix} \frac{1}{5} & \frac{4}{5} \end{bmatrix} = \begin{bmatrix} 0.2 & 0.8 \end{bmatrix}$ 20% will rent, 80% will own

We assume that the trend won't change and that the residents' moving plans are realized.

9. $L = \begin{bmatrix} x & y \end{bmatrix}$, where $x + y = 1$.

$$\begin{bmatrix} x & y \end{bmatrix} \begin{bmatrix} 0.12 & 0.88 \\ 0.31 & 0.69 \end{bmatrix} = \begin{bmatrix} x & y \end{bmatrix}$$

$$\begin{array}{ll} 0.12x + 0.31y & = & x \\ 0.88x + 0.69y & = & y \end{array} \quad \text{becomes} \quad \begin{array}{ll} -0.88x + 0.31y & = & 0 \\ 0.88x - 0.31y & = & 0 \end{array}$$

Replacing one equation with $x + y = 1$:

$$\begin{array}{rl} -0.88x + 0.31y & = & 0 \\ x + y & = & 1 \end{array}$$

Multiplying the second equation by 0.88:

$$\begin{array}{rl} -0.88x + 0.31y & = & 0 \\ 0.88x + 0.88y & = & 0.88 \end{array}$$

$$\begin{array}{rl} 1.19y & = & 0.88 \\ y & = & 0.7395 \end{array}$$

Substituting: $x = 0.2605$ $L = \begin{bmatrix} 0.2605 & 0.7395 \end{bmatrix}$

Sierra Cruiser will eventually control 26% of the market.

11.4 Solving Larger Systems of Equations

1.　Using Row 1 and Row 2 to eliminate the variable z:

Multiplying Row 1 by 2:　　　　Using Row 1 and Row 3 to eliminate z:

$$\begin{array}{rcr} 10x+2y-2z &=& 34 \\ 2x+5y+2z &=& 0 \\ \hline 12x+7y &=& 34 \end{array} \qquad \begin{array}{rcr} 5x+y-z &=& 17 \\ 3x+y+z &=& 11 \\ \hline 8x+2y &=& 28 \end{array}$$

Use these two new equations to eliminate x:

Multiplying the first new equation by 2 and the second new equation by -3:

$$\begin{array}{rcr} 24x+14y &=& 68 \\ -24x-6y &=& -84 \\ \hline 8y &=& -16 \\ y &=& -2 \end{array}$$

Substituting into earlier equations: $x=4$, $z=1$

The solution is $(4,-2,1)$.

3.　Using Row 1 and Row 2 to eliminate z:　　Using Row 1 and Row 3 to eliminate z:

Multiplying Row 2 by -1:　　　　Multiplying Row 1 by -2:

$$\begin{array}{rcr} x+y+z &=& 14 \\ -3x+2y-z &=& -3 \\ \hline -2x+3y &=& 11 \end{array} \qquad \begin{array}{rcr} -2x-2y-2z &=& -28 \\ 5x+y+2z &=& 29 \\ \hline 3x-y &=& 1 \end{array}$$

Use these two new equations to eliminate y:

Multiplying the second new equation by 3:

$$\begin{array}{rcr} -2x+3y &=& 11 \\ 9x-3y &=& 3 \\ \hline 7x &=& 14 \\ x &=& 2 \end{array}$$

Substituting into earlier equations: $y=5$, $z=7$　　　The solution is $(2,5,7)$.

5.   Using Row 1 and Row 3 to eliminate y:

$$
\begin{aligned}
x - y + 4z &= -13 \\
3x + y &= 25 \\
\hline
4x + 4z &= 12 \text{ or } x + z = 3
\end{aligned}
$$

Use this new equation with the original second equation to eliminate z:

$$
\begin{aligned}
x + z &= 3 \\
2x - z &= 12 \\
\hline
3x &= 15 \\
x &= 5
\end{aligned}
$$

Substituting into earlier equations: $z = -2$, $y = 10$. The solution is $(5, 10, -2)$.

7.   Using Row 1 and Row 2 to eliminate x: Using Row 2 and Row 3 to eliminate x:

Multiplying Row 2 by -2:   Multiplying Row 2 by 9 and Row 3 by -4:

$$
\begin{aligned}
8x + 7y - 3z &= 38 \\
-8x + 6y - 4z &= -22 \\
\hline
13y - 7z &= 16
\end{aligned}
\qquad
\begin{aligned}
36x - 27y + 18z &= 99 \\
-36x + 44y - 20z &= -28 \\
\hline
17y - 2z &= 71
\end{aligned}
$$

Use these two new equations to eliminate z:

Multiplying the first new equation by -2 and the second new equation by 7:

$$
\begin{aligned}
-26y + 14z &= -32 \\
119y - 14z &= 497 \\
\hline
93y &= 465 \\
y &= 5
\end{aligned}
$$

Substituting into earlier equations: $x = 3$, $z = 7$.   The solution is $(3, 5, 7)$.

9.   Using Row 1 and Row 2 to eliminate z: Using Row 1 and Row 3 to eliminate z:

                                                            Multiplying Row 1 by -4:

$$
\begin{aligned}
4x + 4y + 2z &= 10 \\
3x - 2y - 2z &= 18 \\
\hline
7x + 2y &= 28
\end{aligned}
\qquad
\begin{aligned}
-16x - 16y - 8z &= -40 \\
7x - y + 8z &= 4 \\
\hline
-9x - 17y &= -36
\end{aligned}
$$

Use these two new equations to eliminate x:

Multiplying the first new equation by 9 and the second new equation by 7:

$$63x + 18y = 252$$
$$-63x - 119y = -252$$

$$-101y = 0$$
$$y = 0$$

Substituting into earlier equations: $x = 4$, $z = -3$. The solution is $(4, 0, -3)$.

11. Using Row 1 and Row 2 to eliminate x:

Multiplying Row 1 by -22 and Row 2 by 47:

$$-1,034x - 1,276y - 814z = 2,486$$
$$1,034x - 1,739y + 1,269z = 15,604$$

$$-3,015y + 455z = 18,090$$

Using Row 1 and Row 3 to eliminate x:

Multiplying Row 3 by -1:

$$47x + 58y + 37z = -113$$
$$-47x - 15y - 52z = -145$$

$$43y - 15z = -258$$

Use these two new equations to eliminate y:

Multiplying the first new equation by 43 and the second new equation by 3015:

$$-129,645y + 19,565z = 777,870$$
$$129,645y - 45,225z = -777,870$$

$$-25,660z = 0$$
$$z = 0$$

Substituting into earlier equations: $x = 5$, $y = -6$. The solution is $(5, -6, 0)$.

13. Enter in matrix form:

$$\begin{bmatrix} 5 & 1 & -1 & 17 \\ 2 & 5 & 2 & 0 \\ 3 & 1 & 1 & 11 \end{bmatrix} \rightarrow \begin{bmatrix} 1 & 0 & 0 & 4 \\ 0 & 1 & 0 & -2 \\ 0 & 0 & 1 & 1 \end{bmatrix}$$

Solution is $(4, -2, 1)$.

15. Enter in matrix form:

$$\begin{bmatrix} 1 & 1 & 1 & 14 \\ 3 & -2 & 1 & 3 \\ 5 & 1 & 2 & 29 \end{bmatrix} \rightarrow \begin{bmatrix} 1 & 0 & 0 & 2 \\ 0 & 1 & 0 & 5 \\ 0 & 0 & 1 & 7 \end{bmatrix}$$

Solution is $(2, 5, 7)$.

17. Enter in matrix form:

$$\begin{bmatrix} 1 & -1 & 4 & -13 \\ 2 & 0 & -1 & 12 \\ 3 & 1 & 0 & 25 \end{bmatrix} \rightarrow \begin{bmatrix} 1 & 0 & 0 & 5 \\ 0 & 1 & 0 & 10 \\ 0 & 0 & 1 & -2 \end{bmatrix}$$ Solution is $(5, 10, -2)$.

19. Enter in matrix form:

$$\begin{bmatrix} 8 & 7 & -3 & 38 \\ 4 & -3 & 2 & 11 \\ 9 & -11 & 5 & 7 \end{bmatrix} \rightarrow \begin{bmatrix} 1 & 0 & 0 & 3 \\ 0 & 1 & 0 & 5 \\ 0 & 0 & 1 & 7 \end{bmatrix}$$ Solution is $(3, 5, 7)$.

21. Enter in matrix form:

$$\begin{bmatrix} 4 & 4 & 2 & 10 \\ 3 & -2 & -2 & 18 \\ 7 & -1 & 8 & 4 \end{bmatrix} \rightarrow \begin{bmatrix} 1 & 0 & 0 & 4 \\ 0 & 1 & 0 & 0 \\ 0 & 0 & 1 & -3 \end{bmatrix}$$ Solution is $(4, 0, -3)$.

23. Enter in matrix form:

$$\begin{bmatrix} 47 & 58 & 37 & -113 \\ 22 & -37 & 27 & 332 \\ 47 & 15 & 52 & 145 \end{bmatrix} \rightarrow \begin{bmatrix} 1 & 0 & 0 & 5 \\ 0 & 1 & 0 & -6 \\ 0 & 0 & 1 & 0 \end{bmatrix}$$ Solution is $(5, -6, 0)$.

25.
$$2x - 3y = 22$$
$$6x + 7y = 2$$

Write in matrix form: $\begin{bmatrix} 2 & -3 & 22 \\ 6 & 7 & 2 \end{bmatrix}$

Multiply Row 1 by –3 and add to Row 2: $\begin{bmatrix} 2 & -3 & 22 \\ 0 & 16 & -64 \end{bmatrix}$

Multiply Row 1 by ½ and multiply Row 2 by 1/16: $\begin{bmatrix} 1 & -\frac{3}{2} & 11 \\ 0 & 1 & -4 \end{bmatrix}$

Multiply Row 2 by 3/2 and add to Row 1: $\begin{bmatrix} 1 & 0 & 5 \\ 0 & 1 & -4 \end{bmatrix}$

The solution is $(5, -4)$.

27.
$$4x - 6y = -10$$
$$2x + 9y = 31$$

Write in matrix form: $\begin{bmatrix} 4 & -6 & -10 \\ 2 & 9 & 31 \end{bmatrix}$

Multiply Row 1 by –½ and add to Row 2: $\begin{bmatrix} 4 & -6 & -10 \\ 0 & 12 & 36 \end{bmatrix}$

Multiply Row 1 by ¼ and Row 2 by 1/12: $\begin{bmatrix} 1 & -\frac{3}{2} & -\frac{5}{2} \\ 0 & 1 & 3 \end{bmatrix}$

Multiply Row 2 by 3/2 and add to Row 1: $\begin{bmatrix} 1 & 0 & 2 \\ 0 & 1 & 3 \end{bmatrix}$

The solution is $(2, 3)$.

11.5 More on Markov Chains

5. Let $L = \begin{bmatrix} x & y & z \end{bmatrix}$, where $x + y + z = 1$

$$\begin{bmatrix} x & y & z \end{bmatrix} \begin{bmatrix} 0.71 & 0.12 & 0.17 \\ 0.32 & 0.34 & 0.34 \\ 0.02 & 0.02 & .096 \end{bmatrix} = \begin{bmatrix} x & y & z \end{bmatrix}$$

$$\begin{aligned} 0.71x + 0.32y + 0.02z &= x \\ 0.12x + 0.34y + 0.02z &= y \qquad \text{becomes} \\ 0.17x + 0.34y + 0.96z &= z \end{aligned} \qquad \begin{aligned} -0.29x + 0.32y + 0.02z &= 0 \\ 0.12x - 0.66y + 0.02z &= 0 \\ 0.17x + 0.34y - 0.04z &= 0 \end{aligned}$$

Multiply by 100 to eliminate the decimals.

$$-29x + 32y + 2z = 0$$
$$12x - 66y + 2z = 0$$
$$17x + 34y - 4z = 0$$

Replace the first equation with $x + y + z = 1$.

$$x + y + z = 1$$
$$12x - 66y + 2z = 0$$
$$17x + 34y - 4z = 0$$

Multiply the first equation by -12 and add to the second equation.

$$-12x - 12y - 12z = -12$$
$$\underline{12x - 66y + 2z = 0}$$
$$-78y - 10z = -12$$

Multiply the first equation by -17 and add to the third equation.

$$-17x - 17y - 17z = -17$$
$$\underline{17x + 34y - 4z = 0}$$
$$17y - 21z = -17$$

Multiply the first new equation by -21 and the second new equation by 10 and add.

$$1638y + 210z = 252$$
$$\underline{170y - 210z = -170}$$
$$1808y = 82$$
$$y = \frac{41}{904}$$

Substituting back into earlier equation, $x = \frac{49}{452}$, $z = \frac{765}{904}$.

Solution is $\left[\frac{49}{452} \quad \frac{41}{904} \quad \frac{765}{904} \right] \approx [0.1084 \quad 0.0454 \quad 0.8462]$.

Silver's 10.8%, Fitness Lab 4.5%, ThinNFit 84.6%

7. Let $L = [x \quad y \quad z]$, where $x + y + z = 1$

$$[x \quad y \quad z] \begin{bmatrix} 0.87 & 0.08 & 0.05 \\ 0.12 & 0.86 & 0.02 \\ 0.13 & 0.10 & 0.77 \end{bmatrix} = [x \quad y \quad z]$$

$$0.87x + 0.12y + 0.13z = x \qquad\qquad -0.13x + 0.12y + 0.13z = 0$$
$$0.08x + 0.86y + 0.10z = y \quad \text{becomes} \quad 0.08x - 0.14y + 0.10z = 0$$
$$0.05x + 0.02y + 0.77z = z \qquad\qquad 0.05x + 0.02y - 0.23z = 0$$

Multiply by 100 to eliminate the decimals.

$$-13x + 12y + 13z = 0$$
$$8x - 14y + 10z = 0$$
$$5x + 2y - 23z = 0$$

Replace the first equation with $x + y + z = 1$.

$$
\begin{aligned}
x + y + z &= 1 \\
8x - 14y + 10z &= 0 \\
5x + 2y - 23z &= 0
\end{aligned}
$$

Multiply the first equation by -8 and add to the second equation.

$$
\begin{aligned}
-8x - 8y - 8z &= -8 \\
\underline{8x - 14y + 10z} &= 0 \\
-22y + 2z &= -8
\end{aligned}
$$

Multiply the first equation by -5 and add to the third equation.

$$
\begin{aligned}
-5x - 5y - 5z &= -5 \\
\underline{5x + 2y - 23z} &= 0 \\
-3y - 28z &= -5
\end{aligned}
$$

Multiply the first new equation by 14 and add to the second new equation.

$$
\begin{aligned}
-308y + 28z &= -112 \\
\underline{-3y - 28z} &= -5 \\
-311y &= -117 \\
y &= \frac{117}{311}
\end{aligned}
$$

Substituting back into earlier equation, $x = \frac{151}{311}$, $z = \frac{43}{311}$.

Solution is $\left[\dfrac{151}{311} \quad \dfrac{117}{311} \quad \dfrac{43}{311} \right] \approx \left[0.4855 \quad 0.3762 \quad 0.1383 \right]$.

Safe Shop 48.6%, PayNEat 37.6%, other markets 13.8%

Chapter 11 Review

1.   Neither; the dimensions are 3×2.   3.   Column; the dimensions are 4×1.

5. $\begin{bmatrix} -3 & -24 \\ -24 & 28 \end{bmatrix}$ 7.   Does not exist   9. $\begin{bmatrix} 35 & -19 & -72 & 24 \\ 12 & 34 & 51 & 24 \end{bmatrix}$

11. $\begin{bmatrix} 5{,}000 & 7{,}000 \\ 5{,}500 & 6{,}200 \end{bmatrix} \begin{bmatrix} 3 \\ 4 \end{bmatrix} = \begin{bmatrix} 43{,}000 \\ 41{,}300 \end{bmatrix}$ NYC: $43,000, DC: $41,300

320

13.   a. Multiplying the first equation by 7 and the second by 5:

$$21x + 35y = -98$$
$$\underline{20x - 35y = 180}$$

$$41x = 82$$
$$x = 2$$
$$y = -4$$

The solutions is $(2, -4)$.

   b. Substituting $x = 2$ and $y = -4$ into the original equations yields true statements.

15.   a. Multiplying the first equation by 4 and the second by 7:

$$20x - 28y = -116$$
$$\underline{14x + 28y = 14}$$

$$34x = -102$$
$$x = -3 \qquad \text{The solution is } (-3, 2).$$
$$y = 2$$

   b. Substituting $x = -3$ and $y = 2$ into the original equations yields true statements.

17.   a. Multiply the second equation by 3 and add to the first equation.

$$7x - 8y + 3z = -43$$
$$\underline{15x + 9y - 3z = 72}$$

$$22x + y = 29$$

Multiply the second equation by 12 and add to the third equation.

$$60x + 36y - 12z = 288$$
$$\underline{11x - 5y + 12z = 0}$$

$$71x + 31y = 288$$

Multiply the first new equation by -31 and add to the second equation.

$$-682x - 31y = -899$$
$$\underline{71x + 31y = 288}$$

$$-611x = -611$$
$$x = 1$$

Substituting back into earlier equations, $y = 7$, $z = 2$. The solution is $(1, 7, 2)$.

   b. Substituting $x = 1$, $y = 7$, and $z = 2$ into the original equations yields true statements.

321

19. a. 0.7 b. 0.37 e.

c. $PT = \begin{bmatrix} 0.63 & 0.37 \end{bmatrix} \begin{bmatrix} 0.3 & 0.7 \\ 0.1 & 0.9 \end{bmatrix} = \begin{bmatrix} 0.226 & 0.774 \end{bmatrix}$

d. $PT^2 = PT \cdot T$

$= \begin{bmatrix} 0.226 & 0.774 \end{bmatrix} \begin{bmatrix} 0.3 & 0.7 \\ 0.1 & 0.9 \end{bmatrix} = \begin{bmatrix} 0.1452 & 0.8548 \end{bmatrix}$

21. a. Let $L = \begin{bmatrix} x & y \end{bmatrix}$, where $x + y = 1$. b. $LT = \begin{bmatrix} 0.4 & 0.6 \end{bmatrix} \begin{bmatrix} 0.7 & 0.3 \\ 0.2 & 0.8 \end{bmatrix} = \begin{bmatrix} 0.4 & 0.6 \end{bmatrix}$

$\begin{bmatrix} x & y \end{bmatrix} \begin{bmatrix} 0.7 & 0.3 \\ 0.2 & 0.8 \end{bmatrix} = \begin{bmatrix} x & y \end{bmatrix}$

$\begin{array}{l} 0.7x + 0.2y = x \\ 0.3x + 0.8y = y \end{array}$ becomes $\begin{array}{l} -0.3x + 0.2y = 0 \\ 0.3x - 0.2y = 0 \end{array}$

Delete the first equation and replace with $x + y = 1$.

$x + y = 1$
$0.3x - 0.2y = 0$

Multiply the first equation by 0.2 and add to the second equation.

$0.2x + 0.2y = 0.2$
$\underline{0.3x - 0.2y = 0}$

$0.5x \quad\quad = 0.2$
$\quad\quad x = 0.4$

Substitute $x = 0.4$ into an original equation, $y = 0.6$. Therefore, $L = \begin{bmatrix} 0.4 & 0.6 \end{bmatrix}$.

23. a. Let $L = \begin{bmatrix} x & y \end{bmatrix}$, where $x + y = 1$. b. $LT = \begin{bmatrix} \frac{1}{7} & \frac{6}{7} \end{bmatrix} \begin{bmatrix} 0.4 & 0.6 \\ 0.1 & 0.9 \end{bmatrix} = \begin{bmatrix} \frac{1}{7} & \frac{6}{7} \end{bmatrix}$

$\begin{bmatrix} x & y \end{bmatrix} \begin{bmatrix} 0.4 & 0.6 \\ 0.1 & 0.9 \end{bmatrix} = \begin{bmatrix} x & y \end{bmatrix}$

$\begin{array}{l} 0.4x + 0.1y = x \\ 0.6x + 0.9y = y \end{array}$ becomes $\begin{array}{l} -0.6x + 0.1y = 0 \\ 0.6x - 0.1y = 0 \end{array}$

Delete the first equation and replace with $x + y = 1$: $\begin{array}{l} x + y = 1 \\ 0.6x - 0.1y = 0 \end{array}$

Multiply the first equation by 0.1 and add to the second equation.

$0.1x + 0.1y = 0.1$

$0.6x - 0.1y = 0$

$0.7x \quad = 0.1$

$x = \dfrac{1}{7}$

Substitute $x = \dfrac{1}{7}$ into an original equation, $y = \dfrac{6}{7}$. Therefore, $L = \begin{bmatrix} \dfrac{1}{7} & \dfrac{6}{7} \end{bmatrix}$.

25.　a.　Let $L = \begin{bmatrix} x & y & z \end{bmatrix}$, where $x + y + z = 1$.

$$\begin{bmatrix} x & y & z \end{bmatrix} \begin{bmatrix} 0.5 & 0.1 & 0.4 \\ 0.2 & 0.2 & 0.6 \\ 0.4 & 0.3 & 0.3 \end{bmatrix} = \begin{bmatrix} x & y & z \end{bmatrix}$$

$0.5x + 0.2y + 0.4z = x$	$-0.5x + 0.2y + 0.4z = 0$
$0.1x + 0.2y + 0.3z = y$　becomes	$0.1x - 0.8y + 0.3z = 0$
$0.4x + 0.6y + 0.3z = z$	$0.4x + 0.6y - 0.7z = 0$

Delete the first equation and replace with $x + y + z = 1$.

$x + y + z = 1$

$0.1x - 0.8y + 0.3z = 0$

$0.4x + 0.6y - 0.7z = 0$

Multiply the first equation by -0.1 and add to the second equation.

$-0.1x - 0.1y - 0.1z = -0.1$

$0.1x - 0.8y + 0.3z = 0$

$-0.9y + 0.2z = -0.1$

Multiply the first equation by -0.4 and add to the third equation.

$-0.4x - 0.4y - 0.4z = -0.4$

$0.4x + 0.6y - 0.7z = 0$

$0.2y - 1.1z = -0.4$

Multiply the first new equation by 0.2 and the second new equation by 0.9, then add.

$$-0.18y + 0.04z = -0.02$$
$$0.18y - 0.99z = -0.36$$

$$\overline{-0.95z = -0.38}$$

$$z = \frac{2}{5}$$

Substituting back in equations, $x = \frac{2}{5}$, $y = \frac{1}{5}$. Therefore, $L = \begin{bmatrix} \frac{2}{5} & \frac{1}{5} & \frac{2}{5} \end{bmatrix}$.

b. $LT = \begin{bmatrix} \frac{2}{5} & \frac{1}{5} & \frac{2}{5} \end{bmatrix} \begin{bmatrix} 0.5 & 0.1 & 0.4 \\ 0.2 & 0.2 & 0.6 \\ 0.4 & 0.3 & 0.3 \end{bmatrix} = \begin{bmatrix} \frac{2}{5} & \frac{1}{5} & \frac{2}{5} \end{bmatrix}$

27. After three years:

$$[0.12 \quad 0.88] \begin{bmatrix} 0.62 & 0.38 \\ 0.16 & 0.84 \end{bmatrix} = [0.2152 \quad 0.7848] \qquad \text{Market share: } 21.5\%$$

29. Let $L = [x \quad y]$, where $x + y = 1$

$$[x \quad y] \begin{bmatrix} 0.62 & 0.38 \\ 0.16 & 0.84 \end{bmatrix} = [x \quad y]$$

$$\begin{array}{ll} 0.62x + 0.16y = x \\ 0.38x + 0.84y = y \end{array} \quad \text{becomes} \quad \begin{array}{ll} -0.38x + 0.16y = 0 \\ 0.38x - 0.16y = 0 \end{array}$$

Delete the first equation and replace it with $x + y = 1$.

$$\begin{array}{rl} x + y &= 1 \\ 0.38x - 0.16y &= 0 \end{array}$$

Multiply Row 1 by 0.16.

$$\begin{array}{rl} 0.16x + 0.16y &= 0.16 \\ 0.38x - 0.16y &= 0 \end{array}$$

$$\overline{\begin{array}{rl} 0.54x &= 0.16 \\ x &= 0.2963 \end{array}}$$

$y = 0.7037$ \qquad Long term prediction 29.6%.

324

31.   In two years:

$$\begin{array}{ccc} A & C & H \end{array}$$

$$\begin{bmatrix} 0.23 & 0.18 & 0.59 \end{bmatrix} \begin{bmatrix} 0.92 & 0.03 & 0.05 \\ 0.01 & 0.88 & 0.11 \\ 0.02 & 0.04 & 0.94 \end{bmatrix} \begin{array}{c} A \\ C \\ H \end{array} = \begin{bmatrix} 0.2252 & 0.1889 & 0.5859 \end{bmatrix}$$

In four years:

$$\begin{array}{ccc} A & C & H \end{array}$$

$$\begin{bmatrix} 0.2252 & 0.1889 & 0.5859 \end{bmatrix} \begin{bmatrix} 0.92 & 0.03 & 0.05 \\ 0.01 & 0.88 & 0.11 \\ 0.02 & 0.04 & 0.94 \end{bmatrix} \begin{array}{c} A \\ C \\ H \end{array} = \begin{bmatrix} 0.220791 & 0.196424 & 0.582785 \end{bmatrix}$$

Apartment: 22.1%, Condo/Townhouse: 19.6%, House: 58.3%

Thus, more condominiums and townhouses will be needed.

33.   Emigration and immigration, also affordability based or economy. We don't know if the prospective buyers can afford houses. It ignores economic conditions.

12.0 Review of Linear Inequalities

1. $y < -3x + 4$

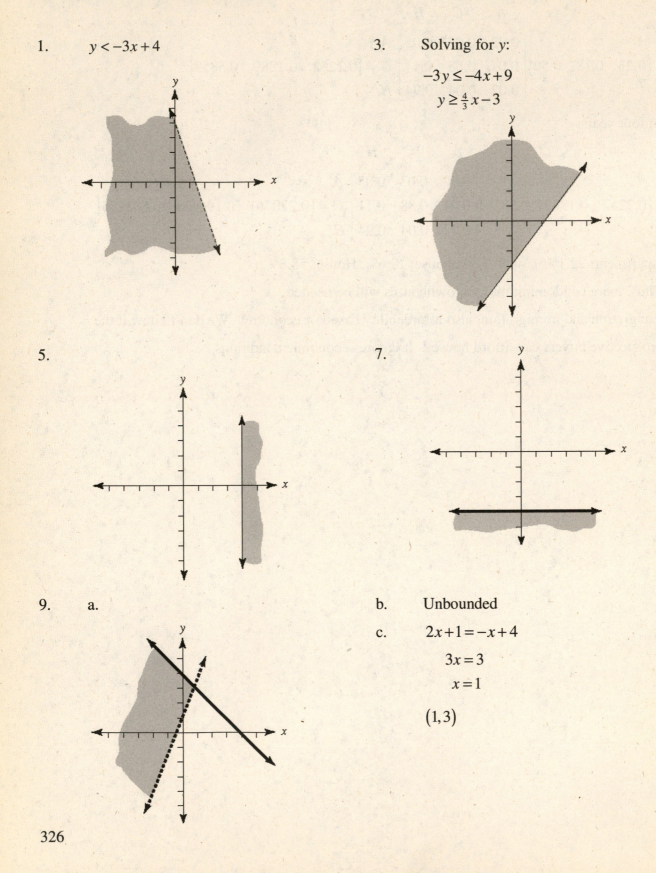

3.   Solving for y:

$$-3y \le -4x + 9$$
$$y \ge \tfrac{4}{3}x - 3$$

5.

7.

9.   a.

b.   Unbounded

c.
$$2x + 1 = -x + 4$$
$$3x = 3$$
$$x = 1$$

$$(1, 3)$$

11. a. Rewriting

$$3y < -2x + 17$$
$$-y \geq -3x - 2$$
becomes
$$y < -\frac{2}{3}x + \frac{17}{3}$$
$$y \leq 3x + 2$$

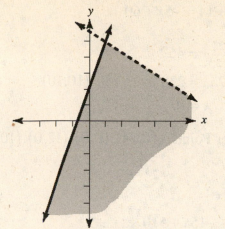

b. Unbounded

c. $-\frac{2}{3}x + \frac{17}{3} = 3x + 2$

$$-2x + 17 = 9x + 6$$
$$-11x = -11 \qquad (1, 5)$$
$$x = 1$$

13. a. Rewriting

$$2y \leq -x + 4$$
$$2y \geq 3x + 12 \quad \text{becomes}$$
$$y > x + 7$$

$$y \leq -\frac{1}{2}x + 2$$
$$y \geq \frac{3}{2}x + 6$$
$$y > x + 7$$

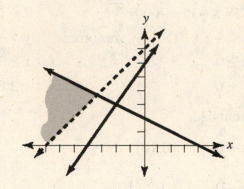

b. Unbounded

c. $-\frac{1}{2}x + 2 = x + 7$

$$-x + 4 = 2x + 14$$
$$-10 = 3x \qquad \left(-3\frac{1}{3}, 3\frac{2}{3}\right)$$
$$x = -\frac{10}{3}$$

15. a.

$$5y \leq -2x + 70$$
$$y \leq -5x + 60$$
$$x \geq 0 \quad \text{becomes}$$
$$y \geq 0$$

$$y \leq -\frac{2}{5}x + 14$$
$$y \leq -5x + 60$$
$$x \geq 0$$
$$y \geq 0$$

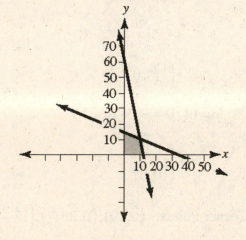

b. Bounded

c. Corner Points:

$$x = 0 \qquad (0, 14)$$
$$x = 0 \qquad (0, 0)$$
$$y = 0 \qquad 0 = -5x + 60$$
$$5x = 60$$
$$x = 12 \qquad (12, 0)$$

327

$$-\frac{2}{5}x + 14 = -5x + 60$$

$$\frac{23}{5}x = 46$$

$$23x = 230 \qquad (10,10)$$

$$x = 10$$

Corner Points: $(0,14),(0,0),(12,0),(10,10)$

17.   a.

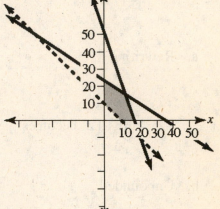

$22y \le -15x + 510$	$y \le -\dfrac{15}{22}x + \dfrac{510}{22}$
$12y \le -35x + 600$	$y \le -\dfrac{35}{12}x + 50$
$y > -x + 10$ becomes	$y > -x + 10$
$x \ge 0$	$x \ge 0$
$y \ge 0$	$y \ge 0$

   b. Bounded

   c. $x = 0 \qquad (0,10)$

$$x = 0 \qquad \left(0, \frac{510}{22}\right)$$

$$y = 0 \qquad 0 = -\frac{35}{12}x + 50$$

$$\frac{35}{12}x = 50$$

$$35x = 600$$

$$x = \frac{120}{7} \qquad\qquad -\frac{15}{22}x + \frac{510}{22} = -\frac{35}{12}x + 50$$

$$\left(\frac{120}{7}, 0\right) \qquad\qquad \frac{295}{132}x = \frac{295}{11}$$

$$y = 0 \quad 0 = -x + 10 \qquad\qquad 295x = 3,540$$

$$x = 10 \qquad\qquad\qquad x = 12$$

$$(10,0) \qquad\qquad\qquad (12,15)$$

Corner Points: $(0,10),\left(0, 23\frac{2}{11}\right),\left(17\frac{1}{7},0\right),(10,0),(12,15)$

19. a.

$$1.30y \le -0.50x + 2.21 \qquad y \le -0.3846x + 1.7$$
$$y \le -6x + 9 \qquad\qquad y \le -6x + 9$$
$$0.6y < -0.7x + 3.00 \quad \text{becomes} \quad y < -1.167x + 5$$
$$x \ge 0 \qquad\qquad x \ge 0$$
$$y \ge 0 \qquad\qquad y \ge 0$$

b. Bounded

c. $x = 0 \qquad (0,0)$

$x = 0 \qquad (0, 1.7)$

$y = 0 \qquad 0 = -6x + 9$

$\qquad\qquad 6x = 9$

$\qquad\qquad x = 1.5$

$\qquad\qquad (1.5, 0)$

$-0.3846x + 1.7 = -6x + 9$

$5.6154x = 7.3$

$x = 1.300$

$(1.3, 1.2)$

Corner Points: $(0,0), (0,1.7), (1.5,0), (1.3,1.2)$

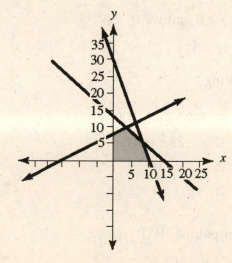

21. a.

$$2y \le x + 16 \qquad y \le \tfrac{1}{2}x + 8$$
$$y \le -3x + 30 \qquad y \le -3x + 30$$
$$y \le -x + 14 \quad \text{becomes} \quad y \le -x + 14$$
$$x \ge 0 \qquad\qquad x \ge 0$$
$$y \ge 0 \qquad\qquad y \ge 0$$

b. Bounded

c. $x = 0 \qquad (0,0)$

$x = 0 \qquad (0,8)$

$$y = 0 \qquad 0 = -3x + 30 \qquad -3x + 30 = -x + 14 \qquad \tfrac{1}{2}x + 8 = -x + 14$$

$$3x = 30 \qquad\qquad 16 = 2x \qquad\qquad \tfrac{3}{2}x = 6$$

$$x = 10 \qquad\qquad x = 8 \qquad\qquad 3x = 12$$

$$x = 4$$

$$(10, 0) \qquad\qquad\qquad (8, 6) \qquad\qquad\qquad (4, 10)$$

Corner Points: $(0,0), (0,8), (10,0), (4,10), (8,6)$

12.1 The Geometry of Linear Programming

1.   Let $x =$ number of shrubs

       $y =$ number of trees

       $1x + 3y \le 100$

3.   Let $x =$ number of hardbacks

       $y =$ number of paperbacks

       $4.50x + 1.25y \ge 5,000$

5.   Let $x =$ number of refrigerators

       $y =$ number of dishwashers

       $63x + 41y \le 1,650$

7.   Let $x =$ number of floor lamps

       $y =$ number of table lamps

       40 hours = 2,400 minutes

       $75x + 50y \le 2,400$ (Labor constraint)

       $25x + 20y \le 900$ (Cost of materials constraint)

       Also, $x \ge 0$ and $y \ge 0$

       $P = 39x + 33y$

       Rewriting:

$$\begin{aligned} 50y &\le -75x + 2,400 \\ 20y &\le -25x + 900 \\ x &\ge 0 \\ y &\ge 0 \end{aligned} \quad \text{becomes} \quad \begin{aligned} y &\le -\tfrac{3}{2}x + 48 \\ y &\le -\tfrac{5}{4}x + 45 \\ x &\ge 0 \\ y &\ge 0 \end{aligned}$$

       Corner points: $(0,0)$

       $x = 0 \qquad (0, 45)$

330

$$y = 0 \qquad 0 = -\frac{3}{2}x + 48 \qquad -\frac{3}{2}x + 48 = -\frac{5}{4}x + 45$$

$$\frac{3}{2}x = 48 \qquad\qquad -\frac{1}{4}x = -3$$
$$3x = 96 \qquad\qquad x = 12$$
$$x = 32$$

$$(32, 0) \qquad\qquad\qquad (12, 30)$$

$$P(0,0) = 0 \qquad\qquad P(0,45) = 1,485$$

$$P(32,0) = 1,248 \qquad\qquad P(12,30) = 1,458$$

She should make 45 table lamps and no floor lamps. The profit is $1,485.

9. Let x = lbs. of Morning Blend

   y = lbs. of South American Blend.

   $\frac{1}{3}x + \frac{2}{3}y \le 100$ (Mexican bean constraint)

   $\frac{2}{3}x + \frac{1}{3}y \le 80$ (Columbian bean constraint)

   Also, $x \ge 0$ and $y \ge 0$

   $P = 3x + 2.5y$

   Rewriting:

$\frac{2}{3}y \le -\frac{1}{3}x + 100$		$y \le -\frac{1}{2}x + 150$
$\frac{1}{3}y \le -\frac{2}{3}x + 80$	becomes	$y \le -2x + 240$
$x \ge 0$		$x \ge 0$
$y \ge 0$		$y \ge 0$

Corner Points: $(0,0)$

$$x = 0 \qquad (0, 150)$$

$$y = 0 \qquad 0 = -2x + 240 \qquad -\frac{1}{2}x + 150 = -2x + 240$$

$$2x = 240 \qquad\qquad \frac{3}{2}x = 90$$
$$x = 120 \qquad\qquad 3x = 180$$
$$\qquad\qquad\qquad x = 60$$

$$(120, 0) \qquad\qquad\qquad (60, 120)$$

331

$$P(0,0)=0 \qquad\qquad P(0,150)=375$$

$$P(120,0)=360 \qquad\qquad P(60,120)=480$$

They should make 60 pounds of Morning Blend and 120 pounds of South American Blend.

The profit is $480.

11. Let x = number of loaves to Shopgood

y = number of loaves to Rollie's

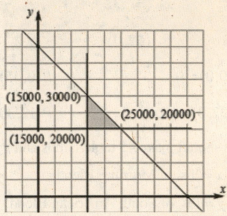

$$x+y \;\le\; 45{,}000 \quad \text{(Total shipment constraint)}$$

$$x \;\ge\; 15{,}000 \qquad \text{(Shopgood constraint)}$$

$$y \;\ge\; 20{,}000 \qquad \text{(Rollie's constraint)}$$

$$C = 0.08x + 0.09y$$

Rewriting:

$$y \;\le\; -x+45{,}000$$

$$x \;\ge\; \quad 15{,}000$$

$$y \;\ge\; \quad 20{,}000$$

Corner Points: $(15{,}000, 20{,}000)$

$$x=15{,}000: \quad y=30{,}000 \quad (15{,}000, 30{,}000)$$

$$y=20{,}000: \quad 20{,}000 = -x + 45{,}000$$

$$x = 25{,}000$$

$$(25{,}000, 20{,}000)$$

$$C(15{,}000, 20{,}000) = 3{,}000$$

$$C(15{,}000, 30{,}000) = 3{,}900$$

$$C(25{,}000, 20{,}000) = 3{,}800$$

They should send 15,000 loaves to Shopgood and 20,000 loaves to Rollie's. The cost is $3,000.

332

13. Let x = number of Orville flights

   y = number of Wilbur flights

   $20x + 80y \geq 1,600$

   $80x + 120y \geq 4,800$

   $x \leq 52$

   $y \leq 30$

   $C = 12,000x + 18,000y$

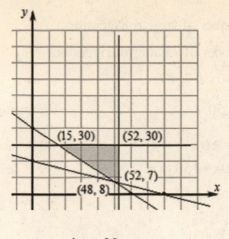

Rewriting,

			becomes			
$80y$	$\geq$	$-20x + 1,600$		y	$\geq$	$-\frac{1}{4}x + 20$
$120y$	$\geq$	$-80x + 4,800$		y	$\geq$	$-\frac{2}{3}x + 40$
x	$\leq$	52		x	$\leq$	52
y	$\leq$	30		y	$\leq$	30

And $x \geq 0$ and $y \geq 0$

Corner Points: $(52, 30)$ $\qquad$ $x = 52$ and $y = 7$

$$(52, 7)$$

$y = 30$

$$30 = -\frac{2}{3}x + 40 \qquad\qquad -\frac{1}{4}x + 20 = -\frac{2}{3}x + 40$$

$$\frac{2}{3}x = 10 \qquad\qquad\qquad \frac{5}{12}x = 20$$

$$2x = 30 \qquad\qquad\qquad\quad 5x = 240$$

$$x = 15 \qquad\qquad\qquad\quad x = 48$$

$\qquad (15, 30) \qquad\qquad\qquad\qquad (48, 8)$

$C(52, 30) = 1,164,000 \qquad\qquad C(52, 7) = 750,000$

$C(15, 30) = 720,000 \qquad\qquad C(48, 8) = 720,000$

Global has many choices, including the following:

   15 Orvilles and 30 Wilbers

   48 Orvilles and 8 Wilburs

Each of these generates a cost of at least $720,000. Another way to express the answer would be any point on the line $y = -\frac{2}{3}x + 40$ produces a minimum of $720,000.

333

15.   Let x = number of large refrigerators

       y = number of smaller refrigerators

       $60x + 40y \le 2,400$

a. $x + y \ge 50$ Rewriting:

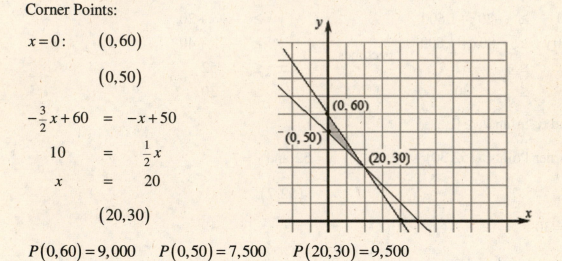

$x \ge 0, y \ge 0$

$P = 250x + 150y$

$$
\begin{aligned}
40y &\le -60x + 2400 \\
y &\ge -x + 50
\end{aligned}
\quad \text{becomes} \quad
\begin{aligned}
y &\le -\tfrac{3}{2}x + 60 \\
y &\ge -x + 50
\end{aligned}
$$

$x \ge 0, y \ge 0$ $x \ge 0, y \ge 0$

Corner Points:

$x = 0$: $(0, 60)$

           $(0, 50)$

$$
\begin{aligned}
-\tfrac{3}{2}x + 60 &= -x + 50 \\
10 &= \tfrac{1}{2}x \\
x &= 20
\end{aligned}
$$

           $(20, 30)$

$P(0, 60) = 9,000$ $P(0, 50) = 7,500$ $P(20, 30) = 9,500$

He should order 20 large and 30 smaller refrigerators. The profit is $9,500.

b. $x + y \ge 40$

       $x \ge 0, y \ge 0$

       $P = 250x + 150y$

       Rewriting:

       $y \le -\tfrac{3}{2}x + 60$

       $y \ge -x + 40$

       $x \ge 0, y \ge 0$

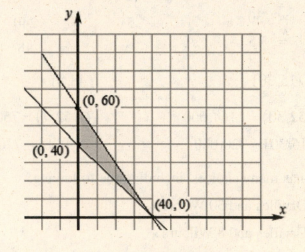

       Corner Points:

       $x = 0$ $(0, 60)$

                   $(0, 40)$

$$-\frac{3}{2}x + 60 \; = \; -x + 40$$

$$20 \; = \; \frac{1}{2}x$$

$$x \; = \; 40$$

$$(40,0)$$

$$P(0,60) = 9,000 \qquad P(0,40) = 6,000 \qquad P(40,0) = 10,000$$

He should order 40 large refrigerators and no small refrigerators. The profit is $10,000.

c. 40

17. 0 Mexican; 0 Columbian

19. 150 minutes unused; $0

21. Rewriting the constraints:

$$y \; \geq \; -3x + 12$$
$$y \; \geq \; -x + 6$$
$$x \; \geq \; 0$$
$$y \; \geq \; 0$$

Corner Points:

$x = 0$: $(0,12)$

$y = 0$: $0 = -x + 6$

$$x = 6$$

$$(6,0)$$

$$-3x + 12 \; = \; -x + 6$$
$$6 \; = \; 2x$$
$$x \; = \; 3$$

$$(3,3)$$

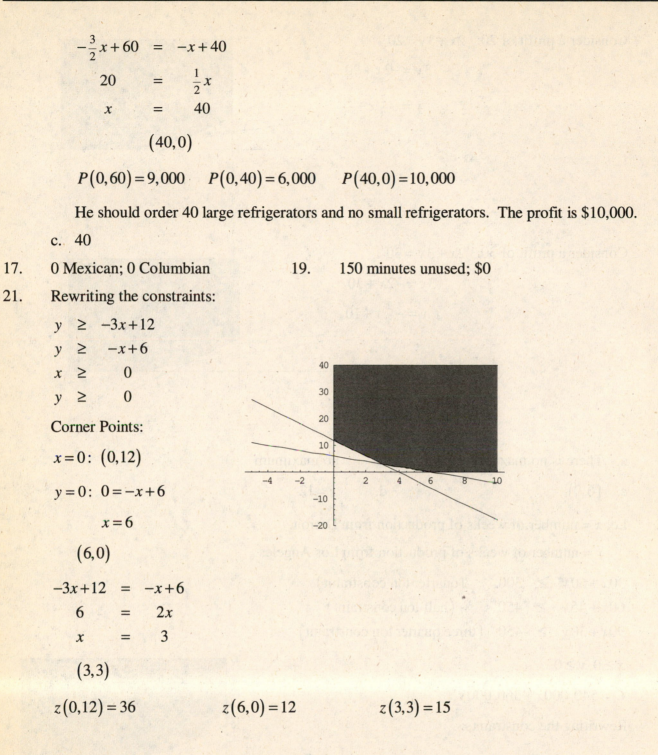

$$z(0,12) = 36 \qquad\qquad z(6,0) = 12 \qquad\qquad z(3,3) = 15$$

Consider a profit of 20: $2x + 3y = 20$

$$3y = -2x + 20$$

$$y = -\frac{2}{3}x + \frac{20}{3}$$

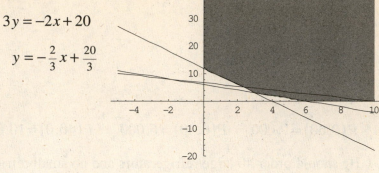

Consider a profit of 30: $2x + 3y = 30$

$$3y = -2x + 30$$

$$y = -\frac{2}{3}x + 10$$

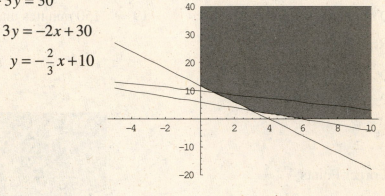

a. There is no maximum.   b.   no maximum

c. $(6,0)$ d. $z = 12$

23.   Let x = number of weeks of production from Detroit.

      y = number of weeks of production from Los Angeles.

$$30x + 60y \geq 300 \quad (\text{quarter ton constraint})$$
$$60x + 45y \geq 450 \quad (\text{half ton constraint})$$
$$90x + 30y \geq 450 \quad (\text{three quarter ton constraint})$$

$$x \geq 0, y \geq 0$$

$$C = 540,000x + 360,000y$$

Rewriting the constraints:

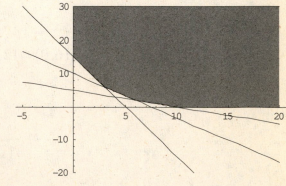

$$60y \geq -30x + 300$$
$$45y \geq -60x + 450 \quad \text{becomes}$$
$$30y \geq -90x + 450$$

$$y \geq -\frac{1}{2}x + 5$$
$$y \geq -\frac{4}{3}x + 10$$
$$y \geq -3x + 15$$

$$x \geq 0, y \geq 0 \qquad\qquad x \geq 0, y \geq 0$$

Corner Points:

$x=0$: $(0,15)$ $y=0$:

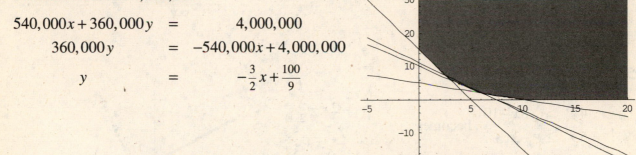

$$0=-\frac{1}{2}x+5 \qquad -\frac{4}{3}x+10 = -3x+15 \qquad -\frac{4}{3}x+10 = -\frac{1}{2}x+5$$

$$\frac{1}{2}x=5 \qquad\qquad \frac{5}{3}x = 5 \qquad\qquad -\frac{5}{6}x = -5$$

$$x=10 \qquad\qquad 5x = 15 \qquad\qquad -5x = -30$$

$$(10,0) \qquad\qquad x = 3 \qquad\qquad x = 6$$

$$(3,6) \qquad\qquad\qquad (6,2)$$

$C(0,15)=5,400,000$ $C(10,0)=5,400,000$

$C(6,2)=3,960,000$ $C(3,6)=3,780,000$

Consider a cost of 4,000,000:

$$540,000x+360,000y = 4,000,000$$

$$360,000y = -540,000x+4,000,000$$

$$y = -\frac{3}{2}x+\frac{100}{9}$$

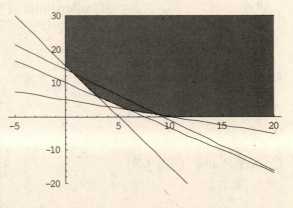

Consider a cost of 5,000,000:

$$360,000y = -540,000x+5,000,000$$

$$y = -\frac{3}{2}x+\frac{125}{9}$$

They should use 3 weeks of production in Detroit and 6 weeks of production in Los Angeles.

The cost is $3,780,000.

Chapter 12 Review

1.   Solving for y:

$$5y \;<\; 4x-7$$
$$y \;<\; \frac{4}{5}x - \frac{7}{5}$$

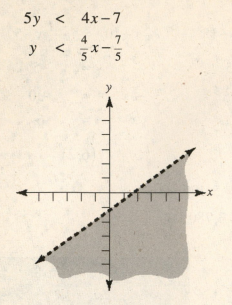

3.   Solving for y:

$$6y \;\leq\; -3x+9$$
$$y \;\leq\; -\frac{1}{2}x + \frac{3}{2}$$

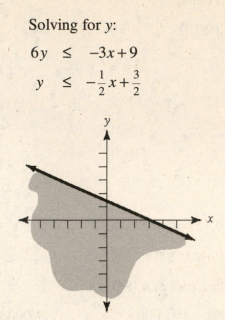

5.   Solving for y:

$$4y \;>\; 8x-10$$
$$5y \;\geq\; -3x+7$$

becomes

$$y \;>\; 2x - \frac{5}{2}$$
$$y \;\geq\; -\frac{3}{5}x + \frac{7}{5}$$

$$2x - \frac{5}{2} \;=\; -\frac{3}{5}x + \frac{7}{5}$$
$$\frac{13}{5}x \;=\; \frac{39}{10}$$
$$x \;=\; \frac{3}{2}$$

Corner Point: $\left(\frac{3}{2}, \frac{1}{2}\right)$

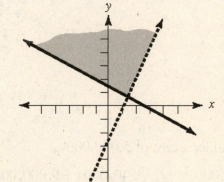

7.   Rewriting:

$$y \;<\; 5x-8$$
$$x \;\geq\; -3$$

Corner Point: $(-3, -23)$

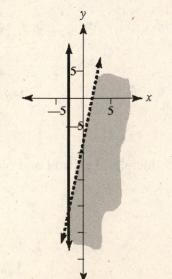

9.   Solving for y:

$$
\begin{aligned}
y &\geq x-7 \\
3y &\leq -5x+9 \\
x &\geq 0 \\
y &\geq 0
\end{aligned}
\qquad \text{becomes} \qquad
\begin{aligned}
y &\geq x-7 \\
y &\leq -\tfrac{5}{3}x+3 \\
x &\geq 0 \\
y &\geq 0
\end{aligned}
$$

$$
\begin{aligned}
-\tfrac{5}{3}x+3 &= 0 \\
\tfrac{5}{3}x &= 3 \\
x &= \tfrac{9}{5}
\end{aligned}
$$

Corner Points: $\left(\tfrac{9}{5},0\right),(0,3)$

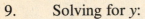

11.   Constraints: Solve for y:

$$
\begin{aligned}
x+y &\leq 20 \\
4x+y &\leq 20 \\
x &\geq 0 \\
y &\geq 0
\end{aligned}
\qquad
\begin{aligned}
y &\leq -x+20 \\
y &\leq -4x+20 \\
x &\geq 0 \\
y &\geq 0
\end{aligned}
$$

Corner Points	$z=4x+3y$
(0, 20)	$z=4(0)+3(20)=60$
(0, 0)	$z=4(0)+3(0)=0$
(5, 0)	$z=4(5)+3(0)=20$

(0, 0) is the minimum.

13.   Constraints: Solve for y:

$$
\begin{aligned}
15x+y &\geq 45 \\
x+6y &\geq 40 \\
8x+7y &\geq 70 \\
x &\geq 0 \\
y &\geq 0
\end{aligned}
\qquad
\begin{aligned}
y &\geq -15x+45 \\
y &\leq -\tfrac{1}{6}x+\tfrac{20}{3} \\
y &\leq -\tfrac{8}{7}x+10 \\
x &\geq 0 \\
y &\geq 0
\end{aligned}
$$

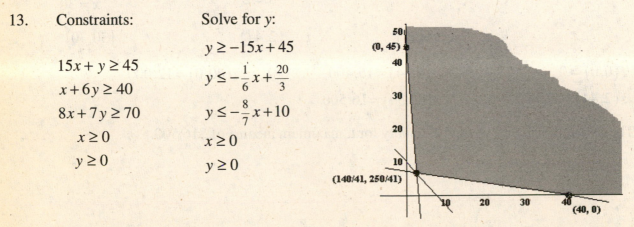

339

Corner Points	$z = 2x + 3y$
$(0, 45)$	$z = 2(0) + 3(45) = 135$
$(40, 0)$	$z = 2(40) + 3(0) = 80$
$\left(\frac{140}{41}, \frac{250}{41}\right)$	$z = 2\left(\frac{140}{41}\right) + 3\left(\frac{250}{41}\right) = 25.12$

$\left(\frac{140}{41}, \frac{250}{41}\right)$ is the minimum.

15. Let x = number of model 110s

$\qquad y$ = number of model 330s

$x + 2y \ \le \ 90$

$x + y \ \le \ 60$

$y \le 44$

$x \ge 0, y \ge 0$

$z = 200x + 350y$

$2y \ \le \ -x + 90 \qquad\qquad y \ \le \ -\frac{1}{2}x + 45$

$y \ \le \ -x + 60 \ \text{ becomes } \ y \ \le \ -x + 60$

$y \ \le \ \quad 44 \qquad\qquad\quad y \ \le \ \quad 44$

Corner Points: $(0,0)$

$x = 0: \ (0, 44) \qquad y = 0: \ 0 = -x + 60 \qquad y = 44: \ 44 = -\frac{1}{2}x + 45 \qquad -\frac{1}{2}x + 45 = -x + 60$

$\qquad\qquad\qquad\qquad\qquad\qquad\qquad x = 60 \qquad\qquad\qquad \frac{1}{2}x = 1 \qquad\qquad\qquad\qquad \frac{1}{2}x = 15$

$\qquad\qquad\qquad\qquad\qquad\qquad\qquad\qquad\qquad\qquad\qquad\qquad x = 2 \qquad\qquad\qquad\qquad\qquad x = 30$

$\qquad\qquad\qquad\qquad\qquad\qquad\qquad (60, 0) \qquad\qquad\qquad (2, 44) \qquad\qquad\qquad\qquad (30, 30)$

$z(0,0) = 0 \qquad\qquad\qquad z(0, 44) = 15,400 \qquad\qquad\qquad z(60, 0) = 12,000$

$z(2, 44) = 15,800 \qquad\quad z(30, 30) = 16,500$

They should make 30 of each assembly for a maximum income of $16,500.

Appendix A

The key strokes may vary depending upon your calculator. Refer to the owner's manual for your calculator.

1. 3 $\boxed{+/-}$ $\boxed{-}$ 5 $\boxed{+/-}$ $\boxed{=}$

3. 4 $\boxed{-}$ 9 $\boxed{+/-}$ $\boxed{=}$

5. 3 $\boxed{+/-}$ $\boxed{-}$ $\boxed{(}$ 5 $\boxed{+/-}$ $\boxed{-}$ 8 $\boxed{+/-}$ $\boxed{)}$ $\boxed{=}$

7. 8 $\boxed{+/-}$ $\boxed{\times}$ 3 $\boxed{+/-}$ $\boxed{\times}$ 2 $\boxed{+/-}$ $\boxed{=}$

9. 3 $\boxed{+/-}$ $\boxed{\times}$ 8 $\boxed{+/-}$ $\boxed{-}$ 9 $\boxed{+/-}$ $\boxed{\times}$ 2 $\boxed{+/-}$ $\boxed{=}$

11. 2 $\boxed{\times}$ 3 $\boxed{-}$ 5 $\boxed{=}$

13. $\boxed{(}$ 4 $\boxed{\times}$ 11 $\boxed{)}$ $\boxed{x^2}$

15. 4 $\boxed{\times}$ $\boxed{(}$ 3 $\boxed{+/-}$ $\boxed{)}$ $\boxed{y^x}$ 3 $\boxed{=}$

17. $\boxed{(}$ 3 $\boxed{+}$ 2 $\boxed{)}$ $\boxed{\div}$ 7 $\boxed{=}$

19. 3 $\boxed{\div}$ $\boxed{(}$ 2 $\boxed{\times}$ 7 $\boxed{)}$ $\boxed{=}$

21. $\boxed{(}$ 3 $\boxed{+}$ 2 $\boxed{)}$ $\boxed{\div}$ $\boxed{(}$ 7 $\boxed{\times}$ 5 $\boxed{)}$ $\boxed{=}$

23. 3 $\boxed{\div}$ $\boxed{(}$ 7 $\boxed{\div}$ 2 $\boxed{)}$ $\boxed{=}$

25. 1.8 $\boxed{x^2}$

27. 47000000 $\boxed{x^2}$

29. $\boxed{(}$ 3.92 $\boxed{+/-}$ $\boxed{)}$ $\boxed{y^x}$ 7 $\boxed{=}$

31. $\boxed{(}$ 3.76 $\boxed{EXP}$ 12 $\boxed{+/-}$ $\boxed{)}$ $\boxed{y^x}$ 5 $\boxed{+/-}$ $\boxed{=}$

33. Using the quadratic equation:

$$x = \frac{-8.3 \pm \sqrt{(8.3)^2 - 4(4.2)(1.1)}}{2(4.2)} = \frac{-8.3 \pm 7.1}{8.4} = -0.1429, -1.8333$$

Using a calculator:

8.3 $\boxed{x^2}$ $\boxed{-}$ 4 $\boxed{\times}$ 4.2 $\boxed{\times}$ 1.1 $\boxed{=}$ $\boxed{\sqrt{x}}$ $\boxed{\text{STO}}$

8.3 $\boxed{+/-}$ $\boxed{+}$ $\boxed{\text{RCL}}$ $\boxed{=}$ $\boxed{\div}$ 2 $\boxed{\div}$ 4.2 $\boxed{=}$

8.3 $\boxed{+/-}$ $\boxed{-}$ $\boxed{\text{RCL}}$ $\boxed{=}$ $\boxed{\div}$ 2 $\boxed{\div}$ 4.2 $\boxed{=}$

35. $\boxed{+}, \boxed{-}, \boxed{\times}, \boxed{\div}$, and $\boxed{y^x}$ because they combine two numbers.

Appendix B

1. $\boxed{(-)}$ 3 $\boxed{-}$ $\boxed{(-)}$ 5 $\boxed{\text{ENTER}}$

3. 4 $\boxed{-}$ $\boxed{(-)}$ 9 $\boxed{\text{ENTER}}$

5. $\boxed{(-)}$ 3 $\boxed{-}$ $\boxed{(}$ $\boxed{(-)}$ 5 $\boxed{-}$ $\boxed{(-)}$ 8 $\boxed{)}$ $\boxed{\text{ENTER}}$

7. $\boxed{(-)}$ 8 $\boxed{\times}$ $\boxed{(-)}$ 3 $\boxed{\times}$ $\boxed{(-)}$ 2 $\boxed{\text{ENTER}}$

9. $\boxed{(-)}$ 3 $\boxed{\times}$ $\boxed{(-)}$ 8 $\boxed{-}$ $\boxed{(-)}$ 9 $\boxed{\times}$ $\boxed{(-)}$ 2 $\boxed{\text{ENTER}}$

11. 2 $\boxed{\times}$ 3 $\boxed{-}$ 5 $\boxed{\text{ENTER}}$

13. $\boxed{(}$ 4 $\boxed{\times}$ 11 $\boxed{)}$ $\boxed{x^2}$ $\boxed{\text{ENTER}}$

15. 4 $\boxed{\times}$ $\boxed{(}$ $\boxed{(-)}$ 3 $\boxed{)}$ $\boxed{\wedge}$ 3 $\boxed{\text{ENTER}}$

17. $\boxed{(}$ 3 $\boxed{+}$ 2 $\boxed{)}$ $\boxed{\div}$ 7 $\boxed{\text{ENTER}}$

19. 3 $\boxed{\div}$ $\boxed{(}$ 2 $\boxed{\times}$ 7 $\boxed{)}$ $\boxed{\text{ENTER}}$

21. $\boxed{(}\ 3\ \boxed{+}\ 5\ \boxed{)}\ \boxed{\div}\ \boxed{(}\ 7\ \boxed{\times}\ 2\ \boxed{)}\ \boxed{\text{ENTER}}$

23. $3\ \boxed{\div}\ \boxed{(}\ 7\ \boxed{\div}\ 2\ \boxed{)}\ \boxed{\text{ENTER}}$

25. $1.8\ \boxed{x^2}\ \boxed{\text{ENTER}}$

27. $47000000\ \boxed{x^2}\ \boxed{\text{ENTER}}$

29. $\boxed{(}\ \boxed{(-)}\ 3.92\ \boxed{)}\ \boxed{\wedge}\ 7\ \boxed{\text{ENTER}}$

31. $3.76\ \boxed{\text{2nd}}\ \boxed{\text{EE}}\ \boxed{(-)}\ 12\ \boxed{\wedge}\ \boxed{(-)}\ 5\ \boxed{\text{ENTER}}$

33. a. $\dfrac{16}{5}$

    b. $\boxed{(}\ 9\ \boxed{-}\ 12\ \boxed{)}\ \boxed{\div}\ 5\ \boxed{\text{STO}}\ \boxed{\text{ALPHA}}\ \boxed{\text{A}}\ \boxed{\text{ENTER}}\ \boxed{(}\ \boxed{\text{ALPHA}}\ \boxed{\text{A}}\ \boxed{+}\ 7\ \boxed{)}\ \boxed{\div}\ 2\ \boxed{\text{ENTER}}$

    c. $\boxed{(}\ \boxed{(}\ 9\ \boxed{-}\ 12\ \boxed{)}\ \boxed{\div}\ 5\ \boxed{+}\ 7\ \boxed{)}\ \boxed{\div}\ 2\ \boxed{\text{ENTER}}$

35. a. $\dfrac{2}{5}$

    b. $\boxed{(}\ 7\ \boxed{+}\ 9\ \boxed{)}\ \boxed{\div}\ 5\ \boxed{\text{STO}}\ \boxed{\text{ALPHA}}\ \boxed{\text{A}}\ \boxed{\text{ENTER}}\ \boxed{(}\ 8\ \boxed{-}\ 14\ \boxed{)}\ \boxed{\div}\ 3\ \boxed{\text{ALPHA}}\ \boxed{\text{B}}\ \boxed{\text{ENTER}}$

       $\boxed{(}\ \boxed{\text{ALPHA}}\ \boxed{\text{A}}\ \boxed{+}\ \boxed{\text{ALPHA}}\ \boxed{\text{B}}\ \boxed{)}\ \boxed{\div}\ 3\ \boxed{\text{ENTER}}$

    c. $\boxed{(}\ \boxed{(}\ 7\ \boxed{+}\ 9\ \boxed{)}\ \boxed{\div}\ 5\ \boxed{+}\ \boxed{(}\ 8\ \boxed{-}\ 14\ \boxed{)}\ \boxed{\div}\ 3\ \boxed{)}\ \boxed{\div}\ 3\ \boxed{\text{ENTER}}$

37. $x = \dfrac{-8.3 \pm \sqrt{(8.3)^2 - 4(4.2)(1.1)}}{2(4.2)} = \dfrac{-8.3 \pm 7.1}{8.4} = -0.1429,\ -1.8333$

    $\boxed{\text{2nd}}\ \boxed{\sqrt{x}}\ \boxed{(}\ 8.3\ \boxed{x^2}\ \boxed{-}\ 4\ \boxed{\times}\ 4.2\ \boxed{\times}\ 1.1\ \boxed{)}\ \boxed{\text{STO}}\ \boxed{\text{ALPHA}}\ \boxed{\text{A}}\ \boxed{\text{ENTER}}$

    First Root: $\boxed{(}\ \boxed{(-)}\ 8.3\ \boxed{+}\ \boxed{\text{ALPHA}}\ \boxed{\text{A}}\ \boxed{)}\ \boxed{\div}\ \boxed{(}\ 2\ \boxed{\times}\ 4.2\ \boxed{)}\ \boxed{\text{STO}}\ \boxed{\text{ALPHA}}\ \boxed{\text{B}}\ \boxed{\text{ENTER}}$

    Second Root: $\boxed{(}\ \boxed{(-)}\ 8.3\ \boxed{-}\ \boxed{\text{ALPHA}}\ \boxed{\text{A}}\ \boxed{)}\ \boxed{\div}\ \boxed{(}\ 2\ \boxed{\times}\ 4.2\ \boxed{)}\ \boxed{\text{STO}}\ \boxed{\text{ALPHA}}\ \boxed{\text{C}}\ \boxed{\text{ENTER}}$

    Check First Root: $4.2\ \boxed{\times}\ \boxed{\text{ALPHA}}\ \boxed{\text{B}}\ \boxed{x^2}\ \boxed{+}\ 8.3\ \boxed{\times}\ \boxed{\text{ALPHA}}\ \boxed{\text{B}}\ \boxed{+}\ 1.1\ \boxed{\text{ENTER}}$

    Check Second Root: $4.2\ \boxed{\times}\ \boxed{\text{ALPHA}}\ \boxed{\text{C}}\ \boxed{x^2}\ \boxed{+}\ 8.3\ \boxed{\times}\ \boxed{\text{ALPHA}}\ \boxed{\text{C}}\ \boxed{+}\ 1.1\ \boxed{\text{ENTER}}$

39. Without the parentheses, the calculator would take the square root of 4.9 only.

343

41. a. $(((5 + 7.1) \div 3 + (2 - 7.1) \div 5) \div 7$ [ENTER]

    Solution: 0.4304761905

    b. Type 2nd, Entry to reproduce the line in part a, then use the ← key to change 7.1 to 7.2, then press Enter

    Solution: 0.4323809524

    c. Type 2nd, Entry to reproduce the line in part b, then use the ← key to change 7.2 to 9.3 and 4.9, then press Enter

    Solution: 0.5980952381

43. a. 81,000

    b. 8.1 E12

    c. The display is not large enough to show the answer to part b without using exponential form.

    d. For the TI 83/84, press [MODE], move cursor over one so that [SCI] is highlighted, and press [ENTER].

Appendix C

1. a. A graph on $[-10,10] \times [-10,10]$

    b. You could set the limits to the same values in WINDOW.

3. The error is because xmin is larger than xmax.

5. a. 0.5

    b. Solving:

$$
\begin{aligned}
2x - 1 &= 0 \\
2x &= 1 \\
x &= 0.5
\end{aligned}
$$

    c. Answers will vary. Possible answer: $(1.9, 2.8)$

    d. $y = 2(1.9) - 1 = 3.8 - 1 = 2.8$

7.   a. The graph is enlarged and centered at that point.

     b. Reset WINDOW.

     c. The graph is reduced and centered at that point.

9.   a. .5

     b. Yes. The numerical routine used by the calculator keeps approximating until it is very precise.

Appendix D

1.   a. $(-1.5, -2.5)$

     b. Substituting:

$$y = 3(-1.5) + 2 = -4.5 + 2 = -2.5$$
$$y = 5(-1.5) + 5 = -7.5 + 5 = -2.5$$

3.   a. $(-12.3333, -112.6667)$

     b. Substituting:

$$y = 8(-12.3333) - 14 = -112.6667$$
$$y = 11(-12.3333) + 23 = -112.6667$$

5.   a. Answers will vary. Possible answer: $(-2.3860, 13.4650),\ (1.8860, 2.7850)$

     b. Substituting:

$$y = (-2.3860)^2 - 2(-2.3860) + 3 = 13.4650$$
$$y = -(-2.3860)^2 - 3(-2.3860) + 12 = 13.4650$$
$$y = (1.8860)^2 - 2(1.8860) + 3 = 2.7850$$
$$y = -(1.8860)^2 - 3(1.8860) + 12 = 2.7850$$

Appendix E

1. a. $\left(12\ \text{ft}\right)\left(\dfrac{1\ \text{yd}}{3\ \text{ft}}\right) = 4\ \text{yd}$ b. $\left(12\ \text{yd}\right)\left(\dfrac{3\ \text{ft}}{1\ \text{yd}}\right) = 36\ \text{ft}$

3. a. $\left(10\ \text{mi}\right)\left(\dfrac{5{,}280\ \text{ft}}{1\ \text{mi}}\right) = 52{,}800\ \text{ft}$ b. $\left(10\ \text{ft}\right)\left(\dfrac{1\ \text{mi}}{5{,}280\ \text{ft}}\right) = 0.0019\ \text{mi}$

5. $\left(2\ \text{mi}\right)\left(\dfrac{5{,}280\ \text{ft}}{1\ \text{mi}}\right)\left(\dfrac{12\ \text{in.}}{1\ \text{ft}}\right) = 126{,}720\ \text{in.}$

7. a. $\left(50\ \text{km}\right)\left(\dfrac{1{,}000\ \text{m}}{1\ \text{km}}\right) = 50{,}000\ \text{m}$ b. $\left(50\ \text{m}\right)\left(\dfrac{1\ \text{km}}{1{,}000\ \text{m}}\right) = 0.05\ \text{km}$

9. a. $\left(2\ \text{cl}\right)\left(\dfrac{1\ \text{l}}{100\ \text{cl}}\right) = 0.02\ \text{l}$ b. $\left(2\ \text{l}\right)\left(\dfrac{100\ \text{cl}}{1\ \text{l}}\right) = 200\ \text{cl}$

11. a. $\left(1\ \text{lb}\right)\left(\dfrac{1\ \text{g}}{0.0022046\ \text{lb}}\right) = 454\ \text{g}$

    b. $\left(8\ \text{lb}\right)\left(\dfrac{1\ \text{g}}{0.0022046\ \text{lb}}\right)\left(\dfrac{1\ \text{kg}}{1{,}000\ \text{g}}\right) = 3.6\ \text{kg}$

13. a. $\left(\dfrac{60\ \text{mi}}{1\ \text{hr}}\right)\left(\dfrac{5{,}280\ \text{ft}}{1\ \text{mi}}\right)\left(\dfrac{1\ \text{hr}}{60\ \text{min}}\right)\left(\dfrac{1\ \text{min}}{60\ \text{sec}}\right) = 88\ \text{ft/sec}$

    b. To cover 80 ft: $\dfrac{80\ \text{ft}}{88\ \text{ft/sec}} = 0.9\ \text{sec}$

15. a. $\left(\dfrac{130\ \text{km}}{1\ \text{hr}}\right)\left(\dfrac{1{,}000\ \text{m}}{1\ \text{km}}\right)\left(\dfrac{39.37\ \text{in.}}{1\ \text{m}}\right)\left(\dfrac{1\ \text{ft}}{12\ \text{in.}}\right)\left(\dfrac{1\ \text{mi}}{5{,}280\ \text{ft}}\right) = 81\ \text{mi/hr}$

    b. $81 - 65 = 16$ more miles every hour.

17. a. $\left(\dfrac{5.75\%}{1\ \text{yr}}\right)\left(\dfrac{1\ \text{yr}}{365\ \text{days}}\right)=0.0157534247\%/\text{day}$

 b. In 30 days: 0.4726027397%

 $10{,}000\cdot 0.004726027397=\$47.26$

 c. In 31 days: 0.4883561644%

 $10{,}000\cdot 0.004883561644=\$48.84$

19. a. $\left(6\ \text{ft}\right)\left(\dfrac{12\ \text{in.}}{1\ \text{ft}}\right)\left(\dfrac{1\ \text{m}}{39.37\ \text{in.}}\right)=1.83\ \text{m}$

 b. $\left(169\ \text{lb}\right)\left(\dfrac{1\ \text{g}}{0.0022046\ \text{lb}}\right)\left(\dfrac{1\ \text{kg}}{1{,}000\ \text{g}}\right)=76.7\ \text{kg}$

 c. $\dfrac{76.7\ \text{kg}}{\left(1.83\ \text{m}\right)^2}=23$; No, he not overweight.

 d. $\left(5.41667\ \text{ft}\right)\left(\dfrac{12\ \text{in.}}{1\ \text{ft}}\right)\left(\dfrac{1\ \text{m}}{39.37\ \text{in.}}\right)=1.65\ \text{m}$

 $\dfrac{76.7\ \text{kg}}{\left(1.65\ \text{m}\right)^2}=28$, so according to this scheme, Fred is overweight.

 e. The World Health Organization is an international organization. Most countries use the metric system.

21. $\left(4\ \text{ft}\right)\left(6\ \text{ft}\right)\left(\dfrac{11}{12}\ \text{ft}\right)\left(\dfrac{7.48\ \text{gal}}{1\ \text{ft}^3}\right)=164.56\ \text{gal}$